BRIAN J. FORD is a renowned biologist who has published hundreds of scientific articles in journals including *Nature*, *New Scientist*, *Scientific American* and the *British Medical Journal*. His revolutionary discoveries have ranged from blood coagulation to the intelligence of microbes, and from human spontaneous combustion to the science of autumn leaves. As a writer he was the first British president of the European Union of Science Journalists' Associations based in Brussels, has written for major newspapers including *The Times*, the *Daily Telegraph* and the *Guardian* and currently contributes his regular column to *The Microscope* in America. He has provided editorial advice for major reference works including *Encyclopaedia Britannica* and *Guinness World Records*, hosted programmes including *Science Now*, *Where are you Taking Us?* and *Computer Challenge* for the BBC and has presented his own series *Jenseits des Kanals* (*Across the Channel*) for German television. He has frequently contributed to programmes including *Today*, *Any Questions?*, *You & Yours*, and news programmes on radio and television including CNN and *Newsnight*. He is a fellow of Cardiff University, former fellow of the Open University, a fellow (and former officer) of both the Linnean Society and of the Institute of Biology, and was appointed visiting professor at Leicester University for his work on distance learning. He was awarded a fellowship by the

National Endowment for Science, Technology and the Arts (NESTA) and was recently presented with honorary fellowship of the Royal Microscopical Society. In Cambridge, he is life fellow of Cambridge Philosophical Society, President Emeritus of the Cambridge Society for the Application of Research, and was for many years a dining member at Gonville and Caius College, Cambridge University. Professor Ford's many books have appeared in over 130 editions around the world, and he is frequently invited to travel extensively overseas to lecture on his work and ideas.

Praise for Brian J. Ford:

'Ford offers his audience a reasonable balance and the chance to improve their perspective' *The Times*

'Important, well-written and entertaining' *Times Literary Supplement*

'Well presented and readable' *New Scientist*

'The author's scholarly familiarity with his subject is evident throughout' *New York Library Journal*

TOO BIG TO WALK

THE NEW SCIENCE OF DINOSAURS

BRIAN J. FORD

WILLIAM COLLINS

William Collins
An imprint of HarperCollins*Publishers*
1 London Bridge Street
London SE1 9GF

WilliamCollinsBooks.com

First published in Great Britain by William Collins in 2018
This fully revised and updated William Collins paperback
edition published in 2019

2020 2022 2021 2019
2 4 6 8 10 9 7 5 3 1

A catalogue record for this book is
available from the British Library

ISBN 978-0-00-821893-5

Set in Minion Pro and Trajan Pro

Printed and bound in Great Britain by
CPI Group (UK) Ltd, Croydon, CR0 4YY

This book is produced from independently certified FSC™ paper
to ensure responsible forest management.

For more information visit: www.harpercollins.co.uk/green

CONTENTS

PREFACE

This is the updated, second edition of a book I didn't want to publish. For decades I deliberated on the giant dinosaurs, pondering as the palæontologists unveiled their findings, and it was obvious that they were getting dinosaurs wrong. I waited for the truth to dawn, but it didn't happen. The scientific evidence is now clear – the way dinosaurs are explained is incorrect. So this book has a bold and irreverent aim, for it sets out to demolish our present-day orthodoxies and to create a radical new view of how dinosaurs developed and the way they lived their lives. I am also launching a startling theory which shows how we have misunderstood the Cretaceous period, that great era when the gigantic dinosaurs held sway. Our current understanding is fundamentally misconstrued: the environment was different; the climate was different; the landscape was different. Dinosaurs were different. Everything we know about the age of the dinosaurs is misconceived, and producing this book has been the only way to revolutionize this entire scientific discipline. It has been a colossal undertaking.

We are going to travel back in time to see how the development of our planet was determined, how fossils were discovered, and how science started to understand evolution and the way the world became the way it is. As we set out on this extraordinary journey, I would like to thank the many dinosaur specialists around the world who have assisted with advice. Truly, I'd like to very much; but I cannot. None of them helped – instead, every dinosaur expert has attacked this new theory whenever it has appeared (or tried to). Those palæontologists around the world are so very antagonistic to every word within, that you may have pebbles thrown at your windows if one of them spies this book in your room. This iconoclastic review has been the target of bitter hostility and the most vitriolic insults, though my inquiries into dinosaurs were never intended to be about controversy, but simply about debating how those massive monsters evolved and how they lived their remarkable lives.

There is clearly a requirement for a detailed explanation of dinosaur research. As Larry Witham has pointed out: 'It is bad news to science museums when four in ten Americans believe humans lived with dinosaurs.'[1] There is certainly a need for a survey of the whole field, dating back to prehistory, looking at the pioneers and the early discoveries, and following how opinions have changed over the years.

Why did this study of such colossal creatures capture my attention, since I am a biologist preoccupied by the smallest microscopic living organisms – single cells? Dinosaurs were the largest animals ever, and should be far from my central interests. Yet there is a link between monstrous dinosaurs and microscopic cells. In 1993, Dippy the *Diplodocus* in the Natural History Museum in London had her tail raised. This long tail had rested on the floor since the skeleton was transferred to the entrance hall in 1979, but research had since shown that the tail could not have been like that in life. And so, in 1993 the massive tail was raised aloft, securely supported by stout steel. People looking at the skeleton found themselves imagining the fossil clothed in muscly flesh, the dinosaur sheathed in warty skin as it snarled at visitors. Not me: I always envisioned a minute microscopic muscle inside the tail, each one endlessly burning glucose to provide metabolic energy at a furious rate as it remained resolutely contracted, struggling to hold the heavy tail up against the downward clutch of gravity. No animal evolves to do this: half the dinosaur's intake of food would be expended by the effort of simply holding the tail up in the air. Try standing erect with your arms held straight out sideways and see how long you manage. That standard view of dinosaurs was impossible. It was the single cell that proved it.

Dinosaurs have long been fantasized about by scientists. Palæontologists have been circulating silly stories about dancing dinosaurs and their complex sex lives and these scientists create complex caricatures of lifestyles that are based on nothing more than wishful thinking or idle guesswork. We need scientific evidence for our statements, and for the present-day theories there is little scientific backing. Every textbook and television documentary ever produced, all the sci-fi movies, newspaper and magazine articles published around the world, and every display in museums and theme parks, all are fundamentally misconstrued. What we have been taught about dinosaurs is wrong.

When I proposed my new theory, it was greeted by a hail of invective. 'Who the hell ...?' demanded one commentator online; 'WTF ...?' said another. The theory is 'a rotting corpse' and 'a silly idea' and reporting this 'dinosaur nonsense' is 'bad science journalism', while 'Brian Ford's wild, ignorant, uninformed speculation' became the target for a petition signed by palæontologists all around the world and sent to the BBC after they broadcast an interview about it all. 'The BBC and everyone else who carried this story should be ashamed,' announced the palæontologists. The BBC carefully considered the petition, and said they felt that 'Brian Ford was unlikely to be put off by the condemnation of the established experts.' On that occasion, the BBC was right.

We like to think that revolutionary scientific theories are seized with open arms, but they are usually crushed by conventional conformity. There is a reason. In science you receive your funds for routine research that has a tried-and-tested track record; there is no academic support for something unexpected. Publications in science are subject to peer review, which means that a paper must proceed through a sequence of checks – carried out by the existing authorities in your field – before it is possible to publish. This is a sensible safeguard against an editor (who may know little of the topic) publishing something that's muddle-headed or wrong. Writers have often said to me that they love the internet, because they can publish whatever they like without the intervention of an editor; believe me, that is why there is so much rubbish on the web. Editors are a scientist's best friend. They can detect the infelicities that your readers would spot in an instant. I write a regular column in America and my editor in Chicago,

Dean Golemis, has an editorial eye eagles would envy. In this book, after all its conventional processing, Golemis corrected dozens of infelicities others had missed. Never edit your own writing!

Yet peer review has a downside. If you are publishing a new theory which says, in essence, that the authorities in your field are heading the wrong way, then obviously they aren't guaranteed to agree. An iconoclastic new theory is likely to be squashed before it gains currency. Establishment academics need to keep things under control or they lose their authority and, worse still, their funding. Although the opinion of our peers may guard against our publishing hastily, it can also conceal crucial new concepts. Peer review has become the single most pervasive obstacle to revolution in science.

It is also being seized upon as the key to success – not for the academics alone, but for the online community of entrepreneurs. Hundreds of newly invented journals with names conjured up to seem prestigious are being set up around the world. They send dignified missives to eminent professors, inviting them to become editorial advisers, and soon establish an editorial board of great names. The typefaces are chosen for their elegant and refined lines, and four-figure fees are demanded from authors for open-access publication. Apart from the time taken to set up and format the website, the running costs are minimal. Whereas an established scientific journal has high costs for paper and production, and for binding and distribution, these online enterprises have negligible outgoings and almost all the work is done by vain volunteers eager to see their names on the editorial page. The profits to the proprietor are immense. Their key to success lies in the extreme gullibility of scientists, to whom publishing peer-reviewed papers matters more than anything else. This emergent form of scientific publishing is a racket and it is exploiting the naïve vanity of academics and dignifying their hollow enterprise with the touchstone of peer review. Millions are being made every day because of this futile faith in a questionable concept.

Part of the problem is the lack of scientific awareness on the part of the media, which allows outrageous flights of fancy to proceed unchallenged. In any other field of endeavour – sport or politics, economics or art – commentators are quick to pounce on any infelicity and argue the toss with the most eminent of authorities. Politicians can hardly get a word in these days. But in the specialist sciences? The interviewers simply trot along lamely, asking anodyne questions, and allowing duplicitous answers to float away like smoke in summer sunshine, so that the scientist is confident they can say what they want. This is why pictures of imaginary planets feature in specialist magazines and newspapers, and on television, which have no basis whatever in reality. Scientists can get away with anything in this ignorant world, and dinosaur palæontologists have exploited that to the full. Most of what they tell us is fake news, stories spun to perpetuate their income and preserve their mystique.

In this book we will discover the facts about the way new knowledge was nudged from the revelations of research. We will see how philosophers came to realize how the Earth had changed, how the climate had altered, and how scientists came to understand that there had been eras populated by mighty, magnificent monsters. We will see how geology and palæontology were born and shall trace their roots from antiquity. When we consider evolution, we will encounter the hero worship of Charles Darwin – but I will also introduce you to a dozen people who came up with 'evolution' long before him. It was not his theory; it will come as a surprise to know that the word 'evolution' did not appear anywhere in Darwin's book when *On the Origin of Species* was published; neither did the phrase 'survival of the fittest' – indeed, that expression was coined by someone else, and not by Charles Darwin. We will look back at the pioneering theories of continental drift proposed by Alfred Wegener and see how the theory of plate tectonics was being rejected in the United States within living memory; and I will surprise you with a dozen investigators who had the idea long before Wegener. The untold stories of the early movies made about dinosaurs also feature in the book, and so do some of the curious novels in which dinosaurs feature prominently. Through all this complex network of developing ideas we can follow the generations of dinosaur hunters and perceive how today's conventions slowly emerged.

It took centuries before the strange fossilized remains found centuries ago on a beach, or dug out by quarrymen, were recognized for what they were. Yet I believe that the view of dinosaurs which those investigators bequeathed to us is wrong. I am going to demonstrate that the conventional conception of those gargantuan monsters pounding across the prairie, fighting like demons and roaring like banshees, is completely fanciful. I am going to show that the theory of birds as dinosaurs is not a new idea at all (it was first proposed in 1888), and I will show that it is absurd. There are animals very close to dinosaurs in our modern world, but they aren't anything like birds, any more than a chihuahua is a kind of kitten. You have been told so many times that an asteroid or a meteorite was the cause of the dinosaurs becoming extinct, but we will discover that even this is wrong. That can never have been the case. If so, then all other reptiles – the crocodiles and tortoises, lizards and alligators, turtles and snakes – would have vanished at the same time.

In this book we will see how the names of dinosaurs were derived, and discover something of the times in which they lived. I have set down imperial and metric measurements, though have used 'ton' throughout. The ton, short ton and metric ton (or tonne) are all within 10 per cent of each other, and all the estimates of the weight of dinosaurs are approximate, some wildly so; therefore sticking to the single unit is sensible.

Why did I not wish to publish these investigations? My assumption was that some sensible palæontologist, somewhere, would draw the same conclusions; yet they didn't. It was only when the global community of palæontologists advised on a series of updated television documentaries perpetuating these myths that I finally felt the time had come to say so. Even then I held back and delayed publishing anything. In the event, I announced my own alternative conclusions in a modest magazine article, only to find all my research universally condemned by the entire world of dinosaur palæontology. So, can an individual single voice revolutionize an entire modern scientific discipline? Could one person, in this modern world, challenge a major branch of science and show that all its protagonists are mistaken? Can a single scientist, in any field, still show that everybody else is wrong?

You tell me.

1

DINOSAURS AND THE ANCIENTS

You cannot escape from dinosaurs. They are everywhere; indeed, you are probably sitting on top of one as you read these words. Fossil dinosaurs are abundant wherever there are Mesozoic sedimentary rocks, and they occur in every continent, even the Antarctic. Everywhere there are books, games, movies and television documentaries; since Sheryl Leach created Barney the Dinosaur in 1992, it has become one of the most famous (or notorious) new cartoon characters, and you can wade through pages of dinosaur toys online. For ten dollars, you can pick up a frightening pair of foam rubber dinosaur claws that, assuming you do not have relatives of a nervous disposition, you may wear like gloves. Dinosaurs dominate the media; indeed, the first popular cartoon character in the history of movies was a dinosaur. It was not Mickey Mouse (who began life as Mortimer), or Betty Boop, or even Felix the Cat. They were all released in the 1920s and 1930s, whereas a dinosaur named Gertie had become the world's first famous cartoon character in a film that was released before World War I. Gertie was a smiling, endearing sauropod, created by Winsor McCay who had enrolled to study art in Chicago in 1899. He released his movie *Gertie the Dinosaur* at the Palace Theater, Chicago, on February 8, 1914, and soon it became a nationwide hit. His cartoon character had been drawn from the skeleton of a brontosaur in the American Museum of Natural History (A.M.N.H). I have lectured in that historic venue and have seen those dramatic dinosaur displays. Truly, they are awesome. Gertie was portrayed with

A dinosaur named Gertie was the world's first popular cartoon character, launched in 1914 by Winsor McCay of Chicago. The backgrounds were drawn by John A. Fitzsimmons, and McCay based the dinosaur on popular accounts of a *Brontosaurus*.

vivid realism, with her tail resting on the ground and dragging along behind her, just as you'd expect, much like a present-day crocodile.

More recently we have seen those stunning digital re-creations in the *Jurassic Park* movies, where huge sauropod dinosaurs rear up for leaves, standing on their huge hindlegs, then hit the ground with their forelimbs to create an earth-shattering crunch. Disney even released a film about a friendly dinosaur, which had facial musculature unlike that of any reptile and which made doe-eyed expressions of maternal devotion as she smiled affectionately at her young. On television, we have become accustomed to dinosaurs throwing up clouds of dust as they pound across the arid scrub, we have seen a planet populated by dinosaurs, and have watched people walking with them. Yet these are all wrong. The evolution of dinosaurs has been misconstrued. In this book, we will take a new look at the latest evidence about dinosaurs and find that they were very different from everything that you have been told before.

The standard books will tell you that the study of dinosaurs began when Sir Richard Owen coined the term in 1842, but dinosaurs were

Conventional portrayals of herbivorous dinosaurs show them in a familiar modern landscape, usually in a landscape of desert or scrub. Scientific evidence shows this cannot be the case: they consumed huge amounts of lush vegetation.

actually discovered thousands of years before that. They even feature in rituals that date back to prehistoric times, and their fossils have been known to scientists for centuries. Although palæontology first became popular in Queen Victoria's England, more than 1,000 different fossil species had already been identified when she was still a young monarch, long before the word 'dinosaur' had even been coined. When the pioneering dinosaur hunters appeared, some were daredevil characters (one of whom is said to have inspired the character of Indiana Jones), others were quiet and modest geologists who hated writing about their discoveries, along with crazy extroverts, thieves and saboteurs, plagiarists and commercial speculators who smuggled skeletons and became wealthy. Many are fantasists, given to wild speculation without a shred of scientific evidence to back it. Since the 1970s dinosaurs have been a hot topic of discussion, and as we shall see, new species are constantly being named. Today, fresh dinosaur sites are being discovered all around the world and we have recorded a total of some 1,850 genera of dinosaurs that lived between 245 and 66 million years ago. Where did this all start? How did we become aware that the Earth had changed – and when did people realize the true nature of fossils?

Palæontologists will tell you that China is a new hotbed of dinosaur discoveries. Research is continually revealing fresh information, and strange unheard-of dinosaurs are being recorded. Of course, this is true; but it is not as new as people say. The Chinese have known about dinosaurs since the Stone Age. The fossilized skeletons were taken as the remains of gigantic monsters, and it is those that gave rise to the legendary depictions of dragons. Dragons were real – they were dinosaurs. Chinese medicine believes in the administration of tinctures from fearsome creatures to heal and invigorate humans. Thus, just as the bones of present-day predators – like brown bears and tigers – have been used for thousands of years in traditional Chinese medicine, so have the bones of mighty dinosaurs. The Chinese for dinosaur is *kǒnglóng* (恐龍), meaning 'terrible dragon', and their existence was written about by Hua Yang Guo Zhi in the Western Jin Dynasty (AD 265–316). Since the fossil remains of dinosaurs showed that they were the most powerful of all creatures, their bones would logically provide the strongest cure. The tradition persists to this day, and many village communities still regard dinosaur skeletons as the remains of real dragons. Reporter Kevin Holden Platt writes that, during a recent palæontological dig in Henan Province headed by Dong Zhiming of the Institute of Vertebrate Paleontology and Paleoanthropology in Beijing, Dong told of villagers who were said to be using dragon bones in their homeopathic medicines. When he investigated, he found that the bones they were using were the petrified remains of gigantic sauropod dinosaurs. The tinctures were used to treat dizziness and cramp, and were also applied to help wounds heal.[1]

In the street markets of today's China, fossilized teeth and claws of dinosaurs are sold in markets where they are described as being those of dragons. Many families own them, sometimes as curios, sometimes as charms, believing them to be from genuine dragons. Not only did this view persist among the rural communities, but such legends still linger among some city-dwellers. Turn to the chinahighlights website and you will see that their section on dragons begins by reassuring readers that dragons are not actually real.[2]

To Western eyes this is as absurd as a written reminder that elves and fairies are merely imaginary. Yet this reminds us how powerful are the age-old legends of dragons in Chinese eyes. Western myths about dragons may also have originated from the discovery of huge fossils in Europe, though it is also possible that the legends spread from China along the ancient Silk Road.

Chinese traditions tell that it was a dragon that sowed the seed of their race. Thousands of years ago, it is said that a tribal leader, Yandi (炎帝), was born out of his mother's telepathic communication with a mighty dragon. Huangdi (皇帝 the yellow Emperor) and the dragon launched the prelude to the Chinese people when Yandi became the Emperor's deputy. So the ancient Chinese took to referring to themselves as originating with Yandi and Huangdi, as well as being descendants of the Chinese dragon. The dragon is first recorded in Chinese archæology in the Xinglongwa culture, which dates back more than 7,000 years. Sites excavated from Liangzhu, and from the Yangshao era in Xi'an, include clay pots bearing dragon motifs, and a

Remarkable dinosaur-like creatures can be seen on an ancient seal carved from jasper 5,500 years in present-day Iraq, now in the Louvre in Paris. Such images have led people wrongly to believe that our predecessors were acquainted with dinosaurs.

Xishuipo burial plot in Puyang from the Yangshao people reveals a skeleton of a human flanked by mosaics made from seashells, with a tiger on one side and a dinosaur dragon on the other. The Chinese were the first to record their impression of dinosaur fossils, though they were not alone.[3]

From ancient Mesopotamia comes an exquisite seal made some 5,500 years ago that seems to show a dinosaur. It is a small cylindrical seal carved from green jasper and shaped like a barrel, which could be rolled across wax or clay to leave an impression. This example was excavated at Uruk in Mesopotamia (now Iraq) and is in the collections of The Louvre in Paris. Uruk was once a large and advanced civilization for its time, with impressive buildings and a complex hierarchical social structure. The creatures that were engraved into the surface of their seals often represent domesticated or wild animals, though this example is unusual. It seems to have the appearance of a sauropod dinosaur, and could perhaps have been inspired by a fossil. We know that fossils were known to the ancients, because of writings that date back 1,000 years. The *Book of Healing* (in Arabic: باتک ءافشلا) was written by the Persian philosopher Ibn Sīnā (Persian: انیس نبا), who is known to present-day Western scholars as Avicenna. His is a remarkable book, covering natural history and mathematics, astronomy and even music, and was written between AD 1010 and 1020. The work is not all his, of course; this was meant as a compendium of current knowledge and contains much information from the ancient Greek writers including Aristotle and Ptolemy, together with the findings of other Persian and Arabic writers. He wrote of fossils as if they were familiar objects, and believed that fossilization occurred when subsidence caused the release of a 'petrifying virtue' that subtly transformed substances into stone. He felt that there was nothing surprising about this; it was no more remarkable than the 'transformation of the waters,' he wrote.

Notions of the movement of the Earth's surface and changes with geological time were part of the common currency in the ancient Middle East, though they did not emerge in European philosophy until they were expounded by Magnus Albertus, a physician and polymath, born in 1200, who became widely regarded as the greatest

German mind of the Middle Ages. He also wrote of fossils, saying that the rocks around Paris were a rich source of 'shells shaped like the moon' that were enveloped by viscous mud and were preserved by the 'dryness of the stone'.[4] For centuries in the West, the occurrence of fossilized seashells on raised ground was taken as evidence of the biblical flood.

Travel now across the world to Cambodia, where the Khmer people live, and at Angkor Wat you will find an ancient image of a dinosaur. There is a stegosaur carved into a wall in the temple of Ta Prohm which was constructed on the orders of the god-king Jayavarman VII and was dedicated in AD 1186. Surviving records show that more than 12,000 people lived in the temple compound at its peak, including 18 high priests and 615 dancers, with another 100,000 villagers dwelling nearby, trading with the temple authorities and providing goods, food and services. Carved into the temple walls are numerous symbolic images, and they were protected by being overgrown with jungle vegetation for centuries so that the building, even though penetrated by massive tree roots, escaped being restored by overeager Europeans. The stegosaur carving has been cited by creationists to show that humans were acquainted with living dinosaurs. The likeness, they say, is anatomically correct – but it isn't. A real *Stegosaurus* had a small head and a pointed, spiked tail; the carving at Ta Prohm has a distinctive, larger, head and there is no sign of the typical stegosaurian spiked tail. The dorsal plates are vividly carved, but a living stegosaur had two rows of plates that were more numerous than in the carving. This temple decoration was emphatically not carved by someone who had a living dinosaur as the reference for the image. They could, however, have seen petrified remains. The fossil of a *Stegosaurus* trapped in limestone strata often reveals only one set of dorsal plates, and it is perfectly reasonable to assume that the head (or the tail) was absent. Stegosaurs had proportionately tiny heads, and the skulls of fossilized dinosaurs are usually missing. Partial fossils are far more abundant than complete dinosaur skeletons, and it is easy to see how an ancient sculptor would have invented a head for his carving. Some present-day amateurs claim that the large plates along the back of this carving 'more closely resemble leaves' and they try to assert that the Ta Prohm carving is 'a boar or rhinoceros against a leafy background'. Like so much scholarly speculation about dinosaurs, this is fanciful. The evidence offers nothing to suggest this is right.[5]

It has even been alleged that there are no stegosaur fossils in Cambodia by which the carving could have been inspired, but this ignores several realities. First, there are stegosaur skeletons all around the world and they are widespread. Political and academic instability for many years led to a failure for palæontology to develop in Cambodia, and important fossil finds are only now being discovered. Cambodia today is not what it once was. The ancient Khmer empire used to occupy part of Thailand and a great swathe of present-day Vietnam and extended up across today's Laos. All this is an area now known to be rich in dinosaur fossils including *Stegosaurus*, and fossils have been recorded at Angkor Wat, near the site of the ancient temple. There are plenty of opportunities for that temple stonecarver to have known a well-preserved fossilized stegosaur skeleton, some 700 years

There is a stegosaur carving at the Ta Prohm temple in Angkor Wat, Cambodia, which opened in 1186 AD. Some commentators have concluded that the plates along the back are 'leaves', but a fossilized skeleton may have been the inspiration.

before that dinosaur was first revealed to scientists in the Western world.

There are other historic representations of a *Stegosaurus*. We can travel back across the globe and find an example in the legacy of Father Carlos Crespi Croci, born in Italy in 1891. He studied anthropology at the University of Milan, entered the priesthood, and in 1923 he was sent as a missionary to the small city of Cuenca in Ecuador. He worked tirelessly for the indigenous people, establishing an orphanage and school and helping the poor. Croci is commemorated at the church of Maria Auxiliadora in Cuenca by a statue of him assisting a little child. He died in 1982, and the local residents who knew him remember him with great affection. He also attracted global publicity with a series of curious relics, some of them allegedly coming from North Africa, and was surrounded by allegations of fakery and subterfuge. He was the subject of books and television documentaries that argued the case, and many of his objects were exhibited in a museum. A disastrous fire destroyed the building (and many of the artifacts) in 1962, and controversy has persisted ever since over the fate of the treasures that were recovered from the smoking ruins. One item in that collection, however, was a stone carving that does not seem to be a fake. It shows a creature with plates along its back like a stegosaur, and could conceivably have been inspired by a fossilized skeleton.

Fossil dinosaurs were known to the ancient people of South and Central America. When the conquistador Hernán Cortés de Monroy y Pizarro Altamirano, Marquis of the Valley of Oaxaca, and his Spanish troops travelled to explore Mexico in 1519, he was presented, not only with gold and precious stones, but with huge fossilized bones. Missionaries who ventured deep into Ecuador and Peru were shown excavations of fossils, and Georges Cuvier (see p. 34) referred to them as evidence of creatures now extinct. When Cortés reached Tlaxcalteca, the elders brought out some huge fossilized limb bones, and Bernal Diaz del Castillo, a captain under Cortés, later wrote about these bones resembling the remains of gigantic humans. He said that one thigh bone (femur) was as tall as a man, which is typical of the femur of a sauropod dinosaur. Diaz recorded that the largest of all the bones was sent in a ship back to Spain for the interest of the king, though there are no current records of where it now might be.

In Mexico you will find a terracotta model reminiscent of an ankylosaur. These tortoise-like dinosaurs lived in the Cretaceous period between 68 and 66 million years ago, and ranged between the size of a tortoise and a small car. Fossils have been found by palæontologists in America and they are very similar to the terracotta model. This curious artifact is approximately 2,000 years old and was made by the artisans of the Jalisco culture, who flourished along the Pacific west coast of Mexico. Stegosaur-like models were also discovered by an expatriate German merchant named Waldemar Julsrud in Acámbaro, Mexico, in July 1944 while out riding his horse. By chance, he spotted some rock carvings and eventually uncovered a range of clay figu-

The Ica Stones of Peru, covered with detailed dinosaur carvings, were collected by a doctor who used to purchase them from a farmer, Basilio Uschuya, believing them to be genuine. It later transpired that Uschuya had carved them all himself.

rines. Julsrud was conversant with Aztec, Toltec, Mayan and Inca artifacts and recorded that these seemed to be different. Julsrud agreed to hire a Mexican farmer, Odilon Tinajero, to dig in the area and offered to pay him one peso (then worth about 12 US cents) for each object he found. Within weeks, he had piles of these figurines.[6]

Among the models were depictions of stegosaurs and some other dinosaurs. Eventually more than 30,000 of the artifacts were excavated. Considerable controversy surrounded these objects, and scientists dated them using thermoluminescence. These results suggested that they were 4,500 years old, though more recent investigations have claimed that the results were inaccurate, and authorities now insist that the objects are fakes dating from the 1930s. These items were given a museum of their own, and it is open to this day. Most people regard the clay figures as curious fakes,[7] though others are convinced that the original dating experiments are valid, and that these are prehistoric relics from an ancient Mexican civilization. There is also a vociferous cohort of enthusiastic supporters of the view that dinosaurs and humans coexisted, and, to these people, the figurines provide the evidence they seek.[8]

There are other intriguing images that depict humans and dinosaurs together, and some of the carvings from Ica, south of Lima in Peru, even show dinosaurs attacking a human hunter. The images are cut into the surface of volcanic rock of what we now call Ica Stones, rounded nodules of andesite measuring less than 1 foot (30 cm) across. Andesite is a stony mineral that can easily be carved, and these stones purport to depict pre-Columbian scenes and bear symbols from the ancient Inca, Paracas, Nazca and Tiwanaku peoples, as well as the Ica. In the 1930s a Peruvian doctor named Darquea began to collect beautifully carved stones from the town of Ica. His son, Javier Cabrera Darquea, was fascinated by the delicate carvings and began a quest for more. He based his collection on 350 Ica Stones purchased from Carlos and Pablo Soldi, brothers who marketed pre-Columbian artifacts to archæologists and collectors. Through them Javier Darquea met a farmer, Basilio Uschuya, who sold him stones at regular intervals until he had amassed a collection of some 11,000 different examples. He published a book on the message of the stones of Ica,

Dinosaurs fired from clay were found by Waldemar Julsrud in Acámbaro, Mexico, in July 1944. He thought they were from the preclassical Chupicuaro Culture (about 2,000 years ago), but they proved to be forgeries made by a farmer, Odilon Tinajero.

and went on to establish a museum to display some of his impressive collection.[9]

The museum is still there, featuring a range of unmistakable dinosaurs, and many visitors leave impressed by his claims that humans are at least 400 million years old. However, any claim that the artifacts might be authentic disappeared when Uschuya finally conceded that the carvings were fakes. He and a farming friend, Irma Gutierrez de Aparcana, admitted that they had forged the images by copying dinosaur pictures from comics and magazines. A BBC television crew were said to have visited the farmers and paid Uschuya to produce an Ica Stone while they filmed. He cut the patterns with a dentist's drill and then gave the rock an authentic-looking patina by baking it in cow dung. Many people wondered why he had owned up, but the Peruvian authorities at the time were beginning to enforce the regulated marketing of pre-Columbian artifacts, and – if Uschuya had truly been selling genuine relics – he could have been arrested and put on trial. Conceding that they were fakes was his guarantee against prosecution. Basilio Uschuya was quoted as saying, 'Carving stones is an easier way of making a living than farming the land,' while Ken Feder, author of a 2010 book on dubious archæology, wrote: 'The Ica Stones are not the most sophisticated of the archæological hoaxes, but they certainly rank up there as the most preposterous.'[10] It would have been an easy matter to resolve: the microscope would show in an instant whether the images had been engraved by a primitive tool or cut with a dentist's drill, and microchemical analysis would as easily detect the presence of cow manure on the surface of a stone.

Some similar examples have not been dismissed as fakes. In 1971, at Girifalco in Calabria, southern Italy, a landslip after a 20-hour deluge revealed a cache of artifacts from of a pre-Greek civilization. A lawyer named Mario Tolone Azzariti reported that he had found some terracotta statues, one of which was a model very like a stegosaur. It measures some 7 inches (18 cm) long and shows solid, strong legs (different from those of a present-day lizard). This seems to be a relic of the Stone Age – though I am not aware that it has ever been dated – and may well be the result of inspiration by a fossilized *Stegosaurus*.

The indigenous peoples of the U.S. knew about fossil dinosaurs for thousands of years. There is an ancient saga among the Delaware people, who inhabited what became New Jersey and Pennsylvania, telling of a party of hunters who returned to their village with a huge, ancient bone which they said had come from a massive monster. This fearsome creature was said to massacre people. A ritual was developed, involving burning tobacco with small fragments of this massive bone, hoping that this would ensure safety from the monster, good hunting for the future, and a long life for all. Fossils found in the area include a range of dinosaurs that we shall encounter later, including *Cœlosaurus*, *Dryptosaurus*, *Ankylosaurus* and *Hadrosaurus*

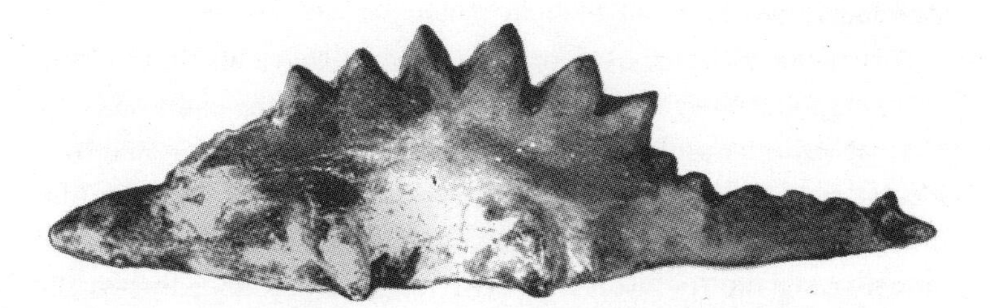

In 1971 in southern Italy, a lawyer named Mario Tolone Azzariti claimed to have found terracotta models, one of which was a like a stegosaur. It measures 7 inches (18 cm) long. Perhaps this was inspired by a fossilized skeleton.

(subsequently, a hadrosaur was to become the first dinosaur ever to be scientifically investigated in America). As we shall discover later, the Cheyenne people taught that a mythical animal named Ahke once lived in the prairies. These were gigantic god-like bison whose remains had been turned into stone. It seems likely that fossils of the horned dinosaur *Triceratops* were known since ancient times, and these gave birth to the legend.

Just as the indigenous inhabitants of North America maintain a culture that dates back to prehistory, Stone-Age traditions are also perpetuated by Australian Aborigines. They too have ancient legends about dinosaurs – though there are no fossilized bones to be found. Their dreamtime stories stem from an abundance of dinosaur footprints, notably in northwestern Australia. There are extensive exposures of sandstone on the Dampier Peninsula in the Kimberley region, where innumerable dinosaur footprints are to be found. There are extensive trackways that stretch from Roebuck Bay near Broome north to Cape Leveque; that's at least 125 miles (200 km). The culture extends inland for at least 100 miles (160 km) across the scrub. So important are these finds that a great swathe of intertidal coastline along the Dampier Peninsula coastline has recently been declared a heritage site.[11]

The 130-million-year-old rocky strata along the beach, known to geologists as the Broome Sandstone, are marked with countless footprints of three-toed dinosaurs, and these clear tracks have played a part in the local culture for thousands, perhaps tens of thousands, of years. The strange footprints in these tracks are clearly recorded in their local legends.[12]

Australian indigenous people everywhere have an oral culture featuring song cycles that trace in chanted refrains the paths taken by gigantic supernatural beings. Everything in nature is denoted by a specific song – features of the landscape, creatures of the wild, life-sustaining and medicinal flowers, and the stars in the sky – all perpetuating the tradition that existence was predicated upon the legends in these songs.[13]

The care of both the song cycles and the land – known as Country – rests in the hands of custodians they call Maja, who are selected from the community not because of possessions or family connections, as is common in the West, but purely because of their wisdom and personality. There are different songlines in different places; along the Dampier Peninsula coast they are called *ululong*, whereas the song cycle that extends east is *dabber dabber goon*, which reaches past Uluru (which we used to call Ayer's Rock) and on to Australia's Pacific coast. At the core of these ancient songs is a mystical being known as Marala, a spirit that laid down the law and established the codes of conduct and morality. The legends about Marala are endless; many of them have been documented by anthropologists, though most remain secret among the tribes and are never vouchsafed to outsiders. One of the songs tells about Marala fighting with Warragunna, the eagle-man, and goes on for hours. Now, it is tempting to dismiss all this as the stuff of legend; as the wild imaginings of a religious people who are now out of their time. Yet their religion has a lot more going for it than any of ours, for they have physical evidence. Marala is a gigantic emu man, and you can see precisely where he walked. There, in the rocks, are the three-toed footprints. There are even some stony giants, weathered in the fierce Sun, that look like petrified monsters. The Aboriginals even know where Marala sat down with Warragunna, because there are the signs of the eagle-man's feathers clearly preserved in the solid rock. We, with our palæontological insights, know that the footprints of the emu man Marala are actually those of a huge three-toed dinosaur named *Megalosauropus*. We can also confidently conclude that the feathery impressions of that gigantic eagle are really the fossilized remains of bennettitalean plants, related to the present-day cycads from which we obtain sago, for those leaves really do look like the feathers of a gigantic eagle. The thought processes of Aboriginal people are very different to ours and are hard to follow. A statement by an Aboriginal elder named Lulu from the Gularabulu tribe of the Nyikina people, an ancient community that still lives in the Stone-Age traditions, stated that: 'The Country now comes from Bugarri-Garriand [dreamtime] and it was made by all the dreamtime ancestors, who left their tracks and statues behind and gave us our law. We still follow that law, which tells us how to look after the Country and how to keep it alive.'[14]

So here we have an entire way of life and an ancient culture that are based on detailed knowledge of dinosaur footprints and the fossilized remains of Cretaceous plants. Australian Aborigines know their Country intimately, and knew about the remains left by prehistoric monstrous beings thousands of years earlier than we did in the West. The authorities now take these remains seriously. In 2000, an Aborigine from Broome named Michael Latham admitted cutting stegosaur footprints from the sandstone strata, when the tide was low, using an angle-grinder. He was jailed for two years.

The first records from Europeans about the Australian dinosaur tracks were written around 1900 when an immigrant from Ireland, Daisy Bates, spent three months on a mission station in Aboriginal territory. She later returned to the Roebuck Plains Station with her husband and son, and devoted herself to a study of the indigenous coastal communities. She observed many of the areas where footprints were visible. Then, in 1935, Catherine Milner and her young twin daughters discovered some on their own. They wrote later that they had come across the footprints one morning when the tide was low, and they had looked as if 'whatever had made them had just passed by, so clear and perfect they were.' Little wonder the tracks were regarded by the Aboriginal people as clear evidence of events.[15]

The only fossilized remains of actual dinosaurs in Australia are of occasional scattered bones and teeth; yet there is now a growing understanding of the variety of the dinosaur population derived from the fossilized trackways set in stone. They are usually left undisturbed for visitors to enjoy, though this inevitably exposes them to damage. Trackways left by theropod dinosaurs were discovered at the eastern side of Australia when palæontologists first discovered the footprints of a theropod dinosaur in tidal strata at Flat Rocks, Victoria, in 2006. Thousands of scattered bones and teeth have been found in the area, and the dinosaur footprints – measuring about 1 foot (30 cm) across – were left intact for visitors; however, there was no sign pointing them out. In December 2017 someone took a hammer and chisel and chopped out the toes, leaving them scattered nearby. The local Park Ranger, Brian Martin, said: 'They would need to know exactly where it is to find it. Most people quite easily walk right past it,' he said. This time the vandals didn't.

We are beginning to understand how huge some of these monsters were. The massive brachiosaurs which appear so frequently in documentaries and pictures about dinosaurs were colossal creatures measuring about 85 feet (26 metres) long and weighing at least 50 tons, their footprints measuring less than 3 feet (90 cm) in length. Similarly, some unprecedently gigantic footprints were discovered in Mongolia in 2016, each measuring 3 feet 6 inches (1.06 metres) long, which caused considerable surprise among palæontologists. There are Australian huge dinosaur footprints that measure 5 feet 7 inches (1.7 metres) from heel to toe. Nothing so vast has ever been discovered elsewhere.[16]

Fossilized shells had been known for centuries and, as we have seen, they were conventionally interpreted as a natural consequence of the biblical flood. They were written about as radical new thinkers appeared on the European scene, and they caught the attention of Leonardo da Vinci around 1508, two centuries before the Enlightenment. Leonardo was not inclined to believe that they were the aftermath of the biblical accounts of the Noahic flood. Instead, he thought that their presence showed that the surface of the Earth had changed over time, and the fossilized remains represented an earlier, watery phase of the Earth's ancient history. A generation later, a French Huguenot hydraulics engineer and ceramicist named Bernard Palissy wrote on the origins of fossils. He too believed that they were not the result of a flood, but had formed naturally in a manner reminiscent of that recorded by Avicenna. Palissy thought that mineral-rich water developed 'congelative properties' that transformed once-living creatures to stone. The first report of fossilized bones in Europe dates from 1605, when a British expatriate theologian named Richard Verstegan (living in Antwerp) became interested in fossils and, for the first time, recognized bones and teeth for what they were.[17]

Verstegan portrayed ichthyosaur vertebræ in a book, though he interpreted them as the remains of fish, which he took as evidence that Britain and mainland Europe were once connected.[18]

Fossils were collected by many enthusiasts during this period, though the first time they were scientifically described was by Robert Hooke in 1665. Hooke was a remarkable polymath and is best known for his role as the founding father of the science of the microscope. In his large folio volume *Micrographia*, published in 1665, Hooke devoted a section to fossils. His microscope showed him that fossilized wood had a structure identical to that of wood taken from a tree nearby, and he described the fossil in terms that fit perfectly with our modern understanding.

Hooke also featured a fine image of the microscopic spheres that comprise limestone. His drawing, captioned 'Kettering-stone', was described in his text: 'This stone is brought from Kettering in Northampton-shire, and digg'd out of a Quarry, as I am inform'd.'[19]

His specimen was demonstrated to the Fellows of the Royal Society on Sunday, April 15, 1663, and attracted much attention. That was an auspicious date; his other demonstration that day was of thin sections of cork. Hooke observed that the specimen (from a wine bottle) showed itself to comprise numerous small boxes, the cell walls of the cork. The room-like nature of each component led Hooke to call them 'cells' – and this is the term that has come down to us today for all the cells that we see in living organisms.[20]

There were other fossils in the Royal Society collections at the time, and examples ranging from fossilized teeth to skeletons of fish were

The Reaſons of all which *Phænomena* ſeem to be,

That this *petrify'd* Wood having lain in ſome place where it was well ſoak'd with *petrifying* water (that is, ſuch a water as is well *impregnated* with ſtony and earthy particles) did by degrees ſeparate, either by ſtraining and *filtration*, or perhaps, by *precipitation*, *cohesion* or *coagulation*, abundance of ſtony particles from the permeating water, which ſtony particles, being by means of the fluid *vehicle* convey'd, not onely into the *Microſcopical* pores, and ſo perfectly ſtoping them up, but alſo into the pores or *interſtitia*, which may, perhaps, be even in the texture or *Schematiſme* of that part of the Wood, which, through the *Microſcope*, appears moſt ſolid, do thereby ſo augment the weight of the Wood, as to make it above three times heavier then water, and perhaps, ſix times as heavie as it was when Wood.

British philosopher Robert Hooke announced the first reasoned account of the process of fossilization in his book *Micrographia*, published in 1665. He thought that organic remains became filled with 'stony particles' and thus became petrified.

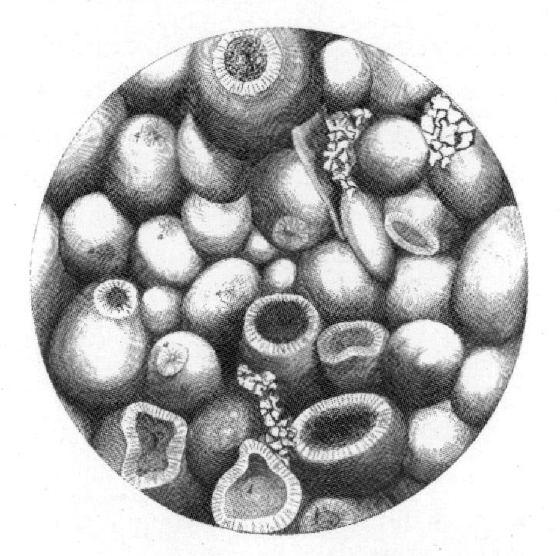

Under his microscope, Hooke could study the fractured surface of this Middle Jurassic oolite rock. He was the first to make detailed studies of rock formations and his conclusions about the processes of fossilization proved to be influential.

included in the Society's catalogue of rarities, with descriptions showing that the petrification of organic remains was understood as a natural process.[21]

The collecting of fossils soon became a popular hobby for the learned classes. In 1695 John Woodward was appointed the first Professor of Geology at Cambridge University. He recognized the widespread occurrence of fossils and taught that they had been laid down by floods to form successive strata. Woodward became an avid collector of fossils and minerals, and eventually amassed over 9,000 specimens. He donated them all to the University, where they became the nucleus of what later became the Sedgwick Museum. In 1699 Edward Lhuyd published accurate engravings of ichthyosaur bones, vertebræ and limb elements in a book along with fossilized shark's teeth and sea urchins, and a variety of petrified seashells and ferns.[22]

Lhuyd was appointed assistant to Robert Plot, a graduate of Magdalen Hall, Oxford, who had been appointed Professor of Chemistry and the first Keeper of the Ashmolean Museum in March 1683. It was Plot who published the first book to feature a picture of a dinosaur bone, *The Natural History of Oxfordshire,*[23] in June 1677. He did not know what it was, and believed it to be a fossilized thigh-bone

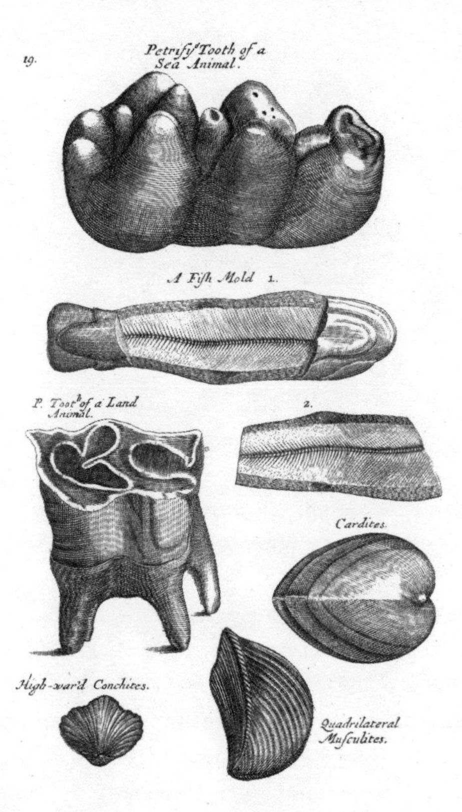

Nehemiah Grew published pictures including 'Animal Bodies Petrify'd' as table 19 in the book *Musæum Regalis Societatis*, or, *A catalogue and description of the Natural and Artificial Rarities* ..., which the Royal Society published in 1681.

from a biblical giant, though it looked like a fossilized scrotum. Lhuyd went on to succeed Plot as Keeper.

By the end of the seventeenth century, popular fossil finds were becoming more familiar in Britain and many had been well documented. In Switzerland, as in England, fossilized bones were always assumed to be human, and so when naturalist Johann Jakob Scheuchzer described two ichthyosaur vertebræ in 1708 he identified them as being the mortal remains of a person who had been drowned in the biblical flood and named them *Homo diluvia tristis testi* ('sad evidence of man in the floods').

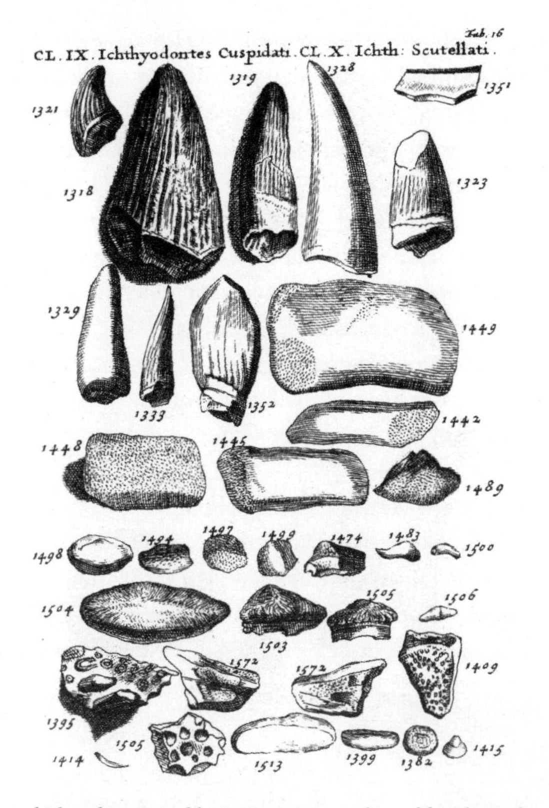

In 1697 Edward Lhuyd engraved his pioneering studies of fossils, and was encouraged by John Ray to publish them. Failing to find a publisher, he solicited subscribers and eventually had enough money to publish *Lithophylacii Brittannici Ichnographia* in 1699.

Then, in 1719, palæontology was quietly born. A clergyman scientist named William Stukeley, a studious man from a legal family, set out for the first time to develop the systematic study of archæological remains. Stukeley came from Holbeach, in Lincolnshire, and was a friend of Isaac Newton. Because of his scientific accomplishments, he was elected a Fellow of the Royal Society and in 1718 became the first Secretary of the Society of Antiquaries in London. The great English prehistoric monuments at Avebury and Stonehenge had long been sites of cultural interest, and indeed Stukeley himself was fascinated by the Druids, but neither site had been academically investigated until Stukeley showed interest in their origins. He examined them systematically, and this is how he came to be regarded as a father of archæology. In 1718 Stukeley visited the Reverend John South, Rector of Elston, near Newark on the boundary with Nottinghamshire, some 50 miles (80 km) from his home. In the nearby quarries of Fulbeck, Blue Lias rock was being mined and a fossil plesiosaur came to light. It was taken home by South, and became known locally as the Elston crocodile. The owner of Elston Hall was Robert Darwin, himself an FRS, and destined to become the great-grandfather of Charles Darwin. It was hoped that Stukeley could examine the ancient skeleton and find out more about it. Stukeley's account of the discovery is meticulous:

> There are Sixteen Vertebræ of the Back and Loyns very plain and distinct, with their Processes and intermediate Cartilages, Nine whole or partial Ribs of the Left-side, the Os Sacrum, Ilium in situ, and two Thigh-Bones displac'd a little, the Beginnings of the Tibia and Fibula of the Right-Leg; on one Corner there seem to be the

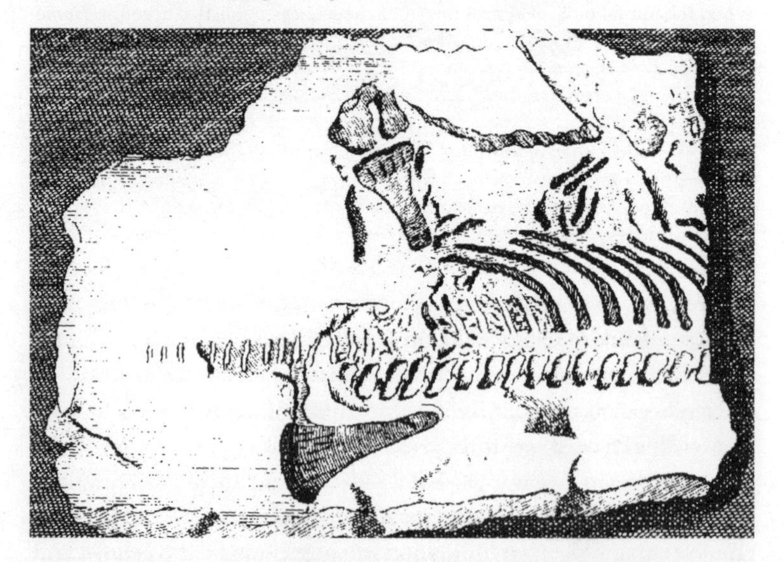

Robert Darwin, ancestor of Charles, discovered this skeleton of *Plesiosaurus dolichodeirus* in 1718. It was displayed in the local vicarage as the bones of a sinner who had died in the great flood, before Darwin recognized it as a fossil.

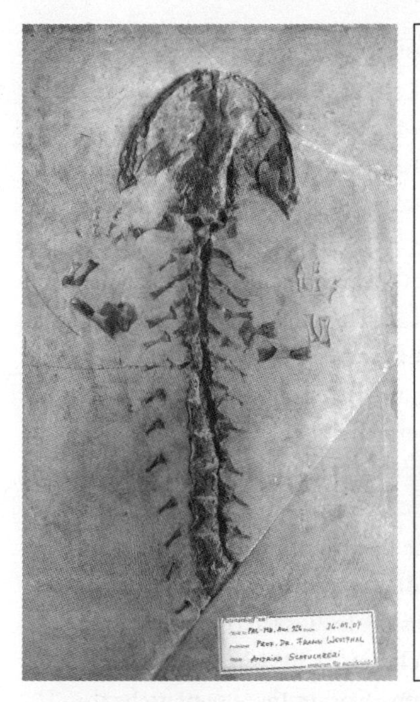

When Johann Jakob Scheuchzer of Zürich was shown this fossilized salamander, he believed it to be the remains of a preserved human victim of the Noahic flood. It is in the collections of the Museum für Naturkunde in Berlin.

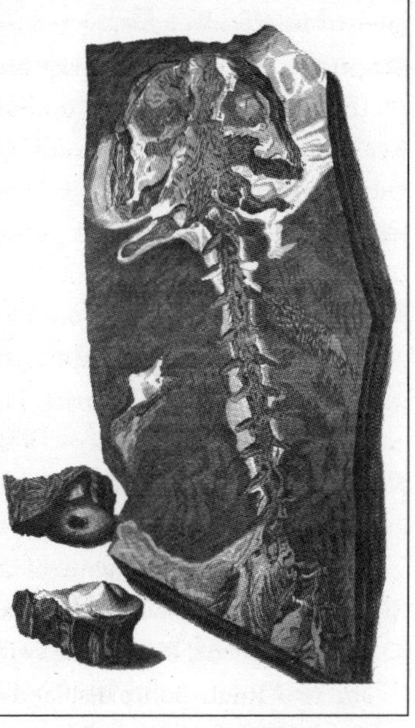

Scheuchzer's published account, *Homo diluvii testis* (A Flood Man), appeared in 1726. Conventional interpretations of these fossils (like Robert Darwin's 'Elston crocodile') were taken as tangible evidence of biblical realities.

> Vestigia of a Foot with four of the five Toes, and a little way off an entire Toe, now left perfect in the Stone – there are no less than Eleven Joints of the Tail, and the Cartilages between them of a White Colour distinguishable from the rest. Sir Hans Sloan has a Fish-Sceleton, amongst his immense Treasure of Curiosities, found near this Place, given by the Duke of Rutland.

Stukeley discussed all this with his Royal Society friends, and his curious specimen was taken to London for inspection by the Fellows. An account of the discovery was presented to a meeting of the Fellows by Robert Darwin on December 11, 1718, and Stukeley was subsequently invited to publish a formal account in the Society's journal. This was the first scientific description of a prehistoric reptile skeleton, which in 1824 was named *Plesiosaurus dolichodeirus*, and today the remains are on display at the Natural History Museum in London. Stukeley's paper was a landmark; the era of palæontology was beginning to dawn.[24]

The swiss naturalist, Johann Scheuchzer, wrote extensively of his travels and also commented on fossils. He was shown a remarkable relic, a clearly preserved skeleton excavated from a quarry in Baden, and published it in 1726. Scheuchzer's interpretation, just as we would expect at the time, was that this represented a human victim of the flood, trapped forever in its stony embrace and preserved for modern man to contemplate his fleeting fate on Earth.[25]

Twenty years later, the first vertebrate fossils to be excavated in the U.S. were discovered by Charles III Le Moyne, the second Baron de Longueuil, when he was exploring the course of the Ohio River in 1739. Le Moyne, who later became Governor of Montreal, had an active military career fighting against both the British and the Iroquois, and did much exploring for the burgeoning fur trade. This part of the Ohio River is swampy, rich with lush vegetation and dotted with mineral pools and hot springs that bubble from the strata seams beneath. Near the river bank Le Moyne discovered some huge bones. Nobody knew what they were, though eventually they were identified as belonging to a mammoth. For decades, the fossils that Le Moyne excavated were known simply as the 'Ohio Animal'. The site in Kentucky is now open to the public as Big Bone Lick State Park, and the welcome signs today proclaim it as the 'Birthplace of American Vertebrate Paleontology'. The bones excavated from this region are far more recent than dinosaurs; those magnificent monsters existed between 250 and 66 million years ago, whereas the mammalian fossils Le Moyne had found are less than 80,000 years old, and some date back only 12,000 years.[26]

These became the first American fossils to be formally described when Jean-Étienne Guettard published a summary of recent discoveries in palæontology. Guettard would become one of the pioneers of this emergent discipline.[27]

In 1780 Guettard went on to publish the first mineralogical map of France, which delineated where valuable resources might be found, and which was the first publication to demonstrate the volcanic origins of the Auvergne district of central France. The book also included maps of sites of interest to fossil collectors and was recognized as a pioneering publication in the emerging science of geology.[28]

Meanwhile, in England in 1755, a naturalist in Oxford named Joshua Platt made a momentous discovery: three 'enormous' vertebræ and then a femur that, when freed from the surrounding rock, weighed 220 pounds (100 kg). Platt had a lifelong interest in fossils, and at the time was studying the nature of belemnites. Colossal bones were not something he had previously experienced, so he sent them on to a Quaker botanist, Peter Collison, who was known for his broad interests in natural history and traded with America. Nothing more was done to investigate them – and the specimens have since disappeared. Fossils were attracting a wider audience and, when his paper on belemnites was published in 1764, Platt began with these words:

> The public hath of late been agreeably entertained with description of many curious Fossils, discovered in different parts of this kingdom; but very little hath been offered with a view to ascertain their origin and formation; a point of much greater importance to a curious mind, than the most accurate descriptions, or the neatest delineations.

This was a key point: describing fossils was all very well, and drawing them, though intrinsically useful, did not reveal what they actually were. For his paper, Platt broke open the belemnites and neatly described what he found within. This was a pioneering attempt to explore the anatomy of fossils, a discipline that became a mainstay of palæontology in the centuries that followed.[29]

Another important discovery was made in England in 1766 when an ichthyosaur jaw bearing teeth was discovered in strata at Weston, near Bath. It was exhibited as the bones of a crocodile by the Society

The huge skull of a mosasaur was excavated in St. Pietersberg, near Maastricht in the Netherlands, and was later given the name of *Mosasaurus hoffmanni*. This fanciful portrayal of the event was engraved in 1799 by G.R. Levillaire.

for Promoting Natural History in 1783; meanwhile, more ichthyosaur fossils were included in the plates for a book by John Walcott published in 1779.[30]

It was in 1764 that European scholars first encountered a dinosaur skull, though nobody realized its significance. This specimen had been dug out of a chalk mine by quarrymen at St Pietersberg, near Maastricht in the Netherlands, and took the form of fragments of jaws with rows of fearsome teeth which were assumed to be the remains of a crocodile. Two years later it was purchased from the miners as a curiosity by Lieutenant Jean Baptiste Drouin; Martinus van Marum, who had opened the Teylers Museum in Haarlem in 1784, subsequently acquired it as a highlight of the museum's displays, and in 1790 he published a description of it as 'a large fish skull'.[31]

A second, similar, skull was excavated a few years later, on nearby land owned by Canon Theodorus Joannes Godding. He displayed it in his home as a curiosity until it was seen by a retired army doctor, Johann Leonard Hoffmann. Hoffmann was a keen fossil collector and decided this must be the skull of a crocodile. A friend of Hoffmann's was the collector Drouin, and together they concurred that the remains were again those of a crocodile – or possibly a whale. French revolutionary forces under Napoleon occupied Maastricht in 1794, and this fossil was traded for 600 bottles of wine

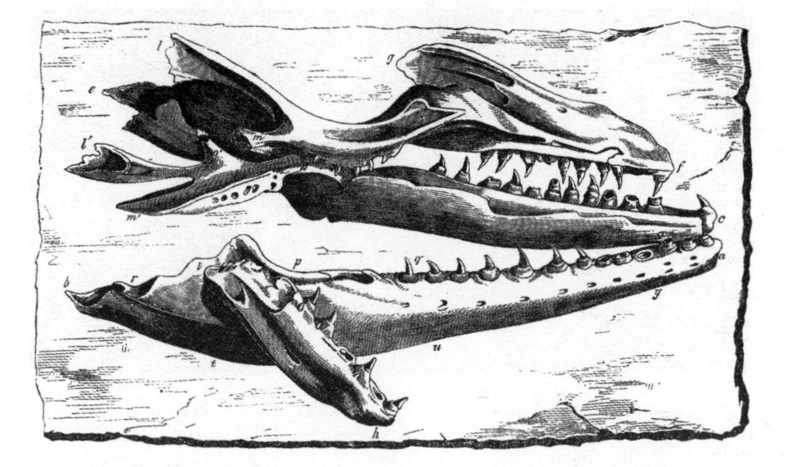

Pieter Harting published this engraving as the 'Jaws of the Mosasaur' in his book *Album der Natuur* in 1866. The original fossil was captured by the French in 1795 and taken to the Muséum National d'Histoire Naturelle in Paris, where it may still be found.

and taken as a trophy straight to Paris. Philosophers conventionally interpreted these petrified remains as coming from familiar creatures, and fossils were widely discussed by learned naturalists. Nobody imagined that fossils might be the remains of long-forgotten creatures, unknown to present-day scholarship, which had roamed the Earth in prehistoric times. It was assumed that the world then was much the same as the world of the past, though there was growing evidence that the Earth had changed over time. As long ago as 300 BC Aristotle's young protégé Theophrastus had written a work entitled *Peri Lithon* ('concerning stones'), which is the earliest work we know of that dealt with rocks. Theophrastus also noted that the draining of coastal swamps had rendered an entire area prone to freezing. He was also a pioneer of climate change, calculating that forest clearance resulted in warmer weather, because more heat could now reach the ground.[32]

The novel notion that the Earth could change over time was slowly emerging. A United Nations report quotes Vitruvius from around 50 BC, who recorded that many ancient settlements along the Anatolian peninsula in the Aegean Sea had been engulfed when changes in the

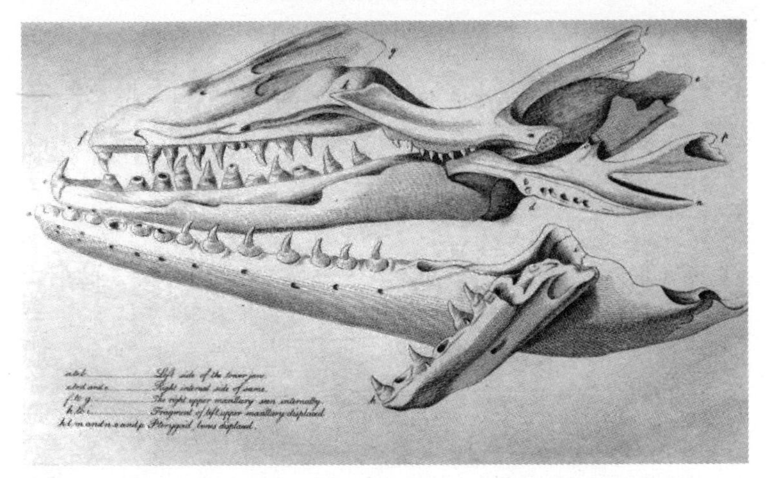

Georges Cuvier published this refined engraving as 'The Mosasaurus of Maastricht, Huge skull found in a quarry at Fort St Peter near Maastricht, Netherlands, in 1780' in his book *The Animal Kingdom* published in London by Whitaker & Co, 1830.

landscape caused the sea to encroach. This is the first known record of major changes in the Earth's surface. We now know that there are many ancient cities at the bottom of the sea – one in the Bay of Cambay, India, dates back 10,000 years, and there is another vast city on the bed of the Yucatán Channel near Cuba.[33]

Another idea that was slowly emerging was that the prehistoric climate had also changed over time. The evidence first emerged from the remote villages of Switzerland, where the local people used to say that the gigantic boulders they found scattered along the valleys were signs that glaciers had once extended further down from the mountains. Pierre Martel, an engineer from Geneva, visited these villages in 1742 and later wrote about these stories. Perhaps the landscape and the climate had once been very different. At the time, this was an unpalatable prospect.[34]

At about the same time, a Swedish mining engineer, Daniel Tilas, was studying the erratic boulders in many of the Scandinavian countries and the nations bordering the Baltic Sea, and he similarly speculated that they were brought to their present positions by glaciers that had existed in the past.[35]

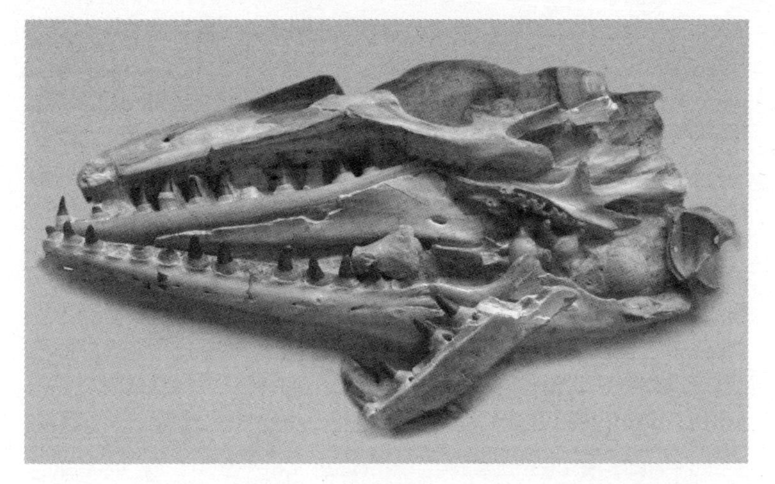

A replica of the original fossil from which the previous illustration was taken is on display in the Natural History Museum, London. The original specimen remains for the time being in the Muséum National d'Histoire Naturelle in Paris.

Matters were about to change with the revolutionary ideas of James Hutton, a Scottish farmer turned naturalist. Hutton had lectured on the structure of the Earth, and in 1785 he presented his findings in a lecture to the Royal Society of Edinburgh. He argued that many rocks were not original, primeval structures, but had been modified and subject to sedimentation and later to weathering. Hutton introduced a basic principle of modern geology, somewhat clumsily termed uniformitarianism, which states that the natural forces that we see operating on the landscape today are essentially the same as those operated in past æons. He concluded that the movement of rocks by prehistoric glaciers, like the progressive erosion of the coast, was the cause of the landscape around us. Geology was suddenly seen as relevant to the study of the present-day Earth.

The term 'geology' had first been used by Ulisse Aldrovandi in his will of 1603, and came from the Greek γῆ (*gē*, earth) and λογία (*logia*, study). Not until Hutton, however, was the subject given proper scrutiny and set on a firm scientific footing. Hutton used his principle of uniformitarianism to argue that many of the features of the present-day rocky landscape had begun at the bottom of the oceans, that many forms of rock had been produced by collections of 'loose or incoherent materials', and that these rocky layers were subject to rising, or falling, over the passage of immense periods of time. This was the first time anyone had put together the ideas fundamental to geology. In 1795, he finally published a book on his conclusions, and the new science of geology was firmly established.[36]

Several years after Hutton's work, a Swiss hunter, Jean-Pierre Perraudin, decided to look further into the massive granite boulders that lay scattered around his home in the Alpine village of Lourtier. Perraudin was a professional hunter of chamois, the pretty mountain goats from which the finest leather is obtained. As he hiked across the mountains, he noticed two key facts: not only were there massive granite rocks scattered across the valley floor, but they were very different from the rocks nearby. They didn't belong there. Secondly, he kept coming across deep longitudinal scratches and grooves carved into the valley sides. He sensibly concluded that this could mean only one thing – just as the villagers had always believed, glaciers had filled the valley in the past and had carried the boulders along, so it was these that had cut the deep grooves into the valley sides. Whenever an explanation for these curiosities had been sought in the past, philosophers said they were the evidence of the biblical flood, but Perraudin dismissed this as nonsense – he was certain that rising flood water could not move granite boulders. In 1815 he went to present his reasoned conclusions to Jean de Charpentier, born in Saxony and an enthusiast for geology, but Charpentier dismissed the idea as absurd. Undeterred, Perraudin raised the matter with a visiting engineer Ignaz Venetz, who lived in the nearby Rhône Valley, and the evidence he presented was so convincing that Venetz himself approached Charpentier and finally convinced him that the theory was right.[37]

Venetz worked further on the theory, and became convinced that – just as the hunter Perraudin had concluded – there had been eras in the past where glaciers had existed, and they had left indelible marks on the present-day landscape. He claimed that this proved that much of Europe had previously been covered with vast glaciers. At last it was becoming acceptable to say that the world's climate truly had changed over time.[38]

Interest in this subject soon came to the attention of a physician and amateur geologist, Jean Louis Rodolphe Agassiz. He was born in 1807 in Môtier, in the French-speaking part of Switzerland, and qualified in Munich, Germany. He was soon to develop an interest in palæontology and he later studied in Paris under Georges Cuvier, who was greatly impressed by Agassiz's growing knowledge of fossil fish. When Cuvier died, Agassiz saw himself as his natural successor, and later wrote an extensive work on fossil fish.[39]

As he developed his spare-time interests in natural history, Agassiz used to discuss his ideas with Karl Friedrich Schimper, a botanist turned poet and four years his senior. Schimper was a firm believer in the idea that prehistoric glaciation had created the landscape they saw around them, and he persuaded Agassiz that the views of Martel and, more recently, Perraudin, provided the evidence. Louis Agassiz devoted much of his free time to tracing changes in the Alps, monitoring snowfall and recording temperatures, and eventually he published an extensive monograph on glaciation. The title page of his book bears a vivid engraving of a glacier towering high above forest trees, and much of the contents comprises a comprehensive collection of detailed climatological tables.[40]

Agassiz owed his inspiration to Schimper, and to those who preceded him. Scholars have drawn attention to the way in which Agassiz not only failed to give them due credit, but sought actively to conceal it. In 1846, confident that his international reputation was assured, and that his mentors had been eclipsed, Agassiz emigrated to the U.S. and spent the rest of his life as a professor at Harvard.[41]

Since the revelations in the Alps that Pierre Martel documented more than 270 years ago, we have come to accept the occurrence of ice ages.

We have not always reached the right conclusions. Just 40 years ago, instead of the warmer world we are currently experiencing, people were heralding a new ice age by the end of the last millennium. Global cooling, not warming, was the threat that the public were warned about. Two of my friends wrote books prophesying an era of intense cold. In 1976 John Gribbin published *Forecasts, Famines and Freezing* and Nigel Calder wrote *The Weather Machine and the Threat of Ice*. Both confidently assured their readers that we were about to plunge into a new ice age and told how in the temperate latitudes we would need to retreat below the surface to create insulated, habitable cities in order to survive an icy future.[42]

Calder's book followed a mammoth BBC television documentary in which he warned us that an ice age was imminent. However, a survey of all the scientific publications on climate in the 1970s shows that only 10 per cent upheld the idea of a new ice age. Even then, most academic opinion was beginning to point to the era we are now experiencing – a time of rising temperatures.[43] Most papers were typified by one in *Science* in 1975, in which Wally Broecker coined the term 'global warming'.[44]

This prediction can be traced back to the Swedish scientist Svante Arrhenius, who discovered the relationship between carbon dioxide in the air and the temperature of the atmosphere as long ago as 1896.[45] Arrhenius followed this in 1908 with a book in which he suggested that we should expect a gradual warming of our world. He calculated that a doubling of the CO_2 in the atmosphere would raise Earth's average temperature by about 10°F (5–6°C) – an estimate still used by many researchers today.[46]

Between 1938 and 1957 a British engineer named Guy Callendar published a series of 26 papers that set out to relate the atmospheric increase in CO_2 to rising temperatures. Then in 1953 Canadian physicist Gilbert Plass reviewed Callendar's results and calculated that, if the concentration of atmospheric CO_2 were to double, the global temperature would rise by about 7°F (3–4°C). This was a meticulous and important paper, but it was ignored by scientists. Plass was a professor at Texas A&M University when writing his paper and, because his university was regarded as undistinguished, his crucial conclusions were ignored. Academic snobbery drowned his prescient proposal.[47]

For those who want to see more of the way in which those pioneering investigations of climate change have matured into today's concerns, the most authoritative book on the subject is the splendid volume *Experimenting on a Small Planet* by William Hay, Professor Emeritus at the University of Colorado in Boulder.[48] The first edition appeared in 2013 and runs to some 1,000 pages. It tells a curious story and is excellent bedtime reading, although, as we know, no book on the future of our climate is destined to have a happy ending.

2

EMERGING FROM THE SHADOWS

Philosophers took a long time to accept that the world had once been a different place. The next step up our ladder of understanding was the dawning realization that huge monsters had once lived on our Earth – creatures whose remains we could find, yet whose nature we could not yet discern. It was the insight of a young French naturalist that introduced the revolutionary idea that there had once been gigantic life forms that had long since ceased to exist. This great step was the brainchild of Jean Léopold Nicolas Frédéric Cuvier. Cuvier – known ever since as Georges – became the greatest naturalist of his age in France, and arguably in Europe. He was born in 1769 in the ancient town of Montbéliard, nestling close to the border with Switzerland, which at the time was ruled by Germany. His father, Jean Georges Cuvier, served in the Swiss Guards, and his mother, Anne Clémence Chatel, was devoted to him and spent much time teaching him. By the time he went to school he showed a fearsome eidetic memory and could recite in detail the chronology of kings and queens, princes and emperors. An enduring fascination for unravelling the past drove his lifetime of study. As a boy, working with his mother, the young Cuvier pored over the antiquated works on natural history by the great French naturalist Georges-Louis Leclerc, Comte de Buffon, and also those by Conrad von Geßner, the influential Swiss zoologist. As a result, by the age of 10 Cuvier was fully conversant with zoology and the classification of all the main animal types.[1]

Throughout his twenties, Cuvier became increasingly interested in fossils. In 1796 – aged just 27 – he was elected a member of the Academy of Sciences in Paris and presented his first paper on the comparisons between Asiatic and African elephants, and the fossilized remains of mammoths. He listed the differences between African and Asiatic elephants and clearly showed that they were distinct species. Cuvier carefully concluded that the fossils must represent creatures that were extinct. Some of the fossils he studied were of the animal still known as the 'Ohio Animal' – the monstrous bones were first excavated in the U.S. in 1739 and were always thought to be the remains of elephants. Cuvier demonstrated that they were anatomically distinct, and he subsequently proposed the name 'mastodon'. Later that year he spoke of a huge skeleton that had been excavated in Paraguay. He compared its skull with the appearance of present-day sloths, and concluded that this was another kind of gigantic ground-dwelling sloth. He decided to name it *Megatherium*. These were two crucial papers – first, they demonstrated the value of comparative anatomy, and they also established the view that there were huge forms of life which no longer existed. The young Cuvier ensured that the idea of extinct giants was formally established, and his two lectures set the science of palæontology onto a firm footing.

In 1784 an Italian philosopher, Cosimo Alessandro Collini, reported on a curious fossil that had been dug out of the smooth, creamy Solnhofen limestone in Bavaria. The fossil was part of the cabinet of curiosities in the palace of Charles Theodore, Elector of Bavaria at Mannheim, and it had perfectly preserved wings. Collini concluded that these were the remains of a sea creature with huge fins, though a French/German naturalist named Johann Herman insisted that the fossil represented some kind of bat. Collini heard of Cuvier's growing interest in fossils, and wrote to him about it, and by 1801 Cuvier had concluded that the fossil represented a flying reptile. He gave it a name we recognize today: ptero-dactyle. Other investigators continued to debate its true nature; it was sometimes claimed to be a bird, a catfish, or a lizard. Not until the 1860s was the nature of a pterodactyl – as a winged reptile – generally agreed.[2]

It was also Cuvier who recognized the true nature of Johann Scheuchzer's fossilized skeleton; in 1812 he correctly identified it as a fossil salamander. Until that time the petrified remains had been accepted by every scholar as evidence of a drowned man from the biblical flood, and it fell to the French philosopher's genius to reveal the truth. The specimen had been purchased by the Teylers Museum in Haarlem, the Netherlands, in 1802 and it remains there on display to this day. It was formally named *Salamandra scheuchzeri* by a German botanist, Friedrich Holl, in 1831.

Cuvier's work was widely disseminated and in 1799 it came to the attention of a physics professor in the Netherlands, Dutch scientist Adriaan Gilles Camper. He had been contacted by both Hoffmann and Drouin about the 'monster of Maastricht' to resolve whether it truly was a fish, a crocodile or a sperm whale. The fossil had remained a mystery and in England it had become a topic of fascination for James Parkinson, a young and enthusiastic physician with a passion for geology. Parkinson didn't enjoy trudging over rocks and was no great enthusiast for fieldwork. He obtained most of his specimens from the London dealers and eventually amassed a collection of over 3,000. Parkinson became Britain's leading palæontologist.

Meanwhile, an enthusiastic student of rocky strata was a young resident of southern England who spent much of his time collecting in the field. This bright young man was Gideon Algernon Mantell, born in Lewes, Sussex, on February 3, 1790, an enthusiastic youngster destined to study medicine and eventually to become a leading obstetrician. Although medicine was to be his profession, his enduring passion was the study of fossils. During Mantell's childhood, words like geology, scientist and palæontology were largely unknown. There was, however, nothing new about the term fossil. It derives from the Latin *fossilis*, the past participle of the verb *fodere*, 'to dig', and once meant anything excavated from the ground. It acquired its modern-day meaning in the 1730s, some 70 years after Robert Hooke was writing about fossils in his book *Micrographia*.

The young Mantell was keen to meet the great James Parkinson, but the enthusiasm was not reciprocated for several years. This was frustrating for a budding enthusiast, for there were few scientific publications that one could study. Parkinson was persuaded that it was time to publish a formal account, and he compiled three majestic folio volumes entitled *Organic Remains of a Former World*, each extensively illustrated. They were published in London in 1804, 1808 and 1811 – the last being the same year in which Gideon Mantell graduated from St Bart's Hospital in London. In later life, the elderly Parkinson was more inclined to meet Mantell, and eventually they became firm friends. Parkinson wrote:

> I am totally ignorant of the science [of fossils] which teaches us their natural history ... I find myself so totally ignorant of their origin, as not even to know in what class of nature's works to place them.[3]

This was an honest reflection of the degree of understanding at the time, and his three volumes represent an impressive collation of what was then known. There are separate sections dealing with fossilized shells and familiar marine creatures, of course; and he went on to describe fossils of what he concluded were whales, crocodiles, elephants, mastodons, and even several rhinoceroses. Petrus Camper, the Dutch physicist, had by this time speculated that perhaps a fossil he had found was the skull of a giant monitor lizard, and in 1808 Cuvier agreed. Cuvier published an engraving of what he called 'the large fossil animal of the quarries of Maastricht' and he decided to name it *Mosasaurus*, after the River Meuse, near where it was found.[4]

Parkinson later reproduced the illustration as Plate XIX Fig. 1 in his own book. The original specimen taken back to Paris by Napoleon's troops is still on display at the Muséum National d'Histoire Naturelle in the Jardin des Plantes, though the Netherlands authorities have recently been demanding its return.

Nobody dwelled on the significance of these early finds; the fossilized remains of seashells continued to be taken by everybody as proof of the biblical story of the flood. Similarly, when three-toed dinosaur footprints were found preserved in rocky strata, they were conventionally regarded as the marks left by the raven that Noah had sent out looking for land. The significance of these fossil remains lay in verifying the Bible. Without an understanding of prehistory, those biblical interpretations were the obvious first point of reference. Quarrymen and miners used to keep fossils, knowing that they might be sold to collectors. In 1676 a curious stone relic was discovered in the Stonesfield quarry in Oxfordshire, a place that was eventually to become a leading source of fossils for the Victorian palæontologists. It was a strange bilobed object and was purchased by Sir Thomas Pennyston, who later agreed to present it to Robert Plot at Oxford. At that time, Plot was still busily setting up the Ashmolean Museum and he published a report on the find in his *Natural History of Oxfordshire* in 1677.

> I have one dug out of a quarry in the Parish of Cornwell, and given me by the ingenious Sir Thomas Pennyston, that has exactly the Figure of the lowermost part of the Thigh-Bone of a Man or at least of some other Animal, with capita femoris inferiora, between which are the anterior … and the large posterior Sinus: and a little above the Sinus, where it seems to have been broken off, shewing the marrow within of a shining Spar-like Substance of its true Colour and Figure, in the hollow of the Bone. In Compass near the capita femoris, just two foot, and at the top above the sinus measures about 15 inches: in weight, though representing so short a part of the Thigh-Bone, almost 20 pounds.[5]

His suggestion that this came from an animal proved to be prescient, and for a time he interpreted the bone as coming from a Roman war elephant, though his later interpretation was that it came from a gigantic human.[6] Philosophers at the time accepted that 10-foot (3-metre) giants had lived in the past, for they were mentioned in the Bible.[7]

What Plot was describing was actually the end of a fossilized long-bone. His published engraving is the first we have of a dinosaur bone, even though nobody at the time realized its significance. Plot himself was an enthusiastic naturalist and collector who met many of the luminaries of his day and carefully cultivated their acquaintance. Plot saw himself as Britain's answer to Pliny the Elder; just as Pliny had written his *Natural History*, so Plot resolved to publish a Natural History of his own that would commemorate his lifetime's work.[8]

Plot's description of the fossil was meticulous, though he did not assign a scientific name to the specimen. His illustration was re-published by Richard Brookes in 1772. Brookes was a physician and naturalist who wrote a great many books on British wildlife, and his desire to categorize species correctly obliged him to find a suitable designation. Considering its appearance, and disregarding Plot's attempt at a detailed description, it seemed to Brookes that he knew what it was. There was only one name that unambiguously summed up its appearance: in the fourth volume of his *New and Accurate System of Natural History* he boldly named it 'Scrotum humanum'. It certainly looked like one.[9]

Observation and the art of seeing were becoming a philosophical preoccupation of the learned classes at this time; it was even the subject of literature for children. A six-volume book entitled *Eyes or no Eyes; or, the Art of Seeing*, written by John Aiken and his married sister Anna Barbauld, was published in 1780. It told the tale of two

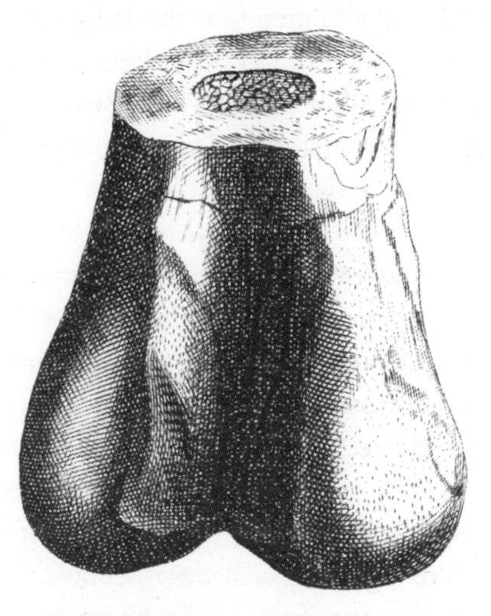

This fossil was found in the Stonesfield quarry near Oxford in 1676 and given to Robert Plot. He identified it as 'Scrotum humanum' but is it actually part of a *Megalosaurus*. His published engraving was the first illustration of a dinosaur bone.

brothers who walked together in the countryside; one finding it a tedious trip, with nothing of interest, while the other was endlessly engaged in the plant species that they encountered, the myriad insects and meadow creatures he could see, and the geology of the landscape – even finding traces of a prehistoric encampment. It was not what you could see that mattered, but what you perceived. The book was so popular it was frequently republished and remained constantly in print for well over a century.[10] So successful was the book that the celebrated W. S. Gilbert and Arthur Sullivan later wrote an opera with the same title.

When Gideon Mantell was growing up in Sussex, the rocky strata around his home were rich in fossils of oyster-like shellfish along with ammonites and belemnites, both of which we now know had swum by jet propulsion like cuttlefish. There was little surprise at the sight of those fossils among the village folk who discovered them. Clearly, they were further evidence substantiating the biblical descriptions of the flood. The shellfish were believed to have been deposited during that inundation, while the coiled shells of ammonites were regarded as serpents that had been turned to stone and the pointed belemnite fossils were taken to be thunderbolts. Collecting these fossilized remains was a popular hobby among youngsters, and young Gideon's enthusiasms were triggered by the discovery of an exquisite ammonite fossil when he was about 12 years old. Even though palæontology was a word yet to be coined, the collecting of fossils now had a term: oryctology. It is now forgotten and absent from most dictionaries (it has no page in Wikipedia), having originated from the Greek *oryktos* meaning 'formed'. And so, by the time Gideon was grown, he was already a seasoned oryctologist.[11]

It was Mantell's desire to become a physician that took him to St Bart's Hospital, where his collecting in the field was replaced with the purchasing of fossils from London dealers including Joseph Stutchbury. Many of the doctors at Bart's were fascinated by fossils, including the celebrated anatomist John Hunter, and many of those doctors simply purchased curiosities from dealers. In 1790 Hunter wrote a revolutionary account of fossils. Wisely, he proposed that the layers of marine fossils he observed had not resulted from the biblical accounts of a flood, and he concluded: 'Many retain some of their form for many thousand years ...'[12]

By this time, the way in which layers of rock were laid down in succession had become a fashionable subject for study in Germany. First to write authoritatively on the subject was a mineralogist born in 1714, Johann Gottlob Lehmann. He studied at Wittenberg and was subsequently invited by the Russian Academy of Sciences to move to St. Petersburg and expand his work. Rocky strata seemed to him amenable to serious scientific study, and he realized that they must have been laid down in strict order. In one mining area he identified more than 20 strata, which he called *Flötzgebirge*, and he soon realized that studying the sequence could perhaps allow prospectors to locate mineral-bearing strata. He concluded that this could be a key to the discovery of vast mineral riches.[13]

The idea was taken up by Abraham Werner, a young mineralogist who had studied at Freiburg, Saxony, and Leipzig. What a curious man was this – sensibly enough, he taught students that rocks were laid down in an orderly fashion, the study of which could help to ascertain where minerals would lie; but, although he never travelled, he confidently concluded that the sequences he observed in Saxony were representative of those everywhere else on Earth, and he decided that volcanoes resulted from the combustion of coal measures deep below the ground. He had a captivating and charming manner. His students hung on every word. He was only 36 when he published a definitive analysis on a classification of mountain ranges that quickly became essential reading for all budding geologists.[14]

One person who bought the book when it appeared was Alexander von Humboldt, a brilliant explorer and naturalist; he was the younger brother of the Prussian linguist and philosopher Wilhelm von Humboldt, and studied mineralogy and geology under Abraham Werner at the School of Mines in Freiburg, Saxony. Alexander von Humboldt bequeathed to us the most familiar geological period of all – the Jurassic. This was the name he gave to an important set of limestone strata that Werner had omitted from his book. This characteristic pale limestone was observed by Humboldt in the Jura mountains, so in 1795 he called it *Jurakalk*. From this, the term 'Jurassic' was soon to emerge. Now we know that this period extended from 201.3 to 145 million years ago, and was an era populated by gigantic sauropod dinosaurs. Alexander von Humboldt became widely admired and internationally famous. He was elected a foreign member of the Royal Society in England and the Royal Academy of Sciences in Sweden, while in the U.S. he was showered with honours, being elected a foreign member of the American Academy of Arts and Sciences, a member of the New York Historical Society, the American Antiquarian Society and the American Ethnological Society. Thomas Jefferson described him as the most important scientist he had ever encountered. Although Humboldt has named for us one of the best-known eras in the whole of palæontology, he did little research in that field and never studied fossil animals. But he did write about prehistoric crocodile tracks in samples of Buntsandstein rock that he discovered in 1834. He thought the footprints had been made by a mammal similar to an opossum, though he hinted that they might alternatively have been made by a primate. The tracks became known casually as the footprints of a 'hand-beast', but at the time the discovery gave rise to no new scientific insights.[15]

The first scientific description of fossil reptiles – which may have included remains of dinosaurs – was published in 1776 by a French zoologist and cleric, Abbé Jacques-François Dicquemare. His primary interest was in sea anemones, but he was fascinated by fossils and he diagnosed his fragmentary fossils as being the petrified remains of fishes and whales. Lurking in his discussion is a crucial concept – he seemed to hint that they might represent creatures that had since become extinct. This was the first suggestion that the remains of prehistoric creatures might possibly be found in the rocky strata.[16]

Meanwhile a French enthusiast, Charles Bacheley (incorrectly identified by Cuvier and every standard scholarly source since as 'Abbé Bachelet'), had developed a passion for fossilized creatures that he found near the pretty coastal town of Honfleur, which nestles close to the mouth of the Seine. I have often retraced his steps. In 1773, he collected fossils that he thought were the remains of a whale. In reality, they comprised cranial and postcranial specimens from two crocodile-like teleosaurs, plus some postcranial vertebræ of a meat-eating theropod dinosaur. Bacheley became acquainted with Jean-Étienne Guettard and sent some of his specimens along to Guettard for him to identify, before publishing an account in 1778.[17]

Recent investigations have confirmed that these fossils were collected at Les Vaches Noires, a zone of coastal rocky strata that would prove to be among the most fossil-rich in Normandy. Three years later, at Le Havre on the opposite bank of the river, Dicquemare reported the find of similar remains, which he interpreted as fossil porpoises and dolphins. Bacheley's collection passed to C. Guersent, who was a geology professor at the museum in Rouen. In 1799 Jacques Claude Beugnot, a local dignitary, ordered that the collections of fossils from Bacheley and Dicquemare should be transported to the Muséum National d'Histoire Naturelle in Paris. The existence of the discoveries was finally published by Cuvier in 1800.[18]

Not until 1808 did Cuvier formally describe the fossils in detail. He had interpreted them as belonging to members of the crocodile family and gave them the name *Streptospondylus*. In fact, as we have seen, the

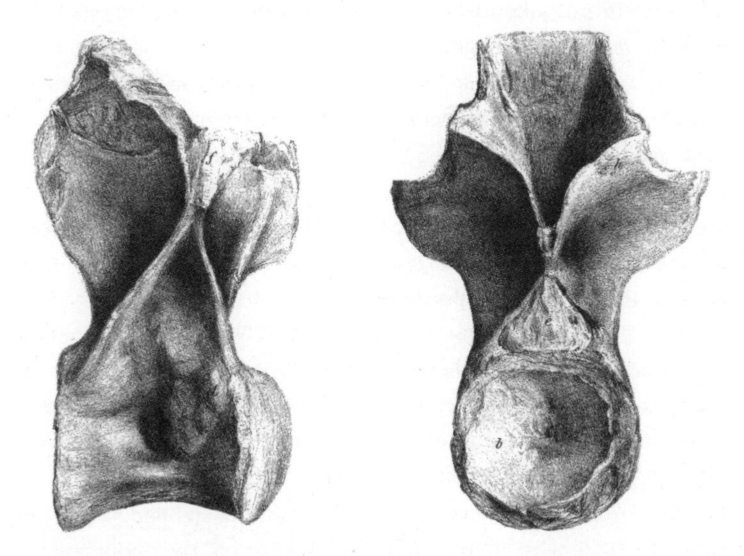

Vertebræ of *Streptospondylus* were discovered in 1778 by a French collector, Charles Bacheley. These were the first known European dinosaur fossils. In 1842 the British palæontologist Richard Owen dubbed the dinosaur *Streptospondylus cuvieri*.

vertebræ came from a medium-sized theropod dinosaur. Even though nobody realized it at the time, these were the first fossils from a meat-eating dinosaur ever to be recorded by science.[19]

In America, the 'Ohio Animal' had continued to attract interest, and in 1796 Thomas Jefferson (who became the president of the United States just five years later) sent a small expedition to look for extinct mastodons and mammoths near the Ohio River in Kentucky. Like most well-educated statesmen of his time, Jefferson liked to keep abreast of discoveries in natural history and science. Indeed, when the White House was first built, it was furnished with a 'Mastodon Room' to house fossil collections. In 1797, Jefferson gave a presentation at the American Philosophical Society in Philadelphia, describing a fossilized giant sloth that now bears his name: *Megalonyx jeffersonii*. When the presentation was printed as an academic paper in the Society's journal, it became one of the first American publications in the developing field of palæontology.[20] More recent American presidents are perhaps less likely to publish academic papers in scholarly journals.

Dinosaur footprints were now being discovered by new arrivals in the United States. The first we know about were unearthed in 1802 by a farm boy named Pliny Moody of South Hadley, Massachusetts. Moody dug up a slab of red sandstone while ploughing. It showed some small, sharp, clear, three-toed footprints. This fine specimen was fixed above the farmhouse door, where a local physician confidently identified them as being the tracks of Noah's raven. The story of the biblical flood was still the conventional explanation at the time, because there was no understanding of fossil footprints left by dinosaurs, so although it seems fanciful to us, this was a popular diagnosis at the time. We are quick to ridicule such early conventions, though a glance at today's religious television channels reminds us that present-day beliefs can be as superstitious and fanciful as anything we have seen in the past.

The Napoleonic Wars had been raging in Europe and they finally ended in 1815, whereupon Georges Cuvier seized the opportunity to visit England. One of his first ports of call was to meet William Buckland in Oxford. Buckland was born on March 12, 1784, in Axminster, Devon, and spent much of his time as a child walking across the countryside with his father, the Rector of Templeton and Trusham. His father used to show him how to collect fossilized shells, including ammonites, from the strata of Jurassic Lias that were exposed in the quarries. The young Buckland went to school in Tiverton, and eventually entered Corpus Christi College at Oxford University, to study for the ministry. He regularly attended lectures on anatomy given by Christopher Pegge, a physician at the Radcliffe Infirmary in Oxford, and he was particularly intrigued by part of a fossilized jawbone that Pegge had purchased for 10s 6d (now about £40 or $55) in October 1797. Buckland also went to the presentations given by John Kidd, Reader in Chemistry at Oxford, on subjects ranging from inorganic chemistry to mineralogy, and discovered that Kidd had himself collected several fragments of huge bones from the Stonesfield Quarry near Witney, some 10 miles (16 km) away. The plot was thickening.

Buckland meanwhile continued searching for fossil shells in his spare time. These he initially took as evidence of the biblical story of the flood, but as the years went by he turned towards more scientific reasoning and abandoned the literal truth of the Old Testament. After graduating, he was made a Fellow of Corpus Christi College, Oxford, in 1809, and he was formally ordained as a church minister. In 1813 Buckland was given the post of Reader in Mineralogy, following John Kidd, and his dynamic and popular lectures began to include a growing emphasis on palæontology. By now he was becoming an experienced collector, and in 1816 he travelled widely in Europe, including Austria, France, Germany, Italy, Poland and Switzerland. He visited Cuvier on several occasions.

Buckland became something of an eccentric. He always wore an academic gown instead of overalls for his fieldwork and claimed to have devoured his way through most of the animal kingdom, serving mice, crocodiles and lions to his guests (and he claimed that the two foods he disliked most were moles and houseflies). On his travels, he was said to have been shown the preserved heart of King Louis XIV nestling in a silver casket and remarked: 'I have eaten many strange things, but have never eaten the heart of a king before,' and so he picked it up and bit into it before anybody could stop him.

Buckland's first major prehistoric discovery was not of a reptile, but a human – the Red Lady of Paviland, a human skeleton found in South Wales. He decided to explore the largely inaccessible Paviland cave on January 18, 1823, only to find this well-preserved skeleton, which he initially took to be the corpse of a local prostitute. He later concluded that the body had been placed there by early residents in pre-Roman times, though more recent tests have shown that it dates back 33,000 years – indeed, it is now accepted as the most ancient human skeleton ever found in Britain. Buckland married an enthusiastic fossil collector and accomplished artist, Mary Morland, in 1825, and thereafter Mary devoted herself to illustrating her husband's palæontological publications with flair and skill.

By the time of Cuvier's visit, Buckland was studying fossil collections, and among the specimens he showed to his French visitor were the *Scrotum humanum* and Pegge's curious specimen of a fossilized jawbone. Cuvier concluded that these were both fragments from gigantic reptiles. William Conybeare, a palæontologist colleague of Buckland's, referred to these specimens as the remains of a 'huge lizard' for the first time in 1821, and the physician and fossil hunter James Parkinson soon announced his intention to call the creature *Megalosaurus* from the Greek μέγας (*megas*, large). Parkinson estimated that this had been a huge land animal measuring 40 feet long and 8 feet tall (12 x 2.5 metres). Parkinson is little known today as a palæontologist, though we all know his name in a different context – he is the physician who correctly identified the degenerative disease known, in his time, as a 'shaking palsy' and which we now call Parkinsonism. Most palæontologists at the time were physicians, and many made discoveries that resonate beyond the world of the fossil collector.

Buckland now faced urgent demands from Cuvier for details to include in his own book, and meanwhile Buckland continued to investigate the fossil remains, while his wife Mary began preparing the detailed drawings of the remains for publication that were to be the basis of the published lithographic plates. Buckland had met Mary while travelling by horse-drawn coach in the West Country. An account records:

> Both were travelling in Dorsetshire and each were reading a new and weighty tome by the French naturalist Georges Cuvier. They got into conversation, the drift of which was so peculiar that Dr. Buckland exclaimed, 'You must be Miss Morland, to whom I am about to deliver a letter of introduction.' He was right, and she soon became Mrs Buckland. She is an admirable fossil geologist, and makes midels in leather of some of the rare discoveries.[21]

They worked together diligently in every spare moment they could find. There was now growing interest in the fossils being found at Lyme Regis on the Dorset coast of southern England. Most people believed these rocky remains to be the fossils of familiar fauna (crocodiles or dolphins). Collectors including Henry de la Beche and William Conybeare carefully examined a range of specimens, and published a joint account in 1821 concluding that they might represent something very different – a new kind of reptile. They mentioned the work of a host of amateur collectors, acknowledging 'Col. Birch, Mr. Bright, Dr. Dyer, Messrs. Miller, Johnson, Braikenridge, Cumberland, and Page of Bristol,' and they now concluded that this new type of reptile formed a bridge between ichthyosaurs and crocodiles, and so they coined a new term for these creatures: plesiosaurs.[22]

Interest in the fossil reptiles started to spread, and in 1822 James Parkinson published a book on his investigations entitled *Outlines of Oryctology*, which, although primarily concerned with seashells and other familiar fossils, also reported the latest investigations of the huge reptile fossils that were now beginning to appear. In this book, *Megalosaurus* was included as 'an animal, approaching the monitor [lizard] in its mode of dentition, &c., and not yet described,' while *Mosasaurus* was defined as 'The saurus of the Meuse, the Maestricht animal of Cuvier.' Parkinson reported that Cuvier and others placed this reptile 'between the *Monitors* and the *Iguanas*. But, as is observed by Cuvier, how enormous is its size compared with all known *Monitors* and *Iguanas*. None of these has a head larger than five inches; and that of this fossil animal approaches to four feet.' Suddenly there was a glimpse of the future – the notion of gigantic prehistoric reptiles was began to emerge.[23]

The Bucklands had by this time assembled a range of fossils carved out from the Stonesfield strata, including a length of lower jaw with a single tooth, a dorsal and an anterior caudal vertebra, five fused sacral vertebræ, two ribs and several sections of the pelvis. Clearly, these did not all come from the same animal, and Buckland's interpretation of some of the bones was incorrect (he thought the ischium was a clavicle). Mary provided perfectly precise pictures of the specimens for

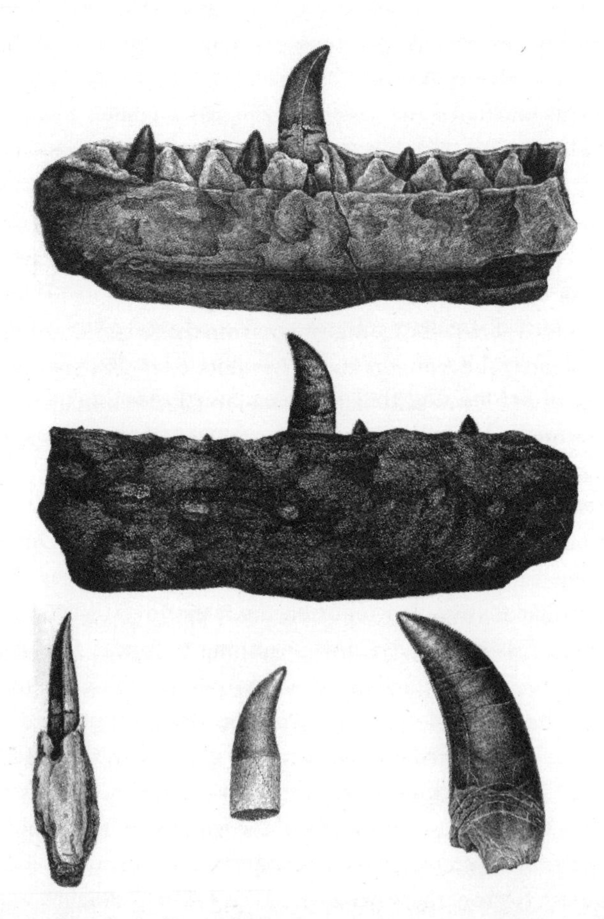

William Buckland asked his wife Mary to prepare these exquisitely detailed drawings of the jawbone found in the Stonesfield quarry, and in February 1824 he announced the name James Parkinson had suggested for this dinosaur: *Megalosaurus.*

the lithographer, and on February 20, 1824, at a meeting of the Geological Society of London, Buckland formally announced the discovery of a new monster reptile bearing the name bestowed upon it by James Parkinson: *Megalosaurus*.[24]

With James Parkinson's pioneering report, and now with William Buckland's formal paper, the world's first dinosaur was formally revealed to the world. It had taken almost 150 years for the true nature of the *Scrotum humanum* specimen to be recognized. No, it was not an ancient gentleman's family jewels, but a monster's elbow. What an extraordinary revelation![25]

Lyme Regis, a coastal village in the English county of Dorset, was emerging as a centre for the study of fossils. The most prominent of the collectors was Richard Anning, a cabinetmaker who had settled in Blandford Forum and married a local girl, Mary Moore (popularly known as Molly), on August 8, 1793. They moved to Lyme and built a house for themselves by the bridge over the River Lym. Storms sometimes struck that shore, roaring in from the Atlantic and devastating the beach. The Annings' home was flooded more than once; on one stormy night it was said that they had to climb out of an upstairs window to escape the rising tide. On Christmas night, 1839, the entire family almost lost their lives. It was after midnight, with everyone in bed and asleep, when there was a mighty roar and the ground suddenly shuddered as a huge slice of the cliff slid into the sea. Witnesses next day said there was a vast chasm where the land had split open for more than half a mile (about 1 km), and a cliff-top field belonging to a farmer slid down 50 feet (about 15 metres) towards the sea. The Bucklands were staying nearby at that time, and Mary used her considerable artistic talents to capture the scene for posterity. Next morning, Boxing Day, the beach and the shattered cliff top were thronged by visitors, eager to see the catastrophic collapse. The landslip caused a huge reef to appear in the sea, towering 40 feet (about 12 metres) tall and enclosing a lagoon at least 25 feet (8 metres) in depth. Within weeks, all this had washed away, and the beach had returned to normal.

Richard and Molly Anning had 10 children. Their first was Mary, who was born in 1794, but tragically died in a fire. The *Bath Chronicle*

In December 1839, Mary Buckland drew the great landslip near Lyme Regis. It was engraved on zinc by George Scharf and printed as a hand-tinted lithograph by Charles Joseph Hullmandel, who studied chemistry under Michael Faraday.

newspaper recorded the incident: 'A child, four years of age of Mr. R. Anning, a cabinetmaker of Lyme, was left by the mother for about five minutes in a room where there were some shavings … The girl's clothes caught fire and she was so dreadfully burnt as to cause her death.' It seems she was trying to rekindle the fire with the slivers of wood.[26]

The distraught parents named their next baby girl Mary in memory of their lost child – and this little girl was destined to become the greatest of all the pioneer fossil hunters. Of the remaining children, only one other, a son named Joseph, survived to adulthood. Infant mortality through this period was around 50 per cent, so the loss of so many infants would not have been regarded as particularly unusual. People at that time lived so close to tragedy. Death was simply a shade of daily life.[27]

The rocky strata near Lyme Regis are marked by numerous layers of Blue Lias, a rock rich in mudstone that was originally the bed of a shallow sea. This form of rock is widely spread across southern England and South Wales and was laid down in late Triassic and early

Jurassic times between 195 and 200 million years ago; it is also known as Lower Lias. The muddy seabed was littered with ammonite shells, and the remains of sea creatures – fish and swimming reptiles – are also abundant. Both youngsters accompanied their parents scouring the rocky shelves exposed after a storm, and they quickly became adept at finding fossils. Then, in 1810, their father Richard Anning suddenly died, and the family was left in penury. Their only possible source of income was now fossil hunting, and the children went out with their mother each day, looking for fossils to sell as souvenirs to visitors. They sold, as they do today, for a present-day value of about £10 ($12). When he was 15, young Joseph discovered part of a remarkably well-preserved ichthyosaur in a rocky shelf and showed his sister where it lay. A year later it was more fully exposed, and Mary had the skill to extricate it from the shore. The skeleton was well preserved, though at the time they could find no skull. Those who saw it concluded that it was some sort of crocodile. Ever since John Walcott's published descriptions in 1779, others had been finding similar specimens. The Blue Lias rock is visible along the coast of South Wales, and as a student I used to find fragments of ichthyosaur skeletons in the smooth strata that storms had exposed. At Welsh St. Donats in 1804, an enthusiast named Edward Donovan discovered an ichthyosaur specimen represented by its jaw, vertebræ, ribs and pectoral girdle. It would have measured 13 feet (4 metres) long and was adjudged to be a gigantic lizard. In the next year two more were found in the same strata on the opposite side of the Bristol Channel, one discovered at Weston by Jacob Wilkinson and the other by the Reverend Peter Hawker. This specimen soon became known as Hawker's Crocodile. In 1810 an ichthyosaur jaw was dug out at Stratford-upon-Avon, but the locals simply put it with some bones from a fossilized plesiosaur to make up a specimen that was more marketable. The name ichthyosaur was becoming popular, derived from the Greek ιχθυς (*ichthys*, meaning fish) and σαυρος (*sauros*, lizard).

By 1811 the time was ripe for a major discovery: in Lyme Regis, along what is now called the Jurassic Coast of Dorset, the first complete ichthyosaur skull was found by Joseph Anning, the brother of Mary, and she soon found the thoracic skeleton of the same animal. Their mother Molly sold the whole piece to Squire Henry Henley for £23 (now about £1,300 or $1,600) and it was later bought by the British Museum for twice the price. It remains on display at the Natural History Museum and is now identified as a specimen of *Temnodontosaurus platyodon*. Both the young Mary and her mother were now adept fossil hunters, while Joseph went on to train as a furniture upholsterer. It was the sight of the young Mary Anning that visitors found so unusual. She did not just collect, but she studied the remains that she found, transcribing lengthy accounts from learned texts and meticulously copying the illustrations of fossils they contained.

By now, Buckland was regularly visiting Dorset to purchase fossils and collect his own specimens, and he became particularly intrigued by the 'bezoar stones' that Mary Anning had been discovering alongside her ichthyosaur skeletons. The bezoar was the name given to an indigestible mass found within human intestines; it was believed that a glass of poison containing a bezoar would be instantly rendered harmless. The word comes from the Persian *pādzahr* (رهزداپ), meaning 'antidote'. Anning discovered that, when those rounded, rough stones were broken open, they always contained the scales and bones of fish and smaller ichthyosaurs. In 1829, Buckland recognized that these stones were present everywhere that fossil reptiles were found, and he suddenly realized what they were. They were fossilized faeces. They really were masses from within the gut. Buckland decided to call them 'coprolites', the term we use to this day. He became devoted to the study of the fossils that Mary Anning had discovered, and his enthusiasms gave rise to an historic painting by Henry de la Beche entitled *Duria Antiquior – a more Ancient Dorset*, which portrayed some of the swimming reptiles Mary Anning had discovered, with some of Cuvier's pterodactyls swooping across the heavens. At last it was becoming clear that there had been an age of strange reptiles that were frighteningly large and had bizarre lifestyles. The age of the dinosaur was steadily coming closer.

A prominent British geologist, Sir Henry Thomas de la Beche, born in London in 1796, had moved to Lyme Regis where he befriended

Mary Anning. He investigated fossil reptiles and wrote extensively on surveying rocky strata. De la Beche was by this time known as one of the most prolific of geologists, and his published works range from the description of fossil marine reptiles to the study of British stratigraphy. He also wrote learned textbooks dealing with the application of geological survey methods. However, he became best known by the public for his talent as a cartoonist. In one of them he satirizes the likes of Buckland and Lyell. This cartoon appeared in 1830, the same year in which Lyell's great, ground-breaking formal book on geology was published in London. In this mighty work Lyell discussed stratigraphy, dealt with the value to commerce of systematic prospecting, arguing that the forces that were acting in nature today were the same as those that had acted in the past, and asserted that they would be the same in the future (the theory that pretentiously became known as uniformitarianism). When Charles Darwin set off on his voyage aboard HMS *Beagle* in 1831, it was Lyell's new book that accompanied him on his geological expeditions.[28]

In 1830 Henry de la Beche, the first director of the Geological Survey of Great Britain, painted this historic watercolour representation of prehistoric life based on Mary Anning's discoveries. He entitled it: *Duria Antiquior – A more Ancient Dorset.*

The French fossils that Cuvier had dismissed as being from a crocodile had meanwhile yet to be properly identified. In June 1793, a zoologist named Étienne Geoffroy Saint-Hilaire had become one of the first 12 professors at the newly opened Muséum National d'Histoire Naturelle in Paris. His principle concern was setting up a zoo, but the following year he had struck up a relationship with Cuvier and they published several joint papers on the classification of animals. Fossils were also discussed. In 1807 Saint-Hilaire was elected to the French Academy of Sciences and concentrated thereafter on the study of the anatomy of invertebrates, corresponding at length with his British friend Robert Edmund Grant. His assistant, a young undergraduate who was particularly interested in barnacles, was a medical school drop-out named Darwin – Charles Darwin. Saint-Hilaire was concerned that Cuvier had too hastily concluded that *Streptospondylus* was a crocodile, and so he examined the fossils again. He decided that they belonged to two species of extinct reptile, and named them *Steneosaurus rostromajor* and *S. rostrominor*. In England, *Megalosaurus* was already officially recognized as a genus, though it still had no species name. It was a German palæontologist, Ferdinand von Ritgen, who gave it the provisional name *Megalosaurus* in 1826. He called the species *Megalosaurus conybeari*, though this name was never formally adopted.[29]

In 1827 Gideon Mantell resolved to include this fossil animal in his geological survey of south-eastern England and felt it appropriate to name it in honour of Buckland. It has been known as *Megalosaurus bucklandii* ever since. This was a crucial step in the history of science: it was the first dinosaur name formally to enter the literature of science. Suddenly, dinosaurs were real.[30]

These new areas of investigation were now attracting increasing attention. The Geological Society of London was inaugurated on November 13, 1807, at the Freemasons Tavern in Great Queen Street, and Buckland was elected their president in 1824–1825 and again in 1840–1841. He had been elected a Fellow of the Royal Society in 1818 and became a member of their Council from 1827 to 1849. Buckland's interest in spreading the word led to his involvement in the newly formed British Association for the Advancement of Science, and in

1832 he was appointed their president and chaired the second conference. By now, Buckland was riding high.

The limestone and chalk quarries at Maastricht continued to provide specimens for collectors, and indeed the final 6 million years of the Cretaceous are known to this day as the Maastrichtian epoch. In spite of his studies of fossil mammals – including extinct species, like mammoths – Cuvier steadfastly refused to accept the concept of evolution. To him, species were immutable, and he substantiated this notion by comparing mummified cats and dogs from ancient Egypt and showing that these creatures were unaltered when compared with present-day specimens. Frozen carcasses of woolly mammoths had first been excavated by explorers in the 1690s, and the first scientifically documented example was discovered in the mouth of the River Lena, Siberia, by a Siberian hunter named Ossip Schumachov in 1799.[31]

Schumachov saw these carcasses as a viable source of tusks that he could sell on to ivory traders, but Johann Friedrich Adam, a Russian explorer who later changed his forename to Michael, went at once to inspect the newly discovered frozen carcass. He found that much had already been devoured by wolves. Even so, it provided the most complete mammoth skeleton ever found and was assembled at the Zoological Institute of the Russian Academy of Sciences, where it was mounted alongside a skeleton of an Indian elephant.

This substantiated that animals could become extinct and that skeletons of preserved carcasses were similar to the fossilized remains that were excavated elsewhere. Cuvier documented his fossil skeletons, and in 1812 he published his research, serving to dignify the study of fossils. The Maastricht fossil was a lizard as big as a crocodile, and the pterodactyl he distinguished from birds or bats, insisting it was a flying reptile. These were exciting conclusions that were already causing consternation and interest among scientists and writers. It was now clear that there had been eras when strange and unfamiliar creatures roamed the Earth[32] and in *Bleak House*, which Charles Dickens started writing in 1852, he said it would be 'wonderful to meet a Megalosaurus, forty feet long or so, waddling like an elephantine lizard up Holborn Hill.'

French palæontologist Louis Figuier published his *La terre avant le deluge* (the world before the flood) in 1863 and it became immensely popular. His illustrator, Edouard Riou, portrayed an *Iguanodon* and a *Megalosaurus* fighting with each other.

The resonances of these fossils in present-day thinking were alluded to by Reverend Charles Kingsley, when he published a serialized story in *MacMillan's Magazine* each month from August 1862 through to March 1863.

> Did not learned men, too, hold, till within the last twenty-five years, that a flying dragon was an impossible monster? And do we not now know that there are hundreds of them found fossil up and down the world? People call them Pterodactyles: but that is only because they are ashamed to call them flying dragons, after denying so long that flying dragons could exist.

The stories were popular, and were brought together into a single volume in 1863. As a classical moralistic tale for children it has remained in print ever since.[33]

What was becoming increasingly apparent was the curious consistency in the way rocks had been laid down. This new science began with Friedrich August von Alberti, born in 1795, when he attended a military academy in Stuttgart and took up geology as a hobby. Later he took up employment in the salt production plant at Rottweil, an exquisite medieval town that has still changed little since the 1500s. (This is the place after which Rottweiler dogs are named; they are believed to be the direct descendants of the military breed that Roman soldiers brought during the invasions by the Caesars.) In 1815 Alberti came to recognize the occurrence of three characteristic strata composed of sedimentary deposits. He regularly found three distinct layers: one of red sandstone, capped by chalk, and succeeded by black shales. They occur throughout Germany and were later found to extend across the whole of northwest Europe. In each he found the same characteristic types of fossils and, from the Latin *trias* (meaning trio), he coined the term Triassic.[34] We can now date this geological period as extending from 252.17 to 201.3 million years ago, for this is when the first dinosaurs appeared. The Jurassic period starts where the Triassic leaves off, and that is when dinosaurs first reached their enormous size.

Shortly after Alberti had coined the term Triassic, Jean Baptiste Julien d'Omalius d'Halloy in Belgium recognized the Cretaceous. Born in 1783 in Liège (now in Belgium, but then in the Austrian Netherlands), d'Halloy was sent by his wealthy family to study literature and the classics in Paris. He wanted none of it – geology became his consuming passion. He attended many lectures given by the distinguished naturalists and geologists and he learned everything he could from Georges Cuvier. He loved studying the rocky structure of France, and being of independent means he could travel as widely as he wished. In 1808 he published a learned paper entitled 'Essai sur la géologie du Nord de la France' in the *Journal des Mines*, an extraordinarily wide-ranging subject for someone working independently. It was in this paper that he recognized the distinct nature of coal-bearing strata, which he named *Terrain Bituminifère*. The name did not gain common currency, but it was the first time that the coal measures were identified as providing a key to an earlier era of the Earth's history. He also recognized the era of chalk formation and drew sections showing its extent; this he named the *Terrain Crétacé*, and it was soon recognized in English as the Cretaceous period. He is estimated to have travelled more than 15,000 miles (24,000 km) across France, and he published an immense range of textbooks on geology, all written with clarity and precision, and including some of the first illustrations of the sequence of geological strata.[35]

Meanwhile, in England, William Coneybeare and William Phillips – both keen amateur geologists – had also concluded that the era of coal-bearing strata must have been confined to one period of time, because of the similar fossil plants that they all contained. Their conclusion substantiated d'Halloy's designation of his *Terrain Bituminifère*. This had been an era of gigantic cycads, massive stands of horse-tail *Equisetum* plants, and huge conifers like today's monkey-puzzle tree *Araucaria*. The two English friends realized that this had been an age of swampy forests and – because of the undeniable fact that this was the era that gave us coal – in 1822 they decided to call it the Carboniferous age, from the Latin *carbō* (coal) and *ferō* (bearing).[36] Today we know this geological period extended from 358.9 to 298.9 million years ago.

Recognizing these great periods – the Jurassic and Triassic, Carboniferous and Cretaceous – had now given palæontologists a greater understanding of the ages through which the Earth had passed. Our present-day countryside reveals the tortured history of churning cliffs and the weathered remains of towering mountains that have resulted from the collisions between continents over millions of years. Now eroded and reduced in stature, the mountain ranges have become rolling hills and the cliffs are cut through like a layered cake so that the æons of prehistory can be seen in strata that stretch back in time. These reveal a timeline of the way that masses of land drifted across the globe, showing us how places like North America and Europe started in the southern Arctic wastes and slowly headed north, where they now extend up towards the northern latitudes of ice and snow. Once, today's western nations sat at the equator, and the sandy deserts from that time are bequeathed to us as sandstone strata. Later, the land was covered with huge swampy stands of conifers and cycad trees as we drifted past the tropics, and we can see them still in the coal measures. The drifting still goes on. As new rocks are spewed up from a massive split in the Earth's crust that runs down the middle of the Atlantic, the mid-Atlantic ridge, continental masses on either side are still being forced apart. America and Europe are moving away from each other at the same speed as your nails grow. Clip a couple of millimetres from your toenail and reflect on the fact that the flying distance between New York and London has increased by precisely the same amount.

Some of the rocky strata tell a powerful story of a prehistoric world dominated by massive monsters and strange landscapes. At Kimmeridge Bay in Dorset, for example, you find oil shales from which crude oil seeps just as it does in the world's greatest petrochemical producing nations. Indeed, there is a nodding donkey oil pump at Kimmeridge, just like those in the American oilfields. Those shale beds formed during the late Jurassic, between 157.3 and 152.1 million years ago, and from them the most beautiful fossils sometimes emerge – bony skeletons of strange swimming plesiosaurs and forbidding dinosaurs. They have been collected as curiosities for centuries. The leading present-day collector is Steve Etches, a plumber who made his living repairing and installing bathroom and kitchen appliances around Kimmeridge. At least, that was in his day job; every moment of his spare time is spent hunting for fossils in the nearby rocky cliffs. He has a keen eye and can spot a fossil when most people see nothing but stone, and his home has been extended to become a private museum. Etches has collected more than 2,000 specimens over the past 35 years. Palæontologists the world over respect his achievements; indeed, in October 2016 the Museum of Jurassic Marine Life was opened a few yards from his home to display the highlights of his collection. It is an exquisite stone building, part of which is the village hall, while the rest is a museum and a laboratory which Etches can enter through a private door and where he can work at will, meeting with the public whenever he wants. This is a unique recognition of his impressive contributions to palæontology, and it was dignified by a formal inauguration ceremony early the next year which my wife and I were privileged to attend.

Steve Etches is the latest in a long line of enthusiasts and, as we have seen, the majority of practitioners of practical palæontology were never formally trained in the discipline. The Dorset shore where

Kimmeridge lies has long been known as the 'Jurassic Coast' and it extends from Exmouth in East Devon to Studland Bay, near Poole in Dorset, a distance of 100 miles (160 km). It is near the middle of this coast that you will find Lyme Regis, known for its beaches and seaside views, and still a popular venue for fossil collectors. For centuries people have taken home fragments of rock with strange skeletal structures embedded in the surface or marked with the remains of peculiar shells. As interest in studying the natural world began to grow in the nineteenth century, families such as the Annings established businesses collecting fossils they could sell on to enthusiasts. The petrified specimens were originally advertised as *thunderbolts* or *devil's fingers* (belemnites), *snake-stones* (ammonites) and *verteberries* (vertebræ). Demand had steadily increased since 1792, as tourism to the south coast of Britain increased when the French revolutionary wars made travel to the continent unsafe. The ancient city of Bath first became a magnet for Georgian tourists, but as visitors were sold the idea of immersion in water rich in minerals, bathing in the sea began to increase in popularity. Bathing machines – little sheds on wheels – sprang up along the coast, from which holidaymakers could demurely emerge and lower themselves into the edge of the ocean. The visitors sought souvenirs, and collecting fossils from the beach became an increasingly popular option, just as enthusiasts had done for centuries.

Among those early collectors was Lieutenant Colonel Thomas James Birch, who used to visit Lyme from his home in Lincolnshire. On a visit in 1820, he became aware that the Annings were in need of money, and he resolved publicly to auction all his fossil collection to help them. Having made no major discoveries for a year, the Annings were at the point of having to sell their furniture to pay the rent. The auction sale at Bullocks auction room in Wareham became a three-day event, with buyers coming from Vienna and Paris, and it raised £400 (over £23,000 or $27,000 today). Mary Anning had become renowned for her expertise and she was mentioned in the media. The *Bristol Mirror* in 1823 reported:

> This persevering female has for years gone daily in search of fossil remains of importance at every tide, for many miles under the hanging cliffs at Lyme, whose fallen masses are her immediate object, as they alone contain these valuable relics of a former world, which must be snatched at the moment of their fall, at the continual risk of being crushed by the half suspended fragments they leave behind, or be left to be destroyed by the returning tide: – to her exertions we owe nearly all the fine specimens of Ichthyosauri of the great collections.

Similarly, the widow of the former Recorder of the City of London, Lady Harriet Silvester, wrote in her diary in 1824:

> The extraordinary thing in this young woman is that she has made herself so thoroughly acquainted with the science that the moment she finds any bones she knows to what tribe they belong. She fixes the bones on a frame with cement and then makes drawings and has them engraved. It is certainly a wonderful instance of divine favour – that this poor, ignorant girl should be so blessed, for by reading and application she has arrived to that degree of knowledge as to be in the habit of writing and talking with professors and other clever men on the subject, and they all acknowledge that she understands more of the science than anyone else in this kingdom.

According to an account in *The Dragon Seekers*, a visiting collector once wrote:

> I once gladly availed myself of a geological excursion with Mary Anning and was not a little surprised at her geological tact and acumen. A single glance at the edge of a fossil peeping from the Blue Lias, revealed to her the nature of the fossil and its name and character were instantly announced.[37]

Mary Anning was a remarkable young woman. It has been claimed that she was the inspiration for the popular tongue-twister: 'She sells seashells by the seashore', though Shelley Emmling, who investigated the legend, says this did not appear in print until Terry Sullivan incorporated it into a lyric he published in 1908, so that widely believed origin may be mistaken.[38]

Mary Anning was not just a fossil collector, or a dealer; she seriously studied what she found and used considerable ingenuity in comparing her fossils with living creatures. She noted that sepia is a brownish ink extracted from present-day cuttlefish, and so – finding fossils of similar animals bearing the traces of fossilized ink-sacs – she made her own ink from the fossils and demonstrated that it could be used in much the same way. When she found fossilized fish, she dissected fresh fish to seek anatomical comparisons. Her diligence and accuracy outshone the work of many professional palæontologists. Among the skeletons that she used to find were the remains of creatures that looked a little like dolphins. These were popularly known as sea dragons or crocodiles. One of her specimens was almost complete and it was inspected by the anatomist and surgeon, Everard Home, an unscrupulous investigator who was responsible for the loss of the Royal Society's collection of microscopes made by the pioneering microbiologist Antony van Leeuwenhoek. Home also took away the anatomical studies written by John Hunter and began publishing them as his own. Home had worked with Edward Jenner on vaccination, and had bribed the burial party of the Irish giant Charles Byrne, who measured 7 feet 7 in (2.31 metres), having them put rocks into Byrne's coffin while taking the corpse for Home to study. Even so, Home was cultivating a reputation for being a leading anatomist and he was more than willing to inspect Anning's latest fossil. Initially, he declared it to be a crocodile; then he changed his mind and decided it was a fish. A year or so later he was saying it was a specimen of a creature that was halfway between fish and crocodiles, and then changed his mind again, deciding it was an amphibian lying between salamanders and lizards. Home was a capricious and devious character, and his personality resonated throughout his work.

By 1826, six years after the Anning family had been rescued from penury by Birch, Mary had saved just enough money to purchase a shop of her own. The family lived in the rooms above, and they named

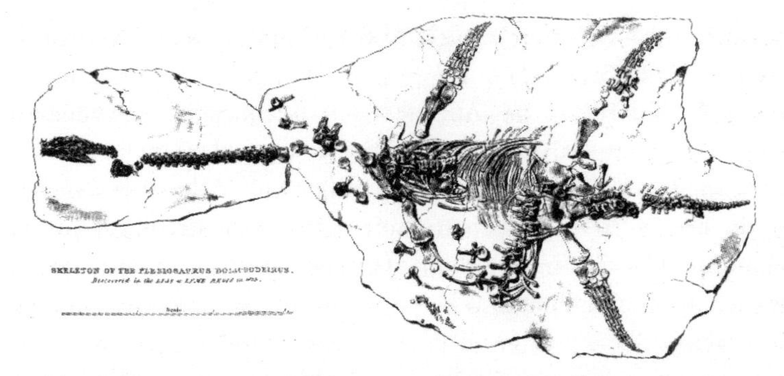

Mary Anning was a student of her subject, and not just a fossil hunter. Many of her finds became popular souvenirs, and this *Plesiosaurus* skeleton, carefully excavated by Anning in 1823, was prepared as a lithograph by Thomas Webster.

the premises 'Anning's Fossil Depot'. Business was soon flourishing, and the local press reported the opening, mentioning that in the middle of the display was a fine ichthyosaur skeleton. Geologists came to buy specimens from her, including collectors like Gideon Mantell, George William Featherstonhaugh (a curious, ancient English surname simply pronounced 'fanshaw') who described her as 'a very clever funny Creature', and even royalty: King Frederick Augustus II of Saxony visited her shop to buy a fossil ichthyosaur skeleton. In her later years Anning lost most of her personal money through a bad investment – sources are uncertain how this occurred – and William Buckland approached the British Association for the Advancement of Science and the Government to propose that she was given funds to continue her work. Mary Anning was granted a modest civil list pension by the royal household, in recognition of her contributions to the new science of palæontology. It brought her a trifling sum of just £25 each year, equivalent to £1,200 or some $1,600 in 2018. The Civil List Act 1837 stipulated that these pensions should be granted 'to such persons only as have just claims on the royal beneficence or who by their personal services to the Crown, or by the performance of duties to the public, or by their useful discoveries in science and attainments in literature and the arts, have merited the gracious consideration of their sovereign and the gratitude of their country.' Artistic civil list pensioners of the early nineteenth century were granted larger sums; the poets Lord Byron and William Wordsworth each received £300.

By the mid-1840s Mary Anning began to acquire a new reputation – her behaviour was changing, and the local people thought she was becoming an alcoholic. They were wrong: she had developed cancer of the breast. To keep the pain under control she drank increasing amounts of laudanum, a solution of opium, which caused her slurred speech and unsteadiness. When the news of her illness spread, the Geological Society of London launched a fund to help with her expenses, and the new Dorset County Museum appointed her an honorary member. In 1847 Anning died, and members of the Geological Society donated money for a stained-glass window in her memory, which was unveiled in St Michael's parish church in Lyme Regis in 1850. An inscription nearby says:

> This window is sacred to the memory of Mary Anning of this parish, who died 9 March AD 1847 and is erected by the vicar and some members of the Geological Society of London in commemoration of her usefulness in furthering the science of geology, as also of her benevolence of heart and integrity of life.

It was a rare distinction, and so richly deserved. Henry de la Beche, president of the Geological Society of London, delivered a eulogy that was published in the society's *Transactions*. No other woman scientist had been similarly commemorated; indeed, the Society did not admit women members until 1904. The president began: 'I cannot close this notice of our losses by death without adverting to that of one, who though not placed among even the easier classes of society, but one who had to earn her daily bread by her labour, yet contributed by her talents and untiring researches in no small degree to our knowledge of the great Enalio-Saurians, and other forms of organic life entombed in the vicinity of Lyme Regis.' The eminent Charles Dickens dedicated an article to her in his literary magazine *All the Year Round*, ending with this tribute: 'The carpenter's daughter has won a name for herself, and has deserved to win it.' More than any other single individual, it was she who launched the study of those curious prehistoric reptiles.

To commemorate Mary Anning's lifetime of devotion to studying and collecting fossils, in 1850 the Geological Society of London funded this stained-glass window for St. Michael's Church, Lyme Regis, showing six religious acts of mercy.

Mary Anning was the first full-time professional palæontologist anywhere in the world.

Since the start of the nineteenth century, natural philosophy had become a common currency for the public. Amateur investigators were everywhere, and collecting curiosities was the perfect pastime for those with social aspirations. The enlightenment had percolated through society, and a new sense of rational thought was replacing traditional superstition. Yet in modern terms, scientific understanding was limited. The term 'scientist' did not exist; that term was not coined until 1834 when William Whewell, a philosopher polymath at Cambridge University, introduced it. Although the existence of now-extinct life forms was widely accepted, there was no understanding of the mechanisms for extinction. Humans were held to be a uniquely gifted form of life, though I have shown elsewhere that a single plant cell can detect stimuli similar to those we know through the classical five senses (sight, hearing, taste, smell and touch). Humans are optimized, but not unique.[39]

As knowledge expanded, the evidence unearthed by geologists was still interpreted in biblical terms. Gravel beds, for instance, simply proved the flood as described in the Old Testament. There was no formal association for oryctologists (as fossil enthusiasts were still known), and most of the fossils in collections were shells and isolated vertebræ. Most wealthy collectors purchased their specimens. Although they enjoyed wandering out and about in nature, picking up items of interest, most lacked the acute levels of perception to identify what was important. Naturalists know that a collector must 'get their eye in', and most fossil enthusiasts were armchair amateurs who found it easier just to buy what they could.

The science of geology was given a boost by the work of an uneducated genius, the son of an Oxfordshire blacksmith. This was William Smith, who became a surveyor and worked on the construction of canals and coal mines across the country. This is when he noticed how the same strata cropped up in disparate parts of Britain, and eventually he collected all his observations together to create the first geological map of the whole of Britain. Because he was of working-class origins he was widely dismissed by educated society; he spent time in a debtor's prison and his work was extensively plagiarized. Eventually, however, his conclusions were published in 1815 as a vast and detailed map, 8 feet 6 inches (2.6 metres) long, the first detailed geological map of an entire country produced anywhere. It was an astonishing achievement. Smith's conviction that rocks could be systematically studied arose from his awareness of the increasing interest in fossils. Whereas most people regarded them as objects of curiosity, Smith saw them as indicators of a hidden reality – the key to unravelling the strata on which Britain was built. Smith had clear sight and a methodical mind, and created lyrical literature. In 1796, he wrote this prescient prose:

> Fossils have been long studied as great curiosities, collected with great pains, treasured with great care and at a great expense, and shown and admired with as much pleasure as a child's hobby-horse is shown and admired by himself and his playfellows, because it is pretty; and this has been done by thousands who have never paid the least regard to that wonderful order and regularity with which nature has disposed of these singular productions, and assigned to each class its peculiar stratum.

Smith fell into debt whan an agreement to provide stone for a customer could not be fulfilled. He raised £700 by selling his fossil

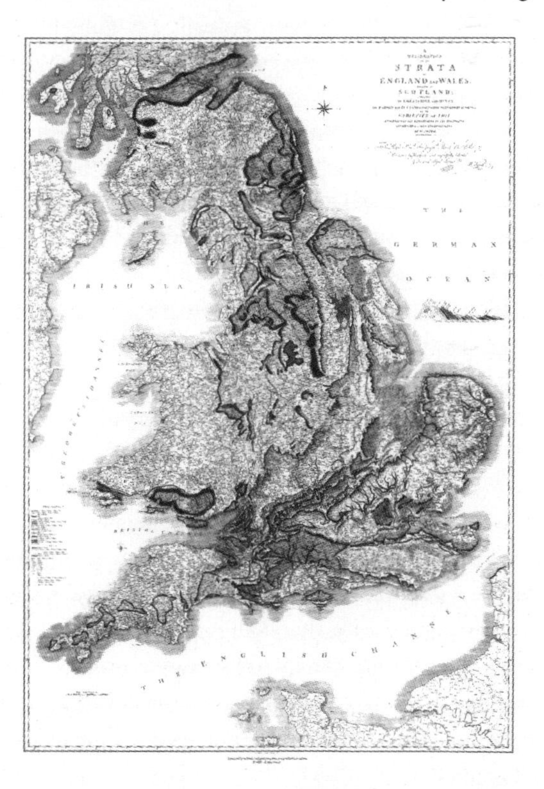

The first geological map of Britain was published by William Smith, a self-educated surveyor, in 1815. It was over 8 feet (2.6 metres) long and was based on an outline map by John Cary. This was the first geological map to cover such a large area.

collection to the British Museum but was confined in the debtor's prison for the remaining £300. After his release in 1819 he worked as surveyor for Sir John Johnstone, who soon appointed him land steward to the family estate in Hackness near Scarborough. Johnstone was astonished at Smith's knowledge, and encouraged him to design the Rotunda, a display of the geology of the Yorkshire coast with strata all shown in their correct order in curved cases. This remarkable building survives to the present day, and is the oldest geological museum anywhere in the world.

Gideon Mantell was also conversant with Smith's map, and Mantell entered into deep discussions with others interested in rocks. For instance, he became friendly with a mining engineer named John Hawkins who was similarly interested in the strata that surrounded them. Other friends were George Bellas Greenough, who was working towards his own map of Britain's rocky strata, and James Sowerby, who encouraged the young Mantell to publish a book that would extend and refine the results for Sussex published in William Smith's pioneering map.

The year 1816 was a crucial watershed for Mantell. He married Mary Ann Woodhouse, a talented young artist, and took on a new post as surgeon at the Royal Artillery Hospital in nearby Ringmer, Sussex. He also published his first scholarly work that year, a book on the minerals in the area around Lewes. Next, he published a magazine article on the rocky strata of southeast Sussex. By this time he was starting to become prominent among the fossil hunters. When Colonel Thomas James Birch decided to auction his entire fossil collection for the benefit of Mary Anning, it was to Mantell that he wrote in March 1820, saying: 'The sale is for the benefit of the poor woman and her son and daughter at Lyme, who have in truth found almost all the fine things which have been submitted to scientific investigation. I may never again possess what I am about to part with, yet in doing it I shall have the satisfaction of knowing that the money will be well applied.'

It was also in 1820 that Mantell heard of a new source of fossilized bones – the quarries in Tilgate Forest near Crawley, Surrey, some 10 miles (16 km) to the north of Cuckfield, from where most of his specimens had come. He travelled by one-horse chaise and stayed at the Talbot inn, which stands to this day, and from there he rode over to the quarries. Pieces of rock had been set out by the quarrymen for him to peruse and perhaps purchase, and he took a selection to study. We remember the men in palæontology much better than the womenfolk, but – just as Mary Anning regularly provided new and exciting specimens for them to study – Mantell later wrote that he owed his greatest leap forward to his wife, when she stumbled across a find that would finally launch the science that we now know as palæontology.

It was the summer of 1822, and Mary Ann was travelling with her husband *en route* to one of his patients at home. By the side of the road, she noticed a chunky fossil on a pile of discarded rubble and showed it to her husband. It was a huge tooth. This was the crucial discovery, and Gideon launched a full-scale investigation of the

A devotee of Smith's geological revelations was Gideon Mantell, an obstetrician and a brilliant amateur fossil hunter, who in 1822 first named a fossil dinosaur – *Iguanodon*. The fossils were originally thought to be those of a rhinoceros. In 1816 Mary Woodhouse married Gideon Mantell and became his co-worker. She was an accomplished artist and prepared many illustrations for publication. It has been said that Mary was the first to discover iguanodon teeth.

Tilgate Forest quarries, looking for more. Mantell did not record until 1827 that he had first been presented with that huge tooth by his wife, but – although they could not know this at the time – this was the tooth of an *Iguanodon*, a dinosaur that we now know measured 43 feet (13 metres) long and weighed some 4 tons. Other specimens were excavated by a quarryman, Mr. Isaac Leney from Cuckfield, and a selection of fossils was soon amassed. Clearly, they could only have come from a huge animal, and Mantell became increasingly excited.[40]

By the end of 1822 Gideon Mantell had at least half a dozen of these specimens, so he travelled to Paris the following year and showed them to Cuvier, who formally identified the specimens that the Mantells had collected as 'the teeth of a gigantic crocodile, the teeth of a rhinoceros [and] bones of an herbivorous animal' – in reality, the 'crocodile' was *Megalosaurus* and the 'herbivorous animal' would eventually prove to be *Iguanodon*. Mary Anning had recently found a virtually complete *Plesiosaurus* skeleton, and this was put on display along with the 'Megalosaurus or great Fossil Lizard of Stonesfield' that Buckland had brought along, and which Parkinson had included in his book. It had been excavated at Stonesfield. Cuvier was less dismissive of this fossil, describing it as 'a monitor [lizard] forty feet long and the size of an elephant.'

Gideon and Mary Ann Mantell worked together on their major volume on the fossils of the South Downs, and it was published at about the same date as Parkinson's book. It was an important book and it launched a more systematic study of fossils. The Mantells' new book received a royal endorsement from King George IV at Carlton House Palace: 'His Majesty is pleased to command that his Name should be placed at the head of the Subscription List for four copies.'[41]

Most of the fine lithographs that Mary Ann meticulously prepared were of shellfish, though towards the end of the plates there were tantalizing glimpses of the scales of the skin of fossil fish, and also a reptilian jawbone, complete with teeth.

It was the large fossilized teeth that Mary Ann and Mr. Leney had collected that continued to fascinate Gideon Mantell. He compared these teeth with those of an iguana in the collection of the Hunterian

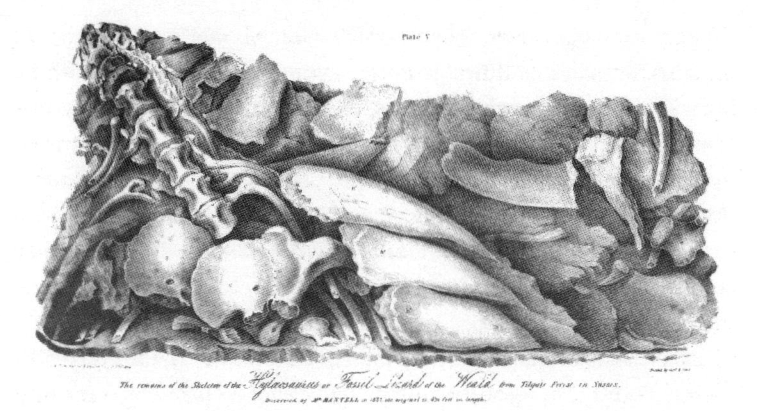

The first engraving of the *Hylæosaurus* fossil that the Mantells unearthed in the Tilgate Forest quarry in 1832 was redrawn & lithographed by F. Pollard for *The Geology of the South East of England*, which was published in London a year later.

Society and became increasingly convinced that his fossils were from a gigantic version of a monitor lizard. He consulted others on his findings, and in 1824 Cuvier wrote again to concede that: 'I believe they belong to the order of reptiles.' Mantell was thrilled by this confirmation from such a well-accepted authority, and thought that he might name the creature *Iguanosaurus*. A colleague, William Daniel Conybeare, by this time dean of Llandaff in Wales and an active amateur collector, suggested that the name be modified. 'The name you propose,' he wrote, 'Iguano Saurus, will hardly do because it is equally applicable to the modern iguana … Iguanodon (having the teeth of an iguana) would be better.' Mantell agreed, and presented a paper to the Royal Society on February 10, 1825, entitled 'Notice on the Iguanodon, a newly discovered fossil reptile, from the sandstone of the Tilgate Forest in Sussex.' His *Iguanodon* was officially acknowledged as a gigantic prehistoric reptile, and he had drawn a sketch showing the bones they had retrieved superimposed on how he envisaged the rest of the creature. He was impressed by a pointed spike that had been unearthed, and had wrongly assumed that it belonged on the snout (like the horn on a rhinoceros). He drew attention to the creature's huge hindlegs, which he compared with those of an elephant, and concluded that their bulk was necessary to fit it for a life on land, writing that 'the legs must have sustained the weight of the body in a manner more nearly resembling those in the pachydermal Mammalia.'

On October 26, 1825, Mantell sent a package of his specimens to Professor Adam Sedgwick at Cambridge University, including 'casts of the best teeth of the Iguanodon in my collection' – the originals he kept for himself. Mantell was writing up his notes for a new book on the geology of Sussex and this was to be a landmark publication. Mantell usually referred to it by part of its subtitle: 'the fossils of the Tilgate Forest'. The specimens were illustrated by his wife Mary Ann

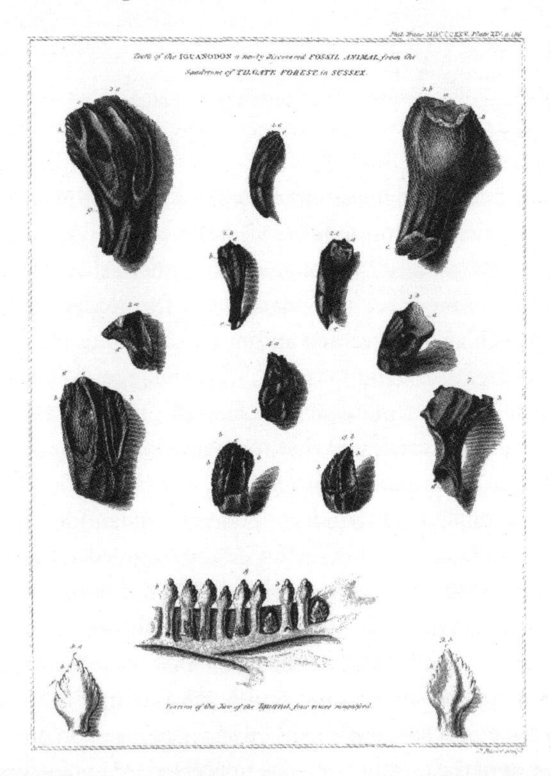

Gideon Mantell illustrated his 1825 paper for the Royal Society with this plate showing the Tilgate Forest iguanadon teeth in comparison with the jaw of an existing iguana. The genus of his fossil has since been reclassified as *Therosaurus*.

in exquisite detail, and have a photographic clarity. For the first time Mantell described the fossil reptiles in detail – the contents page listed them: Crocodiles, Megalosaurus, Iguanodon, Plesiosaurus … these giants of prehistory were now becoming familiar to the academic world. Dinosaurs had at last arrived on the scientific scene.[42]

The book proved to be a landmark, and its level of detail is astonishing. Mantell discusses with precision the strata in which the fossils were found, dissecting each layer meticulously, so that the book stands today as a definitive statement. This is no amateurish essay into the unknown, but a serious scientific study. Yet the workload was almost unendurable: in his dedication of the book to Davies Gilbert, the Member of Parliament for Bodmin in Cornwall, Mantell wrote: 'You are fully aware of the disadvantages under which I have laboured, and will generously make every allowance for the imperfections of a work, composed amidst engagements of the most harassing nature.' Yet he pressed on, and in 1832 discovered the third dinosaur to be scientifically described. It was only half the size of the *Iguanodon* and *Megalosaurus*, measuring less than 30 feet (9 metres) long. It was found, again, in strata at Tilgate Forest, and Mantell wrote: 'I venture to suggest the propriety of referring it to a new genus of saurian … and I propose to distinguish it by the name of *Hylæosaurus*.' In spite of his continued success, life was not easy; in 1833 Mantell moved to the seaside resort town of Brighton but could not sustain his medical practice, and, when he became impoverished, his home was converted by the town council into a museum.

The following year, Mantell received some dramatic news: *Iguanodon* fossils had suddenly been found in a quarry near Maidstone, in Kent. He decided to investigate, but by the time he was able to reach the site, the strata had already been blown up with gunpowder (the standard accounts all say that the rocks were 'dynamited' but that explosive was not invented for another 30 years). Just one fossil-bearing slab survived the demands of the quarrymen, and the owner demanded £25 (in 2018 about £1,200 or some $1,600) before he would release it. Mantell didn't have any spare money, but a group of his friends clubbed together to purchase the rock. It was transported to his home, where it took pride of place in his personal museum. With typical British wit, they called it the Mantell-piece. It proved impossible to separate out the individual bones, so Mantell and his wife worked on reconstructing the appearance of the remains from what they could see. The rock, formally named the Maidstone Slab, can be seen to this day hanging lost and lonely in the dinosaur gallery at the Natural History Museum in London. That is not the new dinosaur gallery; if you follow the signs, you will come to a kind of indoor theme park illustrated with inane cartoons and littered with shops selling plastic souvenirs, fridge magnets and children's clothing. The new dinosaur display in that museum is like a gloomy version of Disney World, and most of the Victorian specimens are hanging on a high wall in a nearby gallery where few people notice them.

Mantell's reconstructions of the dinosaur are largely based on this specimen, and he construed it as a quadruped with the proportions of a bear. Mantell made a sketch on paper which showed the bones they could recognize, placing the horn on the nose; and this remained the conventional interpretation for a century. Mantell began to give public lectures on his work at his home museum, and they proved so popular that in 1838 they were published in a single book devoted to the wonders of geology.[43]

Mantell so liked showing visitors around the collections that he frequently overlooked charging the standard entrance fee, and the

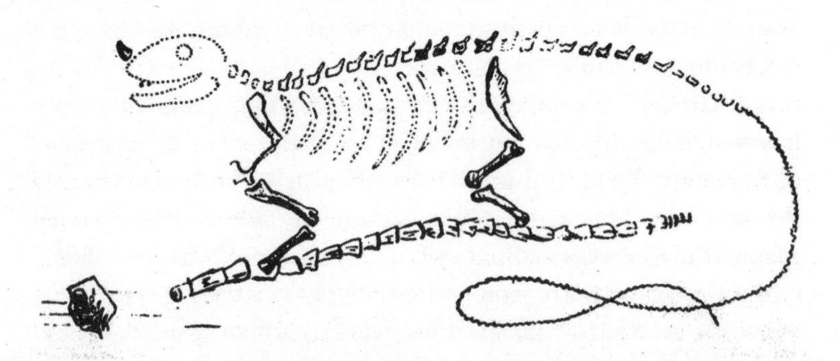

A horn had been found by Mary Mantell among the fossilized iguanodon remains, and when her husband sketched how the animal might have appeared in life, he placed it on the snout. Copies of this incorrect view still appeared 138 years later.

museum soon became insolvent. By now he was becoming desperate, so he offered to sell his entire collection to the British Museum for £5,000 and readily accepted their counter-offer of £4,000 (worth some 50 times as much today: about £200,000 or $250,000). Mantell, who had become the founding father of dinosaurs after his wife's prescient discovery of that first *Iguanodon* tooth, cut back on his fossil lectures and moved to Clapham Common in London to concentrate on practising as a physician. In 1839 Mary Ann left Gideon and emigrated to New Zealand, later sending him some new fossil discoveries, while their daughter Hannah died in 1840. The next year saw a dreadful accident: Mantell was caught in the reins of his horse-drawn carriage and severely damaged his back. From that day onwards his spine was painful and problematic. He moved to Pimlico in 1844 and, confined largely to his home, he continued to write books and papers on his discoveries, deadening the constant pain with laudanum, the solution of opium also taken by Mary Anning.

Mantell was the world's first authority on dinosaurs; four of the five dinosaurs then known had been discovered by him and his wife. Although his life had been enriched with so many discoveries, after his wife left him and he suffered his fall, the pain became increasingly severe. Eventually, on November 10, 1852, Mantell over-dosed on the laudanum, and he died that same afternoon. He was 62. Opinions are divided as to whether it was an accidental overdose or suicide. At autopsy, it was discovered that his vertebral column had curved, a condition now known as scoliosis.[44] After the post-mortem dissection, a section of his spine was removed and preserved for study, and it remained in the pathology museum collection of the Royal College of Surgeons of England until 1969 when it was unceremoniously thrown away, due to a lack of storage space. In 2000, to commemorate Mantell's discovery, a monument was unveiled at Whiteman's Green, Cuckfield, where Mary Ann had found that first tooth. Maidstone, where the celebrated 'Mantell-piece' slab had been retrieved, adopted the iguanodon on its coat of arms. On the town's crest, a lion supports a shield on the right, with an iguanodon on the left. The official designation is written in the ancient blend of Norman French and English beloved of heraldry specialists:

Arms: Or a Fesse wavy Azure between three Torteaux on a Chief Gules a Lion passant guardant Or.
Crest: Issuant from a Mural Crown Or a Horse's Head Argent gorged with a Chaplet of Hops fructed proper, Mantled Azure doubled Or.
Supporters: On the dexter side an Iguanodon proper collared Gules and on the sinister side a Lion Or collared Gules.
Motto: AGRICULTURE AND COMMERCE.

In the Maidstone Museum, there is a stained-glass window with the iguanodon on display. Neither the published version of the iguanodon, nor the one in that window, has that horn on the snout.

By this time, new fossil dinosaurs were being discovered in France. Remains of a large creature were described in 1838 by a French palæontologist, Jacques Amand Eudes-Deslongchamps, who lived in

Maidstone, close to where the Mantells made their discoveries, was granted a coat of arms in 1619. In 1949 the two supporters either side were added, a collared lion (on the right) and an *Iguanodon*. This version shows the correct snout.

Normandy. He didn't have much to go on; bones from the abdomen, a front- and hind-limb, the tail, and three types of ribs. He decided to name it *Poekilopleuron* from the Greek ποίκιλος (*poikilos*, varied) and πλευρών (*pleuron*, rib), and the bones were donated to the Museum of the Faculty of Science at Caen. During World War II an Allied bombing raid demolished the department and all the fossils were destroyed. However, plaster casts had been made, so copies survived in the Muséum National d'Histoire Naturelle in Paris and the Peabody Museum at Yale. This dinosaur was about 23 feet (7 metres) long and weighed about 5 tons. It would have lived 167 million years ago and was similar to *Megalosaurus*.

Although people are taught that the science of dinosaurs began with Richard Owen, the eminent anatomist, he did not begin his serious research until the 1840s and we have seen how far the study of dinosaurs had already advanced. Owen was certainly brilliant, and he was also a tireless investigator. In his early days he had only a few fragmented fossils available for study, yet from this he was able to deduce something of the appearance of an entire dinosaur, and even gain some insights into dinosaur physiology; yet he was also dishonest, vain, vindictive and quarrelsome. Owen frequently claimed the discoveries of others as his own, and was self-righteous in the extreme. He argued repeatedly with Charles Darwin, and refused to accept the theory of evolution. Owen enrolled at the University of Edinburgh Medical School in 1824 aged 20, but – like Darwin, who enrolled at the same school the following year – he was dissatisfied with the quality of instruction and decided instead to study under John Barclay, a convinced anti-materialist. The great debate of the time was the duality of body and soul, and Barclay firmly insisted that life was founded upon a Vital Principle while the individual identity rested in the Soul. Although science was increasingly seen as offering an explanation for living organisms, Barclay adhered rigidly to his non-material views. Neither Darwin nor Owen graduated in Edinburgh, Owen eventually being apprenticed to John Abernathy in London. A philosopher and surgeon, Abernathy was president of the Royal College of Surgeons, and he was instrumental in obtaining membership of the College for Owen in 1826. Charles Darwin, meanwhile, went on his extended holiday as the travelling companion to the captain of HMS *Beagle*.

Owen's first professional task was to assist William Clift, conservator of the Royal College of Surgeons, in cataloguing the collection of 13,000 human and zoological specimens that had been amassed by the eminent surgeon John Hunter. A previous custodian of the papers had been Sir Everard Home, that unscrupulous surgeon who had published many of Hunter's discoveries as his own, and who set fire to the entire archive when an investigation into his conduct loomed near. It fell to Owen to identify what remained in order to rebuild the list. By 1830 he had sorted the documents and identified every anatomical specimen, and was about to publish the full catalogue of the Hunterian Collection. By that time he was accepted as the main authority on Hunter's voluminous work. In July 1835 he married Clift's daughter, Caroline Amelia, and they had one son, William. In 1837 Owen was charged with delivering the first series of Hunterian Public Lectures, and his reputation grew to such an extent that he was appointed to teach natural history to the children of Queen Victoria; but his professional attitude remained obdurate, demanding and unpleasant. Charles Darwin (who attended many of Owen's lectures) wrote later that Owen became his enemy after the *Origin of Species* was published 'not owing to any quarrel between us, but as far as I could judge out of jealousy at its success.'

Fossil dinosaurs were also being discovered elsewhere in England. In the late summer of 1834, the curator of the Bristol Institution, Samuel Stutchbury, accompanied his surgeon friend Henry Riley on an expedition to Clifton. Riley had been intrigued by the news of the newly discovered 'saurian' fossils and took his friend to prospect in the quarry of Durdham Down, on the outskirts of Bristol. There were repeated reports of strange bones being discovered by quarrymen. It seemed that these bones might also be from gigantic prehistoric reptiles, and a short report was published in the U.S. in 1835.[45]

Theirs proved to be a small dinosaur, measuring some 6 feet 6 inches (2 metres) in length and weighing no more than 55 pounds (25 kg). The men were a diligent pair of investigators and they made a significant observation: in lizards, the roots of the teeth merge with the jawbone, but they noted that their dinosaur was different. It possessed tooth sockets, much like those of mammals. In their formal paper the following year they gave their discovery its name: *Thecodontosaurus*, derived from the Greek θήκή (*thēkē*, socket) and οδους (*odous*, tooth) – so here was yet another new dinosaur for scientists to study.[46]

Their searches also turned up teeth of phytosaurian dinosaurs that they named *Paleosaurus cylindrodon* and *P. platyodon*. Although they didn't know it, that generic name had already been thought up by Étienne Geoffroy Saint-Hilaire, so the proposed name was soon abandoned. As *Thecodontosaurus* this became the fifth dinosaur to be academically named, following *Megalosaurus*, *Iguanodon*, *Streptospondylus* and *Hylæosaurus*. The fossil was later provided with an appropriate species name, *T. antiquus*.

Palæontology has long attracted peculiar people, and few were more eccentric than a collector named Thomas Hawkins. Although we imagine the early palæontologists trudging across barren rocks and scrambling into quarries, many of them, as we have seen, adopted a more leisurely approach – they simply purchased specimens dug up by quarrymen, or bought them from specialist dealers like Mary Anning. The young Hawkins was loquacious, brash and bullying, difficult to get on with and given to outbursts of pugnacious prose. He was the son of a wealthy farmer who, rather than encouraging his wayward son to work with him on the family farm in Somerset, gave him a generous allowance to stay away. As a result, the young Hawkins was able to indulge his new passion for fossil collecting and proudly boasted that, by the age of 20 (in 1830), he already had a large and varied collection. He used to wander round the quarries near Walton and Street in Somerset, watching out for discoveries the quarrymen were making. On several occasions, he saw a priceless fossil in a slab of rock that a worker was about to break up and he promptly stepped in with an offer of largesse that could not be refused. He soon gained a reputation for being someone who would pay good money for these fossils whenever they came to light, and so his collection grew, while he had no need to soil his hands by digging. Yet he soon encountered a problem faced by every collector: fossil remains were hardly ever complete. Quite often, valuable fragments were lost as stone was chipped away from the specimen. Dinosaur skeletons usually had limbs missing and often there was no skull (as in the case of Anning's first ichthyosaur, and the *Brontosaurus* later described by Marsh). Early in his career, Hawkins became adept at cleaning up fossils and replaced any parts that were missing with dyed plaster. Today, it is acceptable to create an entire dinosaur skeleton from just a few fossil fragments, but at that time any restoration was frowned upon. Mantell summed up Hawkins perfectly as 'a very young man who has more money than wit'.

Hawkins was a proselytizing Christian and firmly believed in Adam and Eve. Yet he also followed the latest trends in scientific discovery, and formed the view that fossils could reveal what the world was like before humans had been created. The early Earth, he thought, was an alien and hostile place, bathed in murky darkness that sunlight could not penetrate, and peopled by strange monstrous beings that were intent on destruction. On one of his casual visits to a quarry, he found that the tail of a huge ichthyosaur had been laid bare by a workman. The men had agreed to dig out the rest in a few days' time, but Hawkins was insistent that the job should be done at once. It was already dusk – but he made them fetch candles and lamps, and work on through the night. Eventually, the rocky strata bearing the fossil were laid out in pieces on a wagon, ready to be transported to Hawkins' home on the farm. It took him several weeks to chip away the rock to release the whole animal, but in the end he was confronted by an ichthyosaur measuring some 7 feet (2 metres) from nose to tail. The hunks of rock bearing the skeleton were assembled together in a wooden frame, and the result was an entire animal – or almost entire. Whatever was missing, Hawkins created out of plaster that he carefully stained to match the rest of the rock. This gave a convincing result – at least, it did for anybody who wanted to be impressed by the entire creature. For the palæontologists of the time, it posed problems. If you were not certain whether the fossil was entire, it was impossible to tell the real fossil from the replacement plaster, so describing the skeleton accurately would be scientifically invalid.

None of this mattered to the irrepressible and domineering Hawkins. Within a few weeks he had himself forgotten which parts of a fossil were original and which he had created from plaster. The creature looked impressive in its apparent completeness, and that was all that mattered to him. His wish was not to pursue scholarship, but to exhibit monsters from a bygone age. If they had pieces missing, he felt it his duty to bring them back to a state of perfection – only then could their prehistoric magnificence be appreciated. His is a very modern attitude. Present-day palæontologists think nothing of re-creating vast skeletons of imaginary dinosaurs from plastic, when in reality only a very few bones have been discovered. What was considered unprofessional in Hawkins' time is carried out on a grander scale today.

In 1833 Hawkins heard that an ichthyosaur skeleton had emerged on low-lying rocks at Lyme Regis. He travelled to the town, and discovered that it could be accessed only at low tide. Hawkins paid a guinea for the finder to grant him the right to own the fossil (£1 1s, now worth about £60 or $80) and told him to assemble a group of workmen, ready to excavate the entire skeleton. He could not resist telling Mary Anning of his find, and she warned him that the rock in which this fossil lay was likely to crumble as it dried. She said it was marl, and she knew that it was rich in iron pyrites (fool's gold, FeS_2). For the next few days no work took place – storms blew in from the west and the beaches were suddenly inaccessible.

By the time the sky had cleared and the winds had dropped, Hawkins was desperate to extract his new fossil from the beach. At low tide the men were sent to work, and they managed to dig out a large hunk of rock which contained the entire skeleton. It was taken to Hawkins' home where he set to work with his 'magic chisel', and within weeks it was ready for display. Any parts that were missing were created out of the plaster, so the finished result owed as much to Hawkins' creative impulses as it did to nature. By this time, he was spending the winter months in London, mingling whenever he could with the great palæontologists of the day, and he soon managed to become acquainted with William Buckland. Buckland expressed admiration for Hawkins' fossil collection. He was particularly impressed by the pristine cleanliness of the specimens Hawkins had prepared and also by their astonishing completeness. Hawkins was delighted, and carefully cultivated the relationship, talking always of the might of the Creator, the power of nature, and the evil intent of these denizens from the unfathomable past. These fossils represented grim brutes, he insisted, lusting for blood. Encouraged by Buckland, Hawkins soon set about writing up his discoveries for publication. Many of them appeared in his first book, which came out in 1834.[47]

The text was filled with imprecations about the majesty of the creation, and the evil of the monsters that mankind had overcome. He solicited Buckland's support for a proposal to sell his fossil collection to the British Museum for £4,000 (now about £200,000 or $240,000). The management would have none of it, and asked for external opinions as to the real value. Buckland was recommended as a referee by Hawkins, and Mantell was also asked to provide a valuation. Buckland totted up the individual specimens, and provisionally said they were worth between £1,000 and £1,500. Eventually, he decided upon £1,250 as the right price. Mantell separately sent in his own valuation; it came to almost exactly the same sum of money. Doubtless the two men had discussed the total between them, for surely this could not have been coincidence. They wrote to Hawkins, who replied in his oleaginous and glowingly complimentary tone, praising both men for their scholarship and wisdom, and proposing that – as a gesture of his own generosity – the price could perhaps be £2,300. Eventually, after further argument, Hawkins reluctantly accepted £1,250.

No sooner were the specimens in the museum's possession in 1835 than the Keeper of Natural History, Charles König, began to make arrangements for them to go on display. His first choice was a magnificent specimen of an ichthyosaur measuring 25 feet (7.5 metres) long. Everything was immaculate, and the skeleton was perfect in every detail. Naturally, König was concerned; no fossil is ever likely to be entirely complete. Yet nothing seemed to be wrong with the skeleton. Curiously, he looked again at the illustration that Hawkins had printed in his catalogue of the collection – and suddenly he could see something odd. The original lithograph showed that the skeleton had not been complete when Hawkins had first illustrated it. Half the tail was missing, and the right forelimb was shown only as a dotted outline, indicating that it was not present when the fossil had been found. The 'perfect' skeleton had been made up with dyed plaster. Hawkins had protested in his negotiations that his perfect specimens were all genuine, and were testimony to his skill as a conservator, but it was suddenly clear how he had been improving on nature. Buckland and Mantell were both informed immediately, since König now felt that the specimens were worth far less than the museum had paid. He

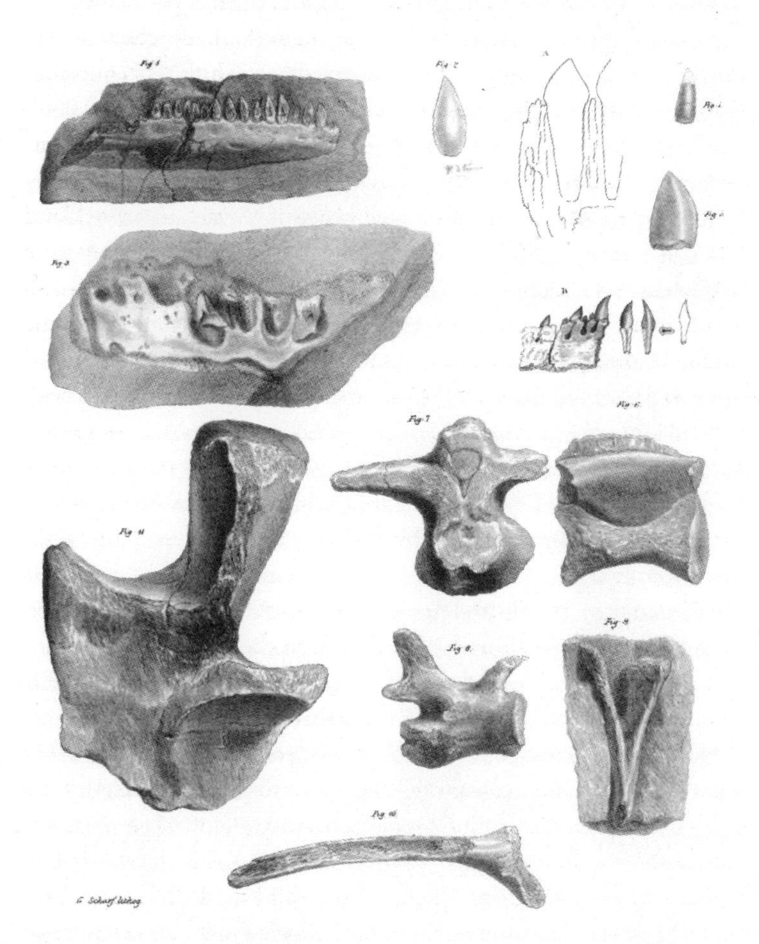

Henry Riley and Samuel Stutchbury found fossils of *Thecodontosaurus* at Clifton near Bristol in 1834, and their paper illustrated the jaw, teeth, part of the ilium, vertebræ and a rib (bottom). The genus was not recognized as a dinosaur until 1870.

wrote to say that the fossils could not be put on display after all, since so much was plaster, and said that he would await further instructions. Buckland, against all expectations, rose to the defence of Hawkins. He insisted that no suggestion had ever been made that the specimens were entirely natural. It was only to be expected, he went on, that repairs here and there were necessary in restoring a skeleton, and there was definitely no hint of 'fraud or collusion' on his part.

None of this debacle helped the museum's reputation, which was already being investigated by an inquiry set up by the House of Commons. When König came before the committee, he was cautious about the whole affair, admitting that the skeletons were less than perfect, and agreeing that the price of £1,250 may have been a little more than the fossil collection was worth. Hawkins railed against König, accusing him of pretending that the specimens were imperfect, when in fact all such specimens had merely been cosmetically improved. Mantell thought that Hawkins had been guilty of double-dealing, but put it down to mental instability. The specimen, along with others from the collection, is in the collections of the Natural History Museum in London, identified as *Temnodontosaurus platyodon*. You can still see tiny indentations all over the specimen. These are the dents left by the point of König's knife, as he probed to distinguish between plaster and stone. All the plaster additions were subsequently painted to be subtly different, and this reveals that – in addition to the forelimb and the tail – many of the ribs, the tips of the hindlimbs, and even a vertebra, had all been constructed out of plaster by Hawkins. In truth, the skeleton is partly faked and was worth considerably less than a perfect specimen.

Not only was Hawkins unreliable as a conveyor of fossils, but in his writing he frequently substituted his own invented Latin names for those already granted to the fossils he found. He often complained that neither Latin nor Greek was good enough for naming fossils – the language in which they should be named, he insisted, was Hebrew. On he rambled, and soon published a second book, with yet more of his startling revelations. The book has an extraordinary style and is virtually unreadable. Here, for example, is a passage from Chapter V:

> The sublime discloses itself only in the silence of which we speak, when, by the most stupendous Efforts of Intellect, by the revivification of Worlds, by the inhabitation thereof of all the Creatures which the labouring Soul can re-articulate, we stand in a Presence which has not, nor ever shall have, one sympathy with ourselves; those Worlds, those antipodal Populations, that Presence passion less, and silent dead; I say the instruments of a few bones verify a Sublimity before which no man can stand unappalled.

And so it drones on, perhaps the most impenetrable prose in the history of science.[48]

When Hawkins heard of the findings of the House of Commons Committee and its report on the part-plaster skeletons, he immediately threatened to sue for defamation. He was given to litigation. Thus, when a visitor at a nearby property casually picked some fruit from his strawberry patch, Hawkins was accused of using 'disproportionate violence' in protesting. He ended up in a legal dispute, meanwhile declaring himself the Earl of Kent.

Richard Owen was becoming increasingly intrigued by the reports of fossilized giant reptiles, and knew of some bones that had been described by an amateur palæontologist, John Kingdon, in a communication to the Geological Society of June 1825. Most learned opinion at the time was still that these were fossils of familiar creatures – porpoises, perhaps, or possibly extinct crocodiles – but Owen was an experienced anatomist and was certain this was wrong. In 1841 he decided the newly discovered fossils represented reptilian animals and he named the genus *Cetiosaurus*. He was only half right – although Owen had correctly determined that these were reptiles, he concluded they were swimming creatures, somewhat like plesiosaurs, which is why he coined the name cetiosaur from the Greek κήτειος (*kèteios*, sea-monster). Owen wanted to learn more, and one day late in 1841 he hastened to 15 Aldersgate Street, near St Paul's Cathedral, to visit his colleague William Devonshire Saull, a businessman and an avid collector of antiquities. Saull had amassed a collection of 20,000 specimens (most of them antiquities from the Middle East) which he had carefully catalogued and labelled. Many were of geological specimens, and some – the important specimens that Owen wanted to inspect – were fossils. Saull was a long-standing friend of Mantell, and they had often exchanged specimens. Among the many relics Saull showed him, Owen picked up a piece of *Iguanodon* bone, and turned it over in his hands. He knew of the various other gigantic specimens that collectors had unearthed – *Megalosaurus*, *Mosasaurus*, his own *Cetiosaurus* – and was suddenly inspired. These were not just creatures from the past, reminders of now-extinct worlds populated by animals like those of the present day; Owen became convinced that these all belonged to a single great family of reptiles. He suddenly realized that they were different from all the other reptiles we knew. Whereas present-day lizards have sprawling legs that splay out either side, these giant reptiles had downward-pointing limbs that functioned like columns to support their weight on dry land. He speculated that dinosaurs might have been warm-blooded, and he noted that they had five vertebræ fused to form the pelvic girdle, which he knew was not the case with other reptiles. Later discoveries would show he wasn't entirely correct (some dinosaurs have different numbers of fused sacral vertebræ), but he was right to recognize these as a new group of huge, extinct monsters. This was a crucial breakthrough, and Owen decided to announce his conclusions at the eleventh annual meeting of the British Association for the Advancement of Science.

The presentation took place on August 2, 1841, on a grey, dank day in London. Owen was a tremendous draw; Cuvier had died in 1832 and Owen had now become Europe's most renowned zoologist. He had lectured on fossil reptiles before, but this time was different – this was to be his announcement of an entire new class of gigantic reptiles. He began by courteously acknowledging the pioneering discoveries made by William Buckland and Gideon Mantell, both of whom he acknowledged with respect. Then he reviewed what was known about crocodiles, and their similarities to plesiosaurs. And then he moved on to the meat of the argument, and began by describing three genera that he was going to analyze in detail: the herbivorous *Iguanodon*, the carnivorous *Megalosaurus* and the armoured *Hylæosaurus*. These, he said, were different from anything alive. These, he told his audience, formed a distinctly new tribe. Gone was the notion that these were merely long-lost forms of animals that were similar to those still in existence; these were a form of life that nobody had ever seen. His audience was spellbound. So many people had accepted that they were ancient forms of crocodiles, or something similar, but Owen was adamant. *Megalosaurus*, he explained, was not a gigantic sprawling lizard, but a huge reptile that stood upright on powerful vertical legs. His new image of *Iguanodon* was of a great monster, standing tall on massive hindlimbs, and towering above the lesser beings that were dotted about its forest landscape. With his majestic prose and his own charismatic powers of oration, Owen had the audience entranced as he radically revised the previous size estimates published by his colleagues. Mantell, he said, had erred in scaling up the size of the limb bones of an iguana to an iguanodon, and reaching an overall length of 75 feet (23 metres). Far better was it to scale up the dimensions of each single vertebra. This posed a problem in knowing how many vertebræ there were in the backbone, for most of the skeletons had a spine that was far from complete. But his calculations worked well: he concluded that an iguanodon would have measured about 28 feet (8.5 metres) from nose to tail, a far more realistic figure that fits well with what we know today. Owen made mistakes of his own in his talk: he described *Thecodontosaurus* as a lizard, and *Cetiosaurus* as a

These bones of a young *Iguanodon* were excavated from Cowleaze Chine on the Isle of Wight and were included by Richard Owen in his monumental work *A History of British Fossil Reptiles* published by Cassells of London between 1849 and 1884.

crocodile, though we now know that both are sauropod dinosaurs. Sometime after the lecture, he coined a new taxonomic name to define the entire group. He resolved to call them Dinosauria, which he said would distinguish the entire 'distinct tribe or sub-order of Saurian Reptiles'. The word came from the Greek δεινός (*deinos*, terrible, awesome) and the familiar σαῦρος (*sauros*, lizard). It is a curious term, in that dinosaurs are definitely reptiles but are certainly not lizards, nor are they descended from them. This new term Dinosauria first appeared in the *Report on British Fossil Reptiles*, published the year following Owen's momentous lecture.[49]

Throughout his speech, Owen adhered to a strictly creationist view. He was convinced that these creatures had been made by divine providence, and the anatomical peculiarities he observed were, he insisted, the sure sign of intelligent design. He remained obdurate in these opinions, and was strongly opposed to any idea of evolutionary progress. He discussed these matters with Charles Darwin on many occasions, and the two became friends for a while. But when Darwin published his *Origin of Species* in 1859, Owen was to write a scathing review that he published anonymously. In later years, there was strong animosity between the two. Others had certainly laid the groundwork for this great revelation of the dinosaurs, even though it was Owen who coined the name. To this day, he is heralded as their great discoverer. As the BBC put it, Richard Owen is 'the man who invented the dinosaur'.

3

THE PUBLIC ERUPTION

Many major discoveries were made by forgotten pioneers. So many of the great dinosaurs we know from today's museums – from *Triceratops* to *Tyrannosaurus* – were first unearthed in the U.S., and it is easy to lose sight of what went before. I have explained that dinosaur fossils were known thousands of years before we usually believe they were discovered. Now we can see that there were ideas, images and sculptures of dinosaurs that are far older than the science of palæontology, and it has become clear that curiosity about these massive monsters dates back long before the word 'dinosaur' was coined. Indeed, it may have surprised you to see that descriptions and images of fossils were being published in learned books back in the 1600s. Alongside the many men whose names we have encountered stand the women who played a crucial role. Remember that the first person systematically to discover and study prehistoric reptile fossils was a woman, as was the first individual to recognize the significance of a dinosaur tooth, and also the first person to draw perfect studies of fossil dinosaurs for publication. The contributions to mainstream science by women have been widely sidestepped in the past; now would be a good time to reinstate their crucial contributions.

Although the dinosaurs are our theme, palæontologists in England were finding the fossilized remains of other plants and animals and recording them in detail long before dinosaurs were recognized. This research was far more extensive than we usually imagine, and it was captured in a book by John Morris that was published in 1845. Morris was born in 1810 in London and had been privately educated to become a pharmaceutical chemist, yet he became increasingly interested in fossils, and began to publish scientific papers on his discoveries. Morris was a man of prodigious energy and had a remarkable memory, but he disliked speaking in public and was not given easily to writing. His strong point, however, was his fastidious facility for cataloguing. In 1845 he published his greatest work, and one which is a forgotten landmark in palæontology: a comprehensive catalogue of all the British fossils. Morris was subsequently appointed Professor of Geology at University College, London, and was elected president of the Geologists' Association in London for 1868–1871 and again between 1877 and 1879. His catalogue is rarely mentioned in present-day books, but it is a remarkable document. It extends over 224 pages and lists well over 1,000 different fossil species that had been formally described.[1]

This may well surprise you; it was only three years after Richard Owen published the notion of a dinosaur, yet already there were hundreds of people pursuing palæontology professionally and there were more than 1,000 known species of fossilized life. Already the burgeoning science of palæontology was becoming well established. The public were increasingly interested in the reality of fossils, and the growth of the railway network in Britain meant that visiting the seashore, and collecting specimens as a hobby, was suddenly available to far more people. During the 1830s, steam railways were inaugurated in England, Ireland, France, Belgium, the Netherlands, Austria, Australia, Cuba, Canada and the U.S.; and by the end of the 1840s seaside holidays had become popular in England. People combed the beaches for shells and the rocky strata for fossils. Many families acquired their own collections and the lure of fossils steadily increased. There was suddenly the perfect opportunity to publicize the latest research into dinosaurs. The Great Exhibition in London of 1851 had caused an upsurge in popular interest for everything scientific, and Owen was asked by Benjamin Waterhouse Hawkins to help design the first life-size models of dinosaurs for public display in the grounds of the Crystal Palace. Hawkins had already mentioned the idea to Mantell, but he had turned it down. With Owen as the chief adviser, teams of artisans set to work, creating the first sculptures of dinosaurs that the world had ever seen. On New Year's Eve 1853, Owen planned a dinner party for 11 prominent academics inside a hollow concrete *Iguanodon*, even though the model was misconstrued. Mantell had realized in 1849 that an *Iguanodon* was not the elephantine monster that Owen was constructing, but was more graceful, with slender forelimbs. However, by now it was too late to change the design. A 30-foot (9-metre) representation of the *Iguanodon* was one of the first of these concrete dinosaurs to be built. To generate publicity, the

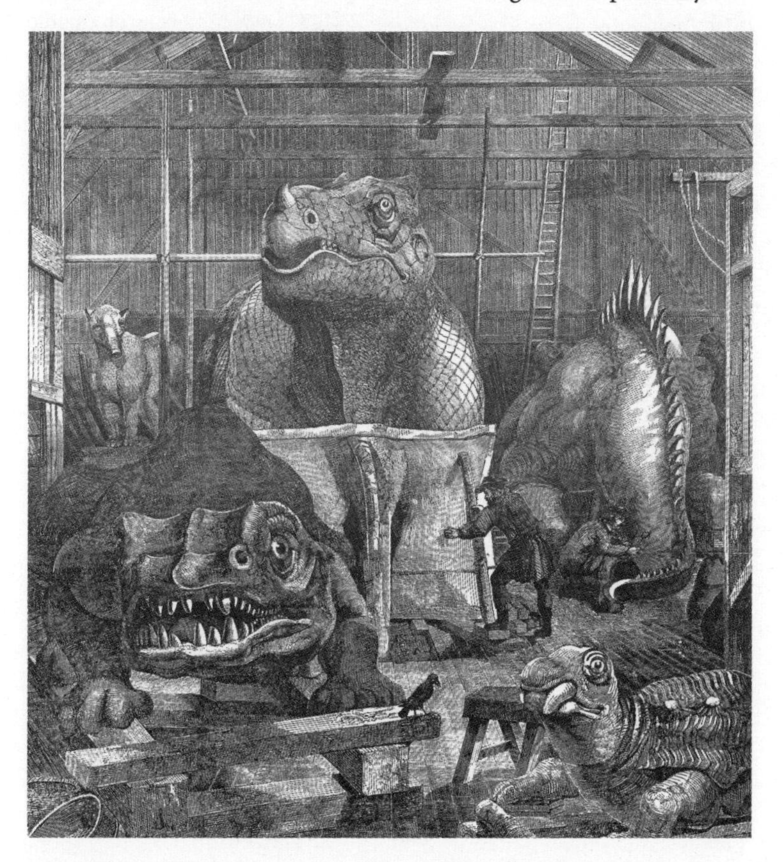

After the Great Exhibition in London of 1851 Sir Richard Owen was asked by Benjamin Waterhouse Hawkins to advise on the first life-size reconstructions of dinosaurs for the Crystal Palace. The workshop was engraved by Philip Henry Delamotte in 1853.

dinner party had been arranged in the open cast of the partly completed sculpture, with Owen sitting at the head of the table opposite Francis Fuller, the managing director of the Crystal Palace, and with nine more seats squeezed into the space. Once the party was over, the top section was added to the sculpture and the world's first life-size dinosaur model was complete. It is one of the original sculptures that can be seen to this day at the Crystal Palace Park, in the London borough of Bromley. These dinosaur models are different to our present-day interpretation, though they are vivid examples of how dinosaurs were first interpreted in Victorian England. Apart from true dinosaurs, the 15 sculptures include plesiosaurs and ichthyosaurs, together with a few prehistoric mammals. They survived in a neglected state until 1973, when they were classified as Grade II listed buildings. In 2002 they were meticulously repaired, and were upgraded to Grade I in 2007. Now that they have been properly restored, they should last forever, or at least as long as London.

The quarries of southern England continued to reveal strange new forms of prehistoric life, and often the excavations began with digging

A unique dinner party took place at 5:00 pm on December 31, 1853, with 11 luminaries seated inside the partly completed *Iguanodon*. Waterhouse Hawkins sent out the invitations, and Sir Richard Owen was seated at the head of the table.

James Harrison, a Dorset quarry manager, discovered this skull of *Scelidosaurus harrisonii* after it was excavated in mudstone destined for the cement furnace. It was purchased by Henry Norris and published by the Palæontological Society in 1861.

into the base of a cliff comprising the desired minerals. One such quarry had been dug out of the cliffs during the 1850s at Black Ven to the east of Lyme Regis. The owner was James Harrison who lived in Charmouth, and who excavated the area for high-quality Charmouth mudstone that dates from the late Sinemurian stage, about 191 million years ago, and was destined for burning into cement. Once in a while, the workmen would retrieve a bone, and these fossils were kept safely as interesting curiosities. Harrison often took them home and displayed them on the mantelpiece or in the hallway for the interest of guests. A surgeon and amateur geologist, Henry Norris, visited Dorset on vacation and became friendly with Harrison. Norris pointed out that these fossils could be important, and even valuable. So, in 1858 the two men sent a parcel containing some broken bones to Owen at the British Museum (Natural History) in London and asked for his opinion. The most conspicuous was a left femur that Owen realized was different from anything previously recorded. He formally described it in 1859, naming the genus *Scelidosaurus*. Owen's intention was to name it from the same Greek word from which the word 'skeleton' is derived, σκέλος (*skelos*, hindlimb), because of the strong femur he had examined, but he confused it instead with σκελίς (*skelis*, rib of beef). He made a mistake: the new dinosaur should have been named *Scelodosaurus*.

Harrison later retrieved a portion of the tibia and fibula of this creature, then a claw, and finally a skull, which Owen formally described in 1861, naming this species *Scelidosaurus harrisonii* in honour of its discoverer. When the rest of the dinosaur had been excavated, it revealed a surprisingly complete skeleton. Although the tip of the animal's snout was missing, the skull and jaws were intact, and the pelvis, ribs, hindlimbs and most of the vertebræ were retrieved. Of the forelimbs (and the end of the tail) there was no sign, but otherwise it was an incredible find. The body of *Scelidosaurus* measured about 13 feet (4 metres) long and was covered with a protective shield of bony scales or scutes, hundreds of which had survived, with many still in roughly the original position. This was the most complete dinosaur skeleton ever found, yet Owen carried out hardly any further investigation. This dinosaur was later described by the prominent American palæontologist Othniel Marsh, who erroneously assumed it had long legs, but not until the 1960s was it further investigated. Acid treatment was used to help release the scutes from their stony matrix, but the entire fossil has yet to be completely recovered. After nearly 160 years, this fascinating fossil is still waiting to be fully described.

These are stories with endless fascination, and they have attracted the attention of innumerable authors and even some movie producers. In 2002 the story of the pioneering British work on dinosaurs became the subject of a television movie produced by National Geographic, *The Dinosaur Hunters*. Henry Ian Cusick played Gideon Mantell and Rachel Shelley played his wife Mary. Alan Cox was Richard Owen, Michelle Bunyan his wife Caroline; Mary Anning was portrayed by Rebecca McClay and William Buckland by Michael Pennington. The movie was well received and remains available online.[2]

Beachcombers were now so abundant in England that they were sometimes teased for their eagerness. A cartoon for *Punch* magazine

Searching for wildlife and fossils on the beach became such a popular pastime in Victorian England that cartoonist John Leech published this portrayal of beachcombers in *Punch* magazine in 1858. Their hooped skirts were reminiscent of giant barnacles.

in 1858 showed a beach scene dotted with bizarre objects that look like barnacles; closer inspection shows they were day-trippers in petticoats, all bending over to search for fossils and seashells. That same year, William Dyce, a leading landscape painter, created his detailed picture *Pegwell Bay, Kent – a Recollection of October 5th, 1858*, which showed an autumnal beach scene with his family gathering specimens from the beach. Dyce was a student of geology and astronomy, and had painted the same bay before. This painting includes finely detailed studies of the chalky cliffs, while, high in the heavens, he captures the faint image of Donati's comet to commemorate the widespread public interest in such phenomena. By this time, people were buying microscopes and telescopes as never before, and the popular understanding of science was burgeoning.

While this excitement was spreading in Britain, new dinosaurs were being discovered in mainland Europe, and in 1859 a German physician and part-time palæontologist, Joseph Oberndorfer, acquired an exquisite little skeleton to add to his collection. He lived in the Riedenburg-Kelheim region of Bavaria, surrounded by quarries where a remarkably smooth and small-grained limestone was

When W.F.A. Zimmerman published *Le monde avant la creation de l'homme* (The World Before the Creation of Man) in 1857, this engraving entitled 'Primitive World' by Adolphe-François Pannemaker was the frontispiece.

obtained. These layers had been laid down some 151 million years ago in a vast shallow lagoon, forming strata with so fine a grain that in 1798 a method was discovered for using the rock to produce flawless lithographic plates for printers. (To this day, lithographers speak of working 'on the stone' even though plastic and metal have long since replaced the German stone slabs.) From time to time, the local quarrymen used to find the remains of creatures trapped in the limestone, and the skeleton of a small dinosaur was perfectly preserved in the thin slab of rock that Oberndorfer obtained. He passed the fossil over to Johann A. Wagner, Professor of Zoology at the University of Munich, who had made extensive studies of mammalian fossils (including mammoths and mastodons) and who was delighted to be able to describe and name a new dinosaur. He decided to call it *Compsognathus longipes* from the Greek κομψός (*kompsos*, delicate) and γνάθος (*gnathos*, jawbone). The specific epithet *longipes* comes from the Latin *longus* (long) and *pes* (foot).[3]

This was a momentous discovery, for it was the first complete skeleton of a carnivorous theropod dinosaur ever to be discovered, and it was also one of the smallest. It measures about 3 feet (90 cm) long and would have been the size of a swan. In 1865 Oberndorfer sold the specimen to the Bavarian State Institute for Paleontology and Historical Geology in Munich, where it is on display to this day. Curiously, we can still see its food: there is the skeleton of a small lizard still visible within the abdomen. When Othniel Marsh examined it in 1881 he concluded that it must have been an embryo within a female *Compsognathus*, though it was later accepted that it was the remains of a meal – lizards would have been a probable prey for a dinosaur like this. At the time, nobody realized that this was a raptor,

An exquisitely preseved specimen of *Compsognathus longipes*, a small theropod the size of a goose, was published by Johann Andreas Wagner in 1861. It had been discovered in limestone deposits from Riedenburg-Kelheim in Bavaria.

in essence one of the tribe to which giants like *Poekilopleuron* and *Megalosaurus* belonged. Nothing more was found of this genus until 1971 when a second, somewhat larger, skeleton was retrieved from the lithographic limestone of Canjuers north of Draguignan, in Provence. This one measures 4 feet (125 cm) long and was quickly given a new species name *Compsognathus corallestris*, though this is almost certainly just a younger example of *C. longipes*. This second specimen was acquired in 1983 by the Muséum National d'Histoire Naturelle in Paris, where it is on public display. This diminutive dinosaur was interpreted as a fierce little hunter, and in 1997 its digital re-creation was given a prominent role in *The Lost World: Jurassic Park*.

In England, Owen was preoccupied with the foundation of a new museum to house all such specimens. He had succeeded in being appointed superintendent of the natural history collections at the British Museum in 1856, whereupon he announced his first mission would be to remove all the biological specimens, not just the dinosaurs, and install them in a new building of their own. One of his strongest supporters was Antonio Panizzi, the museum's librarian, who had never liked the expansion of the museum into natural history, and was eager to see the specimens depart. Panizzi wanted the space. The campaign eventually succeeded, and in 1873 work began on the new museum in South Kensington that we know today. This was the British Museum (Natural History), and it opened in 1881. Not until 1963 did it become a museum in its own right; only since then has it been known as the Natural History Museum. Owen was appointed the first director of the new establishment, and a magnificent statue of him was erected in the main entrance hall. But it didn't last; in 2009 it was summarily removed, only to be replaced by a statue of the single individual that Owen disliked most of all: Charles Darwin. It was the ultimate irony.

Owen had always been a scientific anatomist, given to analysis and discipline, who saw no reason to modify his religious beliefs. Charles Darwin, on the other hand, was a collector and an ardent student of natural history. The publication of Darwin's *Origin of Species* immediately made sense of dinosaurs. Science could now argue that there

We celebrate the life of this man who coined the term 'dinosaur', Sir Richard Owen, founder of the Natural History Museum. He was a brilliant comparative anatomist but was regarded as arrogant, sadistic and deceitful by his peers.

was a steady process of evolution, and there were innumerable extinct species that marked the various stages from the distant past. I have explained that, although we imagine that the study of dinosaurs began with Owen, in fact it dates back much further – and curiously, the same is true of evolution. We regard it as Darwin's great revelation, but in fact the idea goes back to the ancients, and a clear understanding of natural selection was written by others, long before Charles Darwin happened on the scene. We celebrate Darwin as the originator of evolution, indeed 'his' theory is celebrated by everyone as a cornerstone of modern science, but he was not the first to come up with the idea. His pre-eminence in evolution is a myth that has fooled us all. The idea of evolution was understood by the Ancient Greeks; Empedocles wrote about it around 450 BC, as did Lucretius some three centuries later. Aristotle envisaged the progress of life as a Great Chain of Being around 350 years BC – so evolutionary ideas had been around for 2,000 years before Darwin's time. Indeed, he was not even the first in his family to write on the subject. Charles Darwin had a distinguished grandfather named Erasmus who has been lovingly documented by my much-admired friend Desmond King-Hele, himself a distinguished physicist and a specialist on space research. In his spare time, King-Hele has written extensively on Erasmus Darwin, who was a leading physician. He wrote a great work on life entitled *Zoonomia* in two great volumes that embraced many subjects – including evolution. The book was published in 1794, and included these words:

> Since the earth began to exist, perhaps millions of ages before the commencement of the history of mankind, would it be too bold to imagine, that all warm-blooded animals have arisen from one living filament, which the first great cause endued with animality, with the power of acquiring new parts, attended with new propensities, directed by irritations, sensations, volitions, and associations; and thus possessing the faculty of continuing to improve by its own inherent activity, and of delivering down those improvements by generation to its posterity.

There you have it: evolution in a nutshell. Erasmus went further, and even proposed survival of the fittest as the mechanism involved: 'The strongest and most active animal should propagate the species, which should thence become improved,' he wrote back in 1794. Here we have the nature of evolution spelled out decades before Charles began his work. The notion of 'survival of the fittest' was not even mentioned in the papers on evolution by Wallace and Darwin that were presented to the Linnean Society in 1858, yet here we can see that it had been spelled out by Charles's grandfather more than 60 years earlier. When challenged, Charles conceded that he had of course read *Zoonomia*, but insisted that his grandfather's ideas had 'no influence' on his own thoughts, which would be a remarkable example of selective amnesia. In fact, the concept of evolution by natural selection had been current for decades before he wrote his book, and some even predate Erasmus. A French explorer and philosopher named Pierre Louis Maupertuis had written about the idea in his book *Vénus Physique* ('the earthly Venus'), published in 1745. Here it is in the original French:

> Le hasard, dirait-on, avait produit une multitude innombrable d'individus; un petit nombre se trouvait construit de manière que les parties de l'animal pouvaient satisfaire à ses besoins; dans un autre infiniment plus grand, il n'y avait ni convenance, ni ordre: tous ces derniers ont péri.

In English, it reads:

> Chance, you might say, produced an innumerable multitude of individuals; a small number found themselves constructed in such a manner that the parts of the animal were able to satisfy its needs; in another infinitely greater number, there was neither fitness nor order: all of these latter have perished.[4]

There we see 'survival of the fittest' spelled out unambiguously a century before Charles Darwin was active. It is clear from the words of Maupertuis that the idea of natural selection was known far earlier than it is popular to imagine, yet his work too has been largely forgotten. Desmond King-Hele has suggested to me that the disappearance of his work could be due to suppression by Voltaire, to whose lover Maupertuis taught mathematics. She was Gabrielle Émilie Le Tonnelier de Breteuil, the Marquise du Châtelet, and a brilliant mathematician who translated Isaac Newton's *Principia* into French. Little wonder Voltaire regarded Maupertuis with envy.

Four years later the notion of evolution was spelled out by Georges Louis Leclerc, Comte de Buffon, who was the greatest naturalist in France of his generation. He understood what fossils were, and he also recognized that living organisms changed over time, and that they came from simpler, common ancestors. Buffon's books extended to numerous volumes and were highly detailed, while his grounding in practical science was exemplified by the engraving that acted as the opening illustration to the *History of Animals* – it shows him peering intently at a specimen down a microscope.[5]

As we have discovered, James Hutton was a brilliant geologist, and in 1794 he also came up with the essential idea of evolution by natural selection. In a far-reaching book on knowledge and reason, he wrote:

> If an organized body is not in the situation and circumstances best adapted to its sustenance and propagation, then, in conceiving an indefinite variety among the individuals of that species, we must be assured, that, on the one hand, those which depart most from the best adapted constitution, will be the most liable to perish, while, on the other hand, those organized bodies, which most approach to the best constitution for the present circumstances, will be best adapted to continue, in preserving themselves and multiplying the individuals of their race.[6]

In the words 'best adapted' we have 'survival of the fittest' spelled out once more, long before Charles Darwin. Hutton's thoughts might have gained even greater currency, but he had appalling handwriting and many correspondents had difficulty making out what he meant, so his followers relied only on his published works. Here is yet another quotation – still dating from the eighteenth century:

> At length, a discovery was supposed to be made of primitive animalcula, or organic molecula, from which every kind of animal was formed; a shapeless, clumsy, microscopical object. This, by the natural tendency of original propagation to vary to protect the species, produced other better organized. These again produced other more perfect than themselves, till at last appeared the most complete of species, mankind, beyond whose perfection it is impossible for the work of generation to proceed.[7]

These words reiterate the concept of 'survival of the fittest' and were published in a book by Richard Joseph Sullivan in 1794.

This was the same year in which the farmer-cum-physician James Hutton published the idea of natural selection and Erasmus Darwin published *Zoonomia*, embracing similar thoughts. If there was truly a year when 'survival of the fittest' emerged in the literature of science as the principle driving evolution, then in my view it was 1794 – 65 years before Charles Darwin's *Origin of Species*. His grandfather Erasmus revisited the topic of evolution in his poem entitled *The Temple of Nature*, published in 1802:

First forms minute, unseen by spheric glass,
Move on the mud, or pierce the watery mass;
These, as successive generations bloom,
New powers acquire, and larger limbs assume;
Whence countless groups of vegetation spring,
And breathing realms of fin and feet and wing.

Here are the notions that great eras had passed; ages of creatures living in water and aspiring to evolve on land had existed long before our present-day world emerged. By 1810 Jean-Baptiste Lamarck was publishing his views on evolution in France, and came up with a theory that was quickly rejected. Lamarck claimed that organisms evolved because of adaptations made in response to the experiences of successive generations – the reason a giraffe has a long neck, his theory argued, is because successive generations had stretched to reach up for leaves. Survival of the fittest, by contrast, holds that natural selection of longer-necked animals takes place as those with shorter necks were eliminated by an inability to reach up for leaves, and so they would die of starvation. We all know the two versions, and we have been taught to dismiss Lamarck and his views as misguided. Yet Charles Darwin did not do so – for him, the inheritance of acquired characteristics was entirely possible. This surprising view was called Pangenesis by Darwin, and he included many examples of the phenomenon in the last chapter of a book published in 1875 entitled *Variation in Plants and Animals under Domestication*. The argument was that cells within an organism would produce 'gemmules', microscopic particles containing inheritable information that accumulated in the germinal cells, which runs contrary to what is conventionally called Darwinian evolution. This is a remarkable revelation: in some ways, Darwin supported Lamarckism. More recent findings by biologists including Denis Noble at Oxford University have revealed the extent to which genetic change can be passed on through epigenetics – the way gene expression is regulated – so we now know that the extent to which genes are expressed can impose profound controls on evolution. Lamarck was partly right.

Although the theories expounded by Wallace and Darwin were to propose natural selection as an evolutionary principle, that essential idea had recently been published by an experimenter whose name has largely been forgotten. He was an arboriculturist named Patrick Matthew. Like Darwin, he went to Edinburgh University, and like Darwin, he left without a degree. Matthew returned to his family home in Errol, a small Scottish town, where he showed proficiency as a grower of fruit trees. He experimented with cross-breeding, and, notably, with grafting. These activities gave him insights into heredity, and in 1831 he published an important book entitled *On Naval Timber and Arboriculture*. It is in the Appendix that he introduced the crucial concept of natural selection. Wrote Matthew: 'There is a law universal in nature, tending to render every reproductive being the best possibly suited to its condition,' and he continued:

> Nature, in all her modifications of life, has a power of increase far beyond what is needed to supply the place of what falls. Those individuals who possess not the requisite strength, swiftness, hardihood, or cunning, fall prematurely without reproducing … their place being occupied by the more perfect of their own kind.[8]

This was in print, and widely available, 27 years before Charles Darwin's ideas were first presented. When the matter was raised with Darwin, he wrote: 'I freely acknowledge that Mr. Matthew has anticipated by many years the explanation which I have offered on the origin of species under the name of natural selection,' and he promised: 'If another edition of my book is called for, I will insert a notice to the foregoing effect.' He did not keep his word but wrote instead: 'An obscure writer on forest trees clearly anticipated my views … though not a single person ever noticed the scattered passages in his book.' In the fourth edition of the *Origin of Species*, Darwin eventually admitted:

> In 1831, Mr. Patrick Matthew, published his work on Naval Timber and Arboriculture, in which he gives PRECISELY the same view of the origin of species as that ... propounded by Mr. Wallace and myself in the Linnean Journal, and as that enlarged in the present volume.

He then added:

> Unfortunately, the view was given by Mr. Matthew very briefly in scattered passages in an Appendix to a work on a different subject, so that it remained unnoticed until Mr. Matthew himself drew attention to it in the Gardeners' Chronicle, on April 7th, 1860.[9]

This claim was disingenuous. Matthew's views were well known and widely discussed when first they appeared – indeed, many libraries banned his book because of its scandalous allegations that evolution had occurred.

Since I have shown clearly that Charles Darwin was not the first to write a book on evolution, then who was? The first great book on the subject was *Vestiges of the Natural History of Creation*, published in 1844 and written in great secrecy by an anonymous naturalist, who remained unrecognized as the author for many years.[10]

The unnamed author stated that all forms of life had evolved over time, and they had done so according to natural laws and not because of divine intervention. He included in his text a diagram of an evolutionary tree, which was the first ever to appear in print. Although *Vestiges* contained descriptions of the 'progress of organic life upon the globe', the text did not contain the word 'evolution'. The book was enormously successful for its time, selling over 20,000 copies to readers including world leaders such as Queen Victoria and Abraham Lincoln, politicians from William Gladstone to Benjamin Disraeli, scientists like Adam Sedgwick and Thomas Henry Huxley. The book was devoured with enthusiasm by Alfred Russel Wallace. Many years later it was revealed that the author was Robert Chambers, a naturalist who espoused evolutionary theory. His belief in an explanation founded on scientific rationalism went too far when he believed in the claims of W.H. Weekes, who insisted that he had created living mites by passing electricity through a solution of potassium ferrocyanate ($K_4[Fe(CN)_6] \cdot 3H_2O$). Chambers clearly saw biological evolution as steady upward progress, though he felt it was governed by divine laws. Another prior advocate of evolution was the Reverend Baden Powell, Professor of Geometry at the University of Oxford and father of the founder of the Boy Scout movement, Robert Baden Powell. In his *Essays on the Unity of Worlds*, published in 1855, he wrote that all plants and animals had evolved from earlier, simpler forms, through principles that were essentially scientific. Powell also wrote to Darwin, complaining that his own views on evolution should have been cited in his book.

Charles Darwin did admit the influence of Thomas Malthus, who published several editions of *An Essay on the Principle of Population* between 1798 and 1826. In the opinion of Malthus, a leading British scholar, competition was an important factor regulating the growth of societies. Darwin conceded to his readers that his ideas were not original: in the introduction to *The Descent of Man* he emphasized: 'The conclusion that man is the co-descendant with other species of ancient, lower, and extinct forms is not in any degree new.' Darwin knew that; modern scholars, intent on mindless magnification of the man, have forgotten the fact. This is how science is taught to us all. Reality is somewhat different.

Just as we imagine that Richard Owen gave us dinosaurs, we have been taught to hero-worship Charles Darwin as the originator of evolutionary theory. Yet we can now see that evolution was far from being Charles' original idea. Not only had it been summarized by his own grandfather in a previous century, but the essential notion of natural selection was omitted from his early accounts of evolutionary mechanisms, even though it had been published decades earlier by an experimenter whose work Darwin knew. Today, we know Charles Darwin for the crucial concept 'survival of the fittest', and most authorities say that the theory is Darwin's own – yet 'survival of the fittest' was not even his phrase. It was coined by Herbert Spencer in his own book on biology. Wrote Spencer: 'Survival of the fittest, which I have here sought to express in mechanical terms, is that which Mr.

Darwin has called 'natural selection', or the preservation of favoured races in the struggle for life.'[11]

So you can see that evolution by natural selection was thought up long before Darwin began to write about it, and his most famous phrase – survival of the fittest – was coined by somebody else. Indeed the phrase did not enter Darwin's own writings until the *Origin of Species* appeared in its fifth edition. Even by this time, he had not mentioned the word 'evolution' to describe his views, for that term did not appear until the *Origin of Species* was in its sixth edition.

Belief in Darwinian evolution has since become an academic requirement. The question: 'Are you, or are you not, a Darwinist?' is used to mark out real biologists from those beyond the pale. It is rich in resonances of Senator Joseph McCarthy and the famous question: 'Are you now, or have you ever been, a member of the Communist Party?' If you don't espouse Darwinism, then the biological establishment won't want you. Yet we have now seen that Charles Darwin didn't discover evolution. He was not the first to introduce the idea of 'natural selection', and he wasn't even the HMS *Beagle*'s naturalist. When it came to evolution, Charles Darwin was a latecomer on the scientific scene – during his lifetime his book on earthworms outsold the volume on evolution. Modern-day science likes to worship remote figureheads; but much of this tendency is simply science's cult of celebrity. Don't be taken in.

It is clear that the person who triggered Darwin's interests in evolution was a brilliant young explorer named Alfred Russel Wallace. Wallace worked as a watchmaker, a surveyor and then a school teacher before setting out to explore the Amazon in 1848. His intention was to collect specimens for commercial sale to collectors back in Britain, but his ship was destroyed by fire on the way home to England and all the collections went up in flames. You might think that this costly disaster would have ended his career, but Wallace had been fully insured. He claimed for the value of the lost specimens and suddenly was wealthy, without the need to sell everything that he had discovered. With the proceeds safely in his bank, he wrote up his findings on palm trees and on monkeys, and set off again, exploring and collecting in Southeast Asia.

While staying on Borneo, Wallace wrote a paper 'On the Law which has Regulated the Introduction of Species' that was published in the *Annals and Magazine of Natural History* in September 1855. He asserted that 'Every species has come into existence coincident both in space and time with a closely allied species' and noted (as he had done in a book he had written on the Amazon monkeys) that geographical separation seemed to lead to species becoming distinct, a finding that Darwin confirmed with his observations of the Galápagos finches. Wallace then wrote another great paper entitled 'On the Tendency of Varieties to Depart Indefinitely from the Original Type'. Before seeking a publisher, he mailed it to Darwin to ask for his scholarly opinion. Darwin received it on June 18, 1858 and realized that here, set down in writing in detail, was a theory of evolution by natural selection. There was nothing new in the idea. Remember, it had been casually circulating for centuries and the concept had been part of the vernacular currency of biological science since Erasmus Darwin's publications of 1784. For decades, Charles Darwin (along with so many others) had assumed that evolution proceeded through natural selection, and now Wallace had set it out formally in a scientific paper. Evolution had been studied by Wallace as a mechanism operating among wild organisms in nature, whereas Darwin had mostly studied artificial breeding by farmers and horticulturists. This was a suitable time for the theory to be discussed, and Alfred Russel Wallace was the man who triggered the debate. Darwin decided to discuss the paper with his friends. Some of his circle were professional men of science, like Joseph Dalton Hooker; others were struggling to find a position, including Thomas Henry Huxley who had won the gold medal of the Royal Society but still could not find employment. Huxley wrote at the time: 'Science in England does everything but pay. You may earn praise, but not pudding.'[12]

Darwin was also acquainted with a range of independent-minded investigators, including John Tyndall, who was a self-taught physicist; Thomas Hirst, an amateur mathematician; Edwin Lankester, a builder's son who taught himself medical biology; and Arthur Henfrey, who had qualified in medicine but became prominent as a self-taught botanist. Darwin gathered some of these friends together to discuss how he should react to Wallace's detailed article, and it was eventually agreed that a summary of Darwin's ideas could be appended to a formal reading of Wallace's paper. Darwin had set down a few thoughts in a letter to Hooker written in 1847, and there were more in a letter he had written to Asa Gray in 1857. These would make the basis of a submission. A suitable opportunity arose when the Linnean Society suddenly announced the date of a special meeting; there would be time for the reading of new research papers. The Society's president, Robert Brown (the person who named the cell nucleus, and after whom Brownian Motion is named), had suddenly died, and July 1, 1858 was chosen as the date to elect a successor. Since there was time available for additional scientific presentations, it was agreed that the Honorary Secretary would read Wallace's paper on evolution, followed by the extracts from Darwin's letters on the subject. Neither Wallace nor Darwin was present. This launched the theory of evolution for discussion in the world of science, yet it did not catch anyone's attention at the time. The new president of the Society thought it unimportant. In his annual report for 1858, he said that the year 'had not been marked by any of those striking discoveries which at once revolutionize, so to speak, our department of science.' That puts him in the same category as the A & R man at Decca who turned down The Beatles.

Today you can visit Charles Darwin's house as a museum, just 9 miles (15 km) southeast of the Crystal Palace dinosaurs. The building is Down House, and it lies in the village of Downe. Today it looks just the same as it did when Charles Darwin lived there, with the rooms rich in original décor and authentic furnishings – but the back-story is very different. In 1907, the premises were sold and re-fitted to become Downe School for Girls. It remained so until 1922, when another girls' school took over and ran until 1927. By this time, the rooms that Darwin used had all been stripped out, repainted in grey and converted into classrooms. The school building was then purchased in 1929 by a surgeon, Sir George Buckston Browne, with the idea of turning it into a museum. Down House was eventually acquired by English Heritage in 1996 and it reopened to the public after extensive refurbishment in April 1998. I have visited it many times and chaired meetings there. My friend Stephen Jay Gould flew over to speak at a conference I was organizing at the house. Original items from Darwin's time have been returned over the years, and other items that are similar to the original furnishings have been purchased. Walking through the house today, it is hard to imagine it stripped bare and repainted as hordes of pubescent women tramped through its hallowed corridors for so many years – at that time, the interior was unrecognizable. The home now is a latter-day reconstruction, though the visitor may not easily discover the fact.[13]

Another frequent misconception is that Darwin was the official naturalist when he made his famous voyage in the *Beagle*. In a statement in the *Origin of Species* of 1859, Darwin claims to have 'been on board HMS Beagle, as naturalist'. The implication is not correct. The ship already had an officially appointed naturalist, Robert McKormick, who also served as the ship's doctor. Darwin was on board as the travelling companion of the ship's master, Admiral Robert FitzRoy, the person who invented weather forecasts. He had invited Darwin because of his reading theology at Cambridge. FitzRoy was a fervent Christian, and hoped that Charles Darwin could utilize his knowledge to reconcile geology and biology to the teaching of the Bible. Darwin may have claimed to have been the ship's naturalist in his later writings, but at the time he wrote that his appointment was 'not a very regular affair'. We like to imagine that the *Beagle* was an explorers' vessel on a civil voyage of discovery, but she was actually a Royal Navy warship: a Cherokee-class brig. This was a naval expedition, not a journey of discovery. And the captain had an ulterior motive; three captives had been brought to England from Tierra del Fuego; they had been taught English and trained as Christian missionaries. FitzRoy had taken on the major share of financing the voyage, because he wanted to return these men to the country of their birth where they could spread the word of God to this newly discovered land. The Navy organized the voyage on condition that FitzRoy meticulously surveyed the coast and the oceans around South America, so that precise navigational charts could be drawn up. Looking for new forms of life was nowhere in the plans.

The ship set sail on December 27, 1831. Darwin soon became absorbed by the many exotic life forms he encountered on that famous holiday. He was able to go ashore at will to collect and explore to his heart's content. His adventures were comprehensively detailed in a book he wrote describing the voyage of the *Beagle*.[14]

He came to realize how coral atolls were formed, and he published a monograph entitled *The Structure and Distribution of Coral Reefs*. His adventures were many: in Chile he witnessed an earthquake. On the Galápagos Islands he dined on the flesh of the giant tortoises and later described how the different shapes of their shells seemed to match the lifestyle imposed by the environmental situation of the different islands. The Galápagos finches, he concluded, were similar to those on the mainland but had clearly changed over time. He noted: 'Such facts undermine the stability of species.' He then changed it by adding one cautionary word: 'Such facts *would* undermine the stability of species.' There was no implication here that the presumed changeability of species was a novel concept, just that his observations substantiated the accepted view.[15]

The accessible style and exotic nature of the subject brought a wide readership, and suddenly Darwin had a new career – as an author of popular science. To me, his most visible legacy is his list of published books, which represent a remarkable devotion to making science accessible. They are all vividly written. Apart from the *Origin of Species*, he wrote on the geology of South America and on volcanic islands (1844), on the fertilization of orchids (1862), the movements of climbing plants (1865), the effects of cultivation on variation in plants and animals (1868), the *Descent of Man* (1871), insectivorous plants (1875), the effects of cross-fertilization in plants (1876), *The Different Forms of Flowers on Plants of the Same Species* (1877), and finally *The Formation of Vegetable Mould through the Action of Worms, with Observations on their Habits* (1881). This last title was a best-seller. Remember, Darwin's book on worms sold far more copies during his lifetime than the *Origin of Species*.

Curiously, Charles Darwin showed no interest in dinosaurs. He is the person who popularized the theory of evolution, in which dinosaurs would play such an important part, but he did not include them in any of his books. How curious that Robert Darwin, who had presented the first scientific account of a fossil reptile in 1719, had been Charles' great-grandfather. Since Charles' grandfather Erasmus had also written about evolution, it is surprising that although Charles Darwin himself was fascinated by fossils, he had nothing to say about dinosaurs. Although the fact is little discussed these days, Charles Darwin was an expert with the microscope and he became interested in the microscopical structure of fossilized plants. He knew that they had been faithfully preserved in rock, but how much could you discern with the microscope? Was the cellular structure preserved?[16]

Darwin was not the first to speculate thus. As long ago as May 27, 1663, Robert Hooke at the Royal Society of London had looked at fossilized wood under his microscope. As we have seen (p. 18), Hooke had carefully scrutinized his specimen of fossilized wood, and had worked out how it was formed, and he ascertained that the fossil sample showed the same structure as a specimen of fresh wood:

> I found, that the grain, colour, and shape of the Wood, was exactly like this *petrify'd* substance; and with a *Microscope*, I found, that all those *Microscopical* pores, which in sappy or firm and sound Wood are fill'd with the natural or innate juices of those Vegetables, in that they were all empty, like those of *Vegetables charr'd* ...[17]

By 1665 Hooke had recognized that fossil wood was similar to the structure of present-day plants. Darwin made the same observation, but he took it a stage further. Rather than simply inspecting the surface, he resolved to have the rocky fossils ground down with an abrasive paste to produce the thinnest of sections – so fine that light could shine through to reveal the inner structure. He had collected fossilized wood during his sojourn on HMS *Beagle* in 1834 when they called at the Isla Grande de Chiloé, midway along the coast of Chile. He noted at the time that he had found numerous specimens of 'black lignite and silicified and pyritous wood, often embedded close together.' Joseph Dalton Hooker, the founder of geographical botany and the director of Kew Gardens for 20 years, was a close friend of Darwin's and he catalogued the specimens for the British Geological

Survey in 1846. The collections were then lost for 165 years, until Howard Falcon-Lang, of the Department of Earth Sciences at Royal Holloway College of the University of London, investigated some drawers in a cabinet labelled 'unregistered fossil plants' in the vaults of the British Geological Survey near Nottingham. Falcon-Lang reported: 'Inside the drawers were hundreds of beautiful glass slides made by polishing fossil plants into thin translucent sheets, a process [that] allows them to be studied under the microscope. Almost the first slide I picked up was labelled C. Darwin Esq.' This remarkable discovery was a treasure trove, and all the slides have now been digitized and put online for public scrutiny.

Given that the idea of evolution was fundamental to the understanding of dinosaurs in their temporal context, and that Darwin himself was an enthusiastic investigator of fossilized plants, it is surprising to me that he showed little interest in dinosaurs. Yet within two years of his book appearing, a discovery was made that seemed to provide the perfect example of evolutionary theory. This was the discovery in Germany of what seemed to be a creature halfway

Stylised portrayals of an ichthyosaur and plesiosaur were published by Louis Figuier in *La Terre avant le Déluge* (the World before the Flood) in 1863. The drawing, by Édouard Riou, was engraved by Laurent Hotelin and Alexandre Hurel.

between reptile and bird – *Archæopteryx*. The skeletons and feathers have been excavated from the limestone quarries that surround Solnhofen, Germany. First to appear was a lone feather, found in 1860 by Christian Hermann von Meyer and now on display at the Humboldt Museum für Naturkunde in Berlin (see footnote on p. 197). Nobody can be certain it came from *Archæopteryx*, and it may belong to a similar (but different) genus, but the following year a skeleton was found in the same limestone at a quarry in Langenaltheim, 5 miles (8 km) west of Solnhofen. It was donated to a local physician, Karl Häberlein, in lieu of his professional fees. Knowing of the interest in palæontology then spreading across England, Häberlein sold it to the British Museum (Natural History) for the princely sum of £700 (today worth about £45,000 or $52,000). This has long been known as the London Specimen, and it is on display at the Natural History Museum to this day. The skeleton is mostly complete, though it lacks much of the skull and cervical vertebræ. In 1863 Richard Owen formally named it *Archæopteryx macrura*, admitting that it might not be the same species as the one from which the feather found by von Meyer had originated. Darwin was pleased by the find, for it fitted so well with the theories in his book. In the fourth edition, he added a note:

> Now we know, on the authority of professor Owen, that a bird certainly lived during the deposition of the upper greensand; and still more recently, that strange bird, the Archæopteryx, with a long lizard-like tail, bearing a pair of feathers on each joint, and with its wings furnished with two free claws, has been discovered in the oolitic slates of Solnhofen. Hardly any recent discovery shows more forcibly than this how little we as yet know of the former inhabitants of the world.

It was many decades before further specimens of *Archæopteryx* were unearthed. The Eichstätt specimen was discovered in 1951 near Workerszell, Germany, and was not formally described until 1974, when details were published by Peter Wellnhofer. The fossil is on display at the Jura Museum in Eichstätt, Germany, and is of a curiously diminutive form. It has been suggested that it may be a different

Of all the specimens of *Archæopteryx*, this is the only one with a skull. It was found in 1874 by a farmer named Jakob Niemeyer, who sold it to an innkeeper, Johann Dörr, to decorate his bar. It is now in the Museum für Naturkunde, Berlin.

genus, and has been given the alternative name of *Jurapteryx*. The jury is still out on that. More typical of the type is the Maxberg specimen, which was discovered in Germany in 1956 and described in 1959. It was owned by a collector, Eduard Opitsch, who loaned it for exhibition in the Maxberg Museum in Solnhofen. When Opitsch died in 1991 and his estate was catalogued, that fine fossil had vanished. To this day, nobody knows what happened to it. Another fossil from Solnhofen, which had been discovered in 1972, was identified after being classified as an example of *Compsognathus*. This one too is the subject of debate, and some authorities want to classify it as

Wellnhoferia, a cousin to *Archæopteryx*. A further example known as the Munich Specimen was unearthed in August 1992 by quarryman Jürgen Hüttinger who was working for the Solenhofer Aktien-Verein in the limestone quarries of Langenaltheim. Hüttinger duly reported his find to the quarry manager, who again called in Wellnhofer, the specialist palæontologist. The fossil was in fragments, and it was painstakingly reassembled in the State Paläontology Collection workshops. Only then was it realized that the skeleton was almost complete, apart from a single wing-tip. A methodical search through a ton of the nearby strata eventually brought it to light, and a near-perfect skeleton was the result. In April 1993, the finished specimen was formally presented to the press in Solnhofen, and it was put on public exhibition during the 150th anniversary of the Bavarian State Collection in Munich. It then went to the U.S. in 1997, where the Chicago Field Museum in Chicago displayed it as 'Archæopteryx – the bird that rocked the world', as part of the annual meeting of the Society of Vertebrate Paleontology. Eventually, it ended up in Munich's Paläontologisches Museum, who paid 2 million Deutschmark for the fossil (now almost £1 million or $1.3 million).

The best specimen of them all has a mysterious beginning. It was the property of a collector in Switzerland, whose wife – after his death in 2001 – offered it for sale to the Senckenberg Museum in Frankfurt, Germany. They lacked the funds to purchase it until Burkhard Pohl, who founded the Wyoming Dinosaur Center (WDC) in Thermopolis, put them in contact with an anonymous benefactor who came up with the funds. It was put on public display in Frankfurt, and then in 2007 was transferred to Wyoming. German palæontologists were horrified, and began a protest petition. Although no law had been broken by the export of the fossil, it would be unavailable for easy access by German investigators, and would also be passing from a state museum in Frankfurt to a privately owned collection in Wyoming. The directors of the WDC formally issued a statement, saying that the specimen would at all times be freely available for scholarship and study, which mollified the protestors and peace was resumed. This, now known as the Thermopolis Specimen, shows a perfectly preserved skeleton and has been expertly curated. It also retains voluminous sprays of feathers on the body and the wings, and is believed to be the most vivid and perfectly preserved of them all. It has become a key item of evidence in the continuing debate about whether *Archæopteryx* was truly the first bird or was closer to the dinosaur end of the evolutionary spectrum. It was described in 2005 as having 'theropod features' because of the angle of one of its toes; mentioning a connection with meat-eating dinosaurs is a great way to attract maximum attention. Yet nobody knows where it was originally found.[18]

This specimen was subsequently named *Archæopteryx siemensii*, and it is not only the best of them all, but is the only one on display outside Europe. It resides in America, and in that sense it is unique.

4

GREAT AMERICAN DISCOVERIES

It was just when Darwin's book on evolution was published that America entered the dinosaur race. Fossils were already well known in the U.S.; apart from the 'Ohio Animal', innumerable invertebrates had been regularly collected by enthusiasts. A detailed account of animal fossils had been published in the *American Journal of Science and Arts* as early as 1820.[1]

The first dinosaur fossils to be found by geologists in North America were discovered in 1854 by Ferdinand Vandiveer Hayden, a surgeon turned geologist, while he was exploring the Missouri River near its confluence with the Judith River. His team found a few teeth that could not be identified, so they were taken to Joseph Leidy, a knowledgeable geologist who was also Professor of Anatomy at the University of Pennsylvania and a member of the Academy of Natural Sciences of Philadelphia. In 1856 Leidy published a paper that launched North American palæontology, and at the time he was the leading authority on the subject outside of Europe.[2]

The teeth Hayden had found eventually turned out to represent the remains of three dinosaur genera that were named *Trachodon*, *Troodon* and *Deinodon*. Hayden published his findings two years later, in which he briefly reported discovering teeth that Leidy had found to be from 'two or three genera of large Saurians allied to the Iguanodon, Megalosaurus, etc'. It was becoming apparent that dinosaurs were not only to be found in the Old World; they were also relics of prehistoric America. This was a startling revelation to science.

Hayden went on to lead the first major exploration of Yellowstone in 1871, and it was his report that led to the establishment of Yellowstone National Park.[3]

The discovery of the first near-complete dinosaur skeleton in the U.S. would revolutionize palæontology and it was announced by a keen amateur naturalist, William Parker Foulke. He was born in Philadelphia in 1816 and came from a Welsh family of Quakers who had emigrated to America in 1698. Foulke was a successful lawyer and became a prominent campaigner for civil rights. He opposed slavery, fought for prison reform, published political pamphlets, and was a prominent philanthropist. He was also an enthusiast for geology and became a member of the Academy of Natural Sciences. In 1858, just as the papers by Wallace and Darwin on evolution were being read at the Linnean Society of London and two years after Leidy had confirmed the discovery of dinosaur teeth, Foulke was staying in Haddonfield, New Jersey, and was discussing the lie of the land with John Hopkins, who farmed nearby. Hopkins was in the habit of excavating marl from a tributary of the nearby Cooper River for use as a lime-rich soil dressing. He mentioned to Foulke that, some 20 years before, workmen had dug out some huge black bones. He still had a few at his home. Foulke inquired where they came from, but the farmer tersely explained that the purpose of the work was to acquire lime for the farm, not to provide bones for a philosopher. He had thought no more about them. Foulke asked him if the digging could be resumed, to see if further bones could be found, but when they walked across to see the old digging site it had become overgrown with vegetation and much of the marl outcrop had been eroded. One of the workmen was asked to come to look at the site, but he couldn't identify where the bones had been found and for a day or two they excavated in the wrong place. Then the worker had a brainwave – changing position, they dug down about 10 feet (3 metres) and suddenly came across a large bone. It was heavily impregnated with iron and was as black as coal. Careful excavation soon revealed the left side of a large skeleton, including 28 vertebræ, much of the pelvis and almost all the four limbs. As is the curious case with many herbivorous dinosaur fossils, there was no sign of a skull. The bones were in fragments, each of which was carefully cleaned, measured and drawn, before they were packed in straw and transported by horse and cart to Foulke's premises less than a mile away. The skeleton proved to be that of a new type of dinosaur, which was named *Hadrosaurus* from the Greek ἁδρός (*hadros*, large) and σαῦρος (*sauros*, lizard). Analysis of the anatomy showed that, like *Iguanodon*, it seemed to stand erect and walked on its hindlegs.

Joseph Leidy, who had now become known as the nation's leading palæontologist, was informed of the discovery by Foulke. As Leidy arrived on site he went with Foulke to the excavation and the digging continued, though nothing else was found. It was Leidy who decided to name the dinosaur *Hadrosaurus foulkii* to commemorate its discoverer. The skeleton was officially donated to the Academy of Natural Sciences of Philadelphia in December 1858 and 10 years later was put on public display after being meticulously reassembled and mounted by Benjamin Waterhouse Hawkins, the man who had designed the concrete dinosaurs for the Crystal Palace. Hawkins opted not to exhibit the original bones, but to prepare plaster casts that were then stained to look like the real thing. In this way, the fossil remained available for scientific study and there could always be more casts made in future. Not only was this America's first entire dinosaur but it was the first time that any dinosaur skeleton had ever been seen in public anywhere in the world. It was a sensation. Palæontologists, anatomists, zoologists and enthusiasts travelled from far and wide to see the huge display. As the public flocked in to view the spectacular skeleton, the museum staff where soon overwhelmed. So were the exhibits. The thousands of visitors threw up so much dust that other treasures on display were threatened, and the Academy responded at first by limiting the numbers who could attend at any one time. Then they reduced the number of days each week when the exhibit was open; and finally, they began to charge admission to see the skeleton. The museum had been recording some 30,000 people each year, but that number more than doubled when the hadrosaur was put on display, attracting more than 66,000 visitors. The following year, the total topped 100,000 and the administrators launched a public appeal for funding, so that they could obtain premises large enough to

William Foulke's discovery of this *Hadrosaurus* skeleton in 1858 launched the search for dinosaurs in the United States. When the fossil was reconstructed 10 years later, it became the first dinosaur skeleton to be put on public display.

accommodate everyone who wanted to come. Nothing like it had ever been experienced before.

The money they collected was enough to construct a building that was twice the size, and this is the museum that still stands today at 1900 Benjamin Franklin Parkway. It retains much of the original architectural detail, including the galleries with their *fin de siècle* balustrades and the fine brick and stone façade. This is a museum that has blended modernity with tradition, unlike some others (London's Natural History Museum being a case in point) where the dinosaur gallery has become a dimly lit children's theme park. The Philadelphia dinosaur became world-renowned, and it remained the only fossil dinosaur on public exhibition until 1871, when a duplicate plaster copy was erected in Central Park, New York City, as a public attraction. However, the organizers had not reckoned with William Magear Tweed, an influential New York politician. Tweed was a corrupt property developer. Before he was 30, he had been elected to the House of Representatives and, although he never qualified in law, he managed to have himself certified as an attorney and set about extracting protection money from everyone he could. He was appointed to the New York Senate and was repeatedly arrested and freed, sometimes escaping from custody, and was sustained by the support of an adoring public. Tweed became known as 'Boss' and was deeply dishonest – he seemed to have a hand in every commercial deal in the city and soon became the third-largest landowner in New York City. Although his proclaimed wealth was largely an invention, he had himself appointed to the board of a railroad company, a major bank, a luxury hotel and a printing company; little commercial development took place in the city in which his corruption did not play a part (some readers may recognize a pattern here that makes America in 2018 seem tame by comparison). 'Boss' Tweed took a poor view of the new dinosaur display because he was unable to persuade the proprietors to pay protection money, so he secretly ordered the destruction of the plaster skeleton. No sooner had installation work finished than the entire skeleton was smashed to pieces and thrown into the nearby lake by his henchmen. Thus, the second dinosaur skeleton to go on display anywhere in the world fell victim to big-city corruption. Six years later, another of the casts was shipped to Scotland for display in the National Museum in Edinburgh. Fortunately, it was left unscathed and remains there to this day – the first dinosaur skeleton ever put on public display outside of the United States.[4]

A cast of the dinosaur skeleton from Haddonfield stood at the heart of the centennial exhibition of scientific and industrial wonders held in 1876 in Fairmount Park, Philadelphia, where it shared the stage with the world's largest steam engine and the torch that had been manufactured in readiness for installation on the Statue of Liberty. Once again, it was a sensation. Two years later, another of the casts was purchased by the Smithsonian Institution, which displayed the world-famous skeleton outside its headquarters. Another cast was bought by Princeton University and displayed in its Nassau Hall. *Hadrosaurus foulkii* remained the only dinosaur put on public exhibition anywhere in the world until 1883, when a Belgian *Iguanodon* skeleton went on display in Brussels. This was to correct a mistake Mantell had made when the first fossils were excavated – the horn he had assumed went on the snout was actually situated on the wrist, somewhat like a spiked thumb. Artists using the standard drawings as a reference for their own impressions of dinosaurs took longer to adapt; as we shall see, there was still a horn appearing on the snout of an iguanodon as late as the 1960s.

A problem soon emerged. These skeletons were all installed in an imposing upright position, with the tail resting on the ground, as a standard, terrestrial creature. But over the following decades, as palæontologists considered their findings, it was realized that the upright position was impossible – although innumerable footprints of dinosaurs like this were being discovered, there was never any sign of an impression left by the tail. Clearly, the tail of a dinosaur never touched the ground. And so, since the 1990s, dinosaur skeletons have been reconfigured with the tails held aloft. There are still some in the former, upright stance (there is one example in the State Museum in Trenton, New Jersey, and another in the Sedgwick Museum of Cambridge University), but the fossil evidence has proved that this upright stance is impossible. The hadrosaur that Foulkes discovered remains the only one anybody has found. Even though other duck-billed dinosaurs (collectively known as hadrosaurs) have since been found, there are no other specimens of the same strange species that he discovered. Furthermore, nobody knows quite what it looked like, for the skull was never located (the skeleton casts on display in museums use a manufactured skull shaped rather like that of an *Iguanodon*). The team at the Academy of Natural Sciences in Philadelphia had made history by displaying this unique discovery. Just as the concrete sculptures of dinosaurs at the Crystal Palace had caught international attention, so had their exhibit of the first skeleton ever put on display.

This was an exciting project, and one of the team, a young zoologist who had been a child prodigy, was destined to become their curator. He was 18 years old when the skeleton arrived at the Academy, and its majesty transfixed him. This teenager was Edward Drinker Cope, and he became one of the greatest dinosaur hunters of all time. Edward was the son of Alfred and Hanna Cope, a wealthy Quaker family who ran a shipping company and had emigrated from Germany. They were resident near Philadelphia when the young

Cope was born on July 28, 1840. His mother died when he was 3 years old, and Rebecca Biddle became his stepmother. They looked to the future, and Edward was taught to read and write from an early age. The family made extensive visits to parks, zoological collections and museums, for this was how a youngster was prepared for a full and satisfying life in those days. Some of Edward's childhood notebooks survive and they show a wide-ranging interest in natural history. He was also a capable artist. At the age of 12, he was sent to boarding school at West Chester, Pennsylvania, where he studied algebra and astronomy, chemistry and physiology, scripture, grammar and Latin. Looking back at those days, we can see that earlier generations had a tradition of learning instilled at an early age – yet Edward Cope would not study subjects he disliked. Penmanship was a lesson he found difficult, and indeed his handwriting throughout his adult life was often illegible. By 1855 he was back at home and became increasingly preoccupied by natural history. He had often been taken to the Academy of Natural Sciences, and by 1858 he was working there as a part-time assistant, cataloguing and classifying specimens in the collections, when Foulkes' dinosaur skeleton arrived. This sowed a seed that would germinate in Cope's later years. He was captivated by the prospect of investigating magnificent monsters from long-lost worlds.

Later that year, far younger than you would expect a scientist to start publishing, Cope's first paper appeared. It was on salamanders. Meanwhile, his father was not convinced that a life in the museum world was appropriate, and he believed that agriculture would be more suitable, so he purchased a farm for the young Cope. With considerable enterprise, Edward decided to rent out the property to a tenant farmer and he used the regular income to support him in his private interests. In 1860 he enrolled in the University of Pennsylvania, but he found their approach too pedestrian so he did not stay to graduate. (I have to admit that I had a similar experience at Cardiff University; settling down to study a syllabus with which I was already familiar did not appeal to me either, when there was a great wide world awaiting discovery.) Cope always kept the admission slip for the academic year 1861–1862, and – since these forms were always

Edward Drinker Cope (left), born in Philadelphia in 1840, followed Lamarck's theory of acquired characteristics. In a 22-year period he discovered more than 1,000 new fossil species, including 56 new dinosaurs. The co-star of the 'bone wars' of late nineteenth-century America was Othniel Charles Marsh (right), born in 1831 in Lockport, NY. He discovered about 500 new fossil species, including 80 dinosaurs. There was much bitter rivalry between Marsh and Cope.

collected on the first day of the academic year – we can assume that he left university late in 1861. One of his teachers of comparative anatomy was Joseph Leidy, and it was he who encouraged the young Cope to join the staff of the Academy. Because of his position in the Academy of Natural Sciences, Cope was able to join the American Philosophical Society, and it was this august body that provided him with an outlet for many of his pioneering papers on palæontology.

Two years later, the American Civil War erupted, and, to avoid the draft, Cope quit America for Europe. He travelled to England, France, Germany, Ireland, Austria and Italy, taking full advantage of the time to visit seats of learning, museums, and some of the distinguished scientists of the day. But he was unhappy for much of the time, and became depressed. He even set fire to his voluminous diaries and papers until he was prevailed upon by friends to keep some intact – but most of his early documents were lost. Cope was cheered by a visit to Berlin in 1863, where he met a fellow enthusiast who was equally fascinated by fossils. The two frequently went out together and soon became friends. This new acquaintance was Othniel Charles Marsh, who, at 32, was 9 years his senior. They were contrasting young men: whereas Cope lacked formal education after school, Marsh held two university degrees. While the eager Cope had published some 37 scientific papers, Marsh still had only two published papers to his name. Marsh took Cope out to explore the city and its museums, and the two became confidants. As they parted they agreed to correspond, and to exchange specimens. Both seemed destined for greatness. Palæontology was blossoming, and a new world awaited them.

Back in Philadelphia as the Civil War was ending, Cope's father arranged for him to secure a teaching post at Haverford College, a private school with which the family was associated. Cope was still unqualified, so the college awarded him an honorary degree. He married a young Quaker named Annie Pim, aiming less at a romantic liaison than a practical arrangement. He was less interested in poetic passion than in the ability to manage a household, he told his father. They were married in 1865 and next year had a daughter, Julia Biddle Cope. Edward Cope continued to publish papers on anatomy, and also wrote his first paper on palæontology – it described a small carboniferous amphibian fossil *Amphibamus grandiceps* from Grundy County, Illinois, which he said was 'discovered in a bed belonging to the lower part of the coal measures … imbedded in a concretion of brown limestone.'[5]

Cope travelled extensively in the U.S., always searching for specimens, and – although he enjoyed teaching – he complained that he had no time for his serious studies. The work on the *Hadrosaurus* skeleton continued to captivate him, so he sold the farm that his father had given him and moved to Haddonfield, which is where Foulke had found the fossil. Cope now started digging in the marl beds and was soon rewarded by a spectacular find. He unearthed the remains of a new dinosaur that measured 25 feet (8 metres) in length and could have weighed about 2 tons in life. He named it *Laelaps aquilunguis* but later discovered that the genus *Laelaps* had already been assigned to a small mite, so the name was changed. Today we know it as *Dryptosaurus aquilunguis*. He also obtained a fossil that had been discovered in Kansas and excavated by a military surgeon, Theophilus Turner. This was a 30-foot (10-metre) plesiosaur weighing about 4 tons and named *Elasmosaurus platyurus*. He described it as having an elongated tail and a short neck, and, always rushing to publish, he prepared an illustration that showed his newly discovered dinosaur perched on a sandbank as plesiosaurs frolicked in the nearby water.

The family lived comfortably and entertained house guests. Cope was a man of enormous energy and boundless enthusiasm. Henry Weed Fowler, a zoologist friend, said he was 'of medium height and build, but always impressive with his great energy and activity.' Cope was a friendly and open character, and people found him approachable and kind. If a passing youngster drifted from the street into the museum where he was at work, Cope would chat animatedly about the work he was doing. Many modern accounts portray him as an avuncular and warm individual with high moral values and integrity, though some of his associates recorded that his language was foul, he had a bad temper that erupted without warning – and he was a habitual womanizer. A one-time friend, the artist Charles R. Knight, claimed that: 'In his heyday, no woman was safe within five miles of

Cope placed the skull on the tail end of an elasmosaur in the foreground of this illustration published in *American Naturalist* in 1869. Once Marsh had pointed out this mistake, Cope tried to have every incorrect version destroyed.

him.' In an era of machismo, some colleagues even admired this in Cope. One American palæontologist, Alfred Romer, commented that: 'His little slips from virtue were those we might make ourselves, were we bolder.' If Cope was anything, he was bold.[6]

Because of his Quaker upbringing, Cope never challenged his family in their belief in the literal truth of the Bible. Although he maintained a gulf between his life as a palæontologist and his views as a family man, he continued to proclaim his Christian beliefs, and many of his colleagues found his religious fervour to be at odds with his work. Although he looked to the future of science, he remained a traditionalist in family matters and was strongly opposed to women's emancipation. Suffrage, he thought, was irrelevant to a woman whose husband alone could care for her interests and he was certain that it would be pointless for a weak woman, for she would surely be better served by conforming to her husband's beliefs. As a self-educated man, Cope disliked academic orthodoxy, so he turned his back on administrative conformity, and was annoyed when formality threatened to limit his work. He was rebellious by nature, and saw himself on a quest for truth.

When Othniel Charles Marsh returned to America he soon called in to pay his respects. Cope took him excitedly around his excavations. He showed him some protruding bones that he next planned to excavate, and proudly showed the elasmosaur skeleton neatly reassembled and laid out. Marsh considered it all. He was a self-righteous man, but had an analytical mind; suddenly, he realized that Cope had made a fundamental mistake. He had laid out all the vertebræ the wrong way round! Cope was deeply affronted and demanded that someone authoritative should examine the bones to resolve the matter, so the curator of the Academy, Joseph Leidy, was asked to decide who was right. Gravely, Leidy inspected the fossil bones neatly laid out by Cope; then without a word he picked up the skull, strolled right along the skeleton, and placed it where it belonged – at the other end. This was not a creature with a stubby neck and a graceful tail, as Cope had thought. Instead, it had an elongated neck, for it was the tail that was short. Marsh laughed aloud. He ridiculed Cope for his elementary error, and an enduring state of animosity was established between the two. Leidy was right, of course; Cope tried to recover all the copies of the publication in which he had incorrectly portrayed the dinosaur and wanted them all destroyed. The controversy came up at the next meeting of the Academy of Natural Sciences where Leidy was asked to explain. Not only did he account for the error, but he amusingly recounted Cope's elaborate attempt at covering up his mistake. Marsh was delighted. He had become irritated by Cope's easy way of publishing results so rapidly and became convinced that Cope lacked the necessary objectivity to report academic findings in a considered and balanced fashion. Before he left the excavation, he had taken one of Cope's workmen to one side and bribed him to send over any startling new discoveries. Not only did this whole episode mark the end of their friendship, but it launched a bitter battle that was to last a lifetime.

Othniel Charles Marsh was born on October 29, 1831, to Caleb Marsh and Mary Gaines Peabody in Lockport, New York. His mother Mary died of cholera before his third birthday, so her brother, the philanthropist George Peabody, set aside a fund to support the family. As a youth, Marsh developed a close friendship with Colonel Ezekiel Jewett, a remarkable soldier who had travelled the world and fought in many military campaigns. Jewett has been largely forgotten (there is no page for him in Wikipedia, though there should be) and he developed a passion for palæontology. He held classes where he taught geology, and among the young men he mentored was Charles Doolittle Walcott, who was to become a distinguished Secretary of the Smithsonian Institution. The young Othniel Marsh became fascinated by everything Jewett could teach him, for he enjoyed collecting specimens, and through Jewett he learned about geology and the nature of minerals – and soon became fascinated by fossils. Jewett has been described as America's finest field palæontologist of his age, and he found an eager student in Marsh. There were exquisite brachiopod and trilobite fossils to be found near his home, and fine examples of crinoids. On his 21st birthday, Marsh inherited the legacy bequeathed to his mother by Peabody and his life changed forever. The money funded a good education and he was such a promising student that his wealthy uncle agreed to pay for him to attend Yale College. In

1860, aged 28, he finally graduated with a Bachelor of Arts degree. When the American Civil War broke out the following year, Marsh did not try to dodge the draft and was keen to enlist as a soldier, but was rejected because he was so short-sighted. Instead, with his uncle's financial support, Marsh continued postgraduate studies in mineralogy and geology at the Sheffield Scientific School at Yale, and wrote his first scientific paper. It was entitled 'Description of the Remains of a New Enaliosaurian (*Eosaurus acadianus*)' and in July 1862 he sent a copy to Sir Charles Lyell in London. Lyell was the foremost British geologist of the day and had often been at the core of controversy.

In November 1862 Othniel Marsh set sail for England. He was enthralled by the Great London Exposition, an international exhibition held that year, and he wanted to study the dinosaur fossils at the British Museum. Marsh made an appointment to meet Lyell, who was impressed by the young palæontologist and his knowledge of rocks and minerals. Lyell read a paper that Marsh had written, and delivered

Sir Henry Thomas de la Beche was a prolific geologist and artist. This cartoon was drawn by de la Beche in 1830 and it satirizes the views of Sir Charles Lyell. It was published in Francis Buckland's *Curiosities of Natural History* (1857).

it at a meeting of the Geological Society. He then arranged for its publication in their journal, and even proposed Marsh as a new member of the Society, an unusual distinction for someone so new to the field. Early the following year Marsh moved to Berlin to study mineralogy and chemistry, and he then travelled on to Heidelberg and Breslau. His ailing uncle visited Wiesbaden that spring, and when he told Marsh that he was making plans for the allocation of his wealth to worthy causes, Marsh suggested that he should fund a new building at Yale. George Peabody gave a donation of $150,000 (now worth about £5 million, or $6 million) and the Peabody Natural History Museum was the result. Switzerland was next on Marsh's itinerary, and he later returned to Berlin, which is where he first encountered Cope.

These were crucial times for Marsh. New discoveries of fossil dinosaurs were being reported everywhere and – buoyed up by his time spent with Cope – he became determined to devote his life to this fashionable branch of science. In 1866, when he returned to Yale University, it was to become their Professor of Vertebrate Palæontology. This was the first such post in American history. Marsh was appointed a curator of the Peabody Natural History Museum in 1867, and the museum soon became one of the world's leading institutes for dinosaur research. Two years later Peabody died, leaving his nephew a generous inheritance. Marsh was now able to devote himself full-time to palæontology, and built his family a large home in which he entertained luminaries including Alfred Russel Wallace, and in which he could display a vast collection of specimens. The building still exists on Prospect Hill in New Haven, Connecticut. It is now administered by the School of Forestry & Environmental Studies of Yale University.

Marsh could now relish his position as the most eminent palæontologist in the U.S. He stood tall for the time, measuring 5 feet 10 inches (1.7 metres), with a florid complexion and wide-set piercing blue eyes. He was a bulky man with a round face, and he had a reddish beard and sandy hair, the hairline riding high on his forehead. Marsh struck visitors as confident, though at first hesitant and suspicious; his manner was shrewd and, some said, cunning. He had enormous energy and unlimited enthusiasm for palæontology, and was known as a dominant character who fully exploited his position as the leading figure in his field. He wanted to be the best, and indeed, he knew that he was. The British palæontologist Sir Arthur Smith Woodward, whose father Henry Bolingbroke Woodward was one of Marsh's closest friends, said that the family found Marsh to be jealous by nature and desperately keen to attract as much adulation as he could. Nothing pleased him more than the attention of others, and he found it difficult to cooperate with rivals, always seeing them instead as competitors to be vanquished. This attitude fuelled his enduring rivalry with Cope. It was a state of open warfare between the two that led to bitter exchanges and, curiously enough, to the greatest explosion of research into fossil dinosaurs. Had these men not been driven by the passion of animosity, it is unlikely that the astonishing rate of discovery that marked this era in American palæontology would ever have taken place.[7]

These days we look at the Midwest of the U.S. as a vast plain, leading up to the towering Rocky Mountains, and interspersed with genteel cities and scattered settlements. Yet, just a few decades ago, it was a very different place. Older readers may remember the start of *Muffin the Mule* (in England) or *The Ed Sullivan Show* (in America) on television. Well, they were launched some 70 years ago and if we go back just another 70 years earlier, we are in 1878 when the repercussions of the Battle of the Little Bighorn and Custer's Last Stand were still resounding – truly, it is not such a long time ago. The great battle that resulted in Lieutenant Colonel George Armstrong Custer's resounding defeat occurred on June 26, 1876, along the Little Bighorn River in Eastern Montana. The government in Washington had sent out the 7th Cavalry Regiment of the U.S. Army to engage with the Lakota, Cheyenne and Arapaho tribes as the occupying European settlers sought to move west. This battle was the culmination of the resulting Great Sioux War and it was an overwhelming victory for the indigenous people. Of the 12 army companies, 5 were completely annihilated. Lieutenant Colonel George Custer died, along with 2 of his brothers, a brother-in-law and his nephew. Of the 700 soldiers and scouts who had gone out to wage war, 268 were slaughtered outright in the battle and 6 more later died from their wounds. It was a complete victory for the Native American tribespeople but it did not last; after the battle, immigrant Europeans continued to spread across the prairies and their superior arms soon gave them domination of the territory.

The vast central plains of North America extend over 800 miles (1,300 km) west of the Mississippi River to end abruptly against the Front Range of the Rocky Mountains that runs roughly north to south. The first time I travelled that route by road was 40 years ago. The barrenness of such vast areas of wilderness is overwhelming; occasional gateposts mark out the entrance to a ranch, and you know that the residents have to travel many miles to reach any other sign of civilization. Retracing the routes of the early pioneers, imagining their trekking ever westward with their dwindling supplies to reach the gold fields of California, it is hard to imagine the incredible hardship they had to face. Life for the indigenous tribes, who had existed in these desolate areas for tens of thousands of years, is unimaginable. There are other areas of today's U.S. with limpid streams and clear

The digging at Como Bluff in the late 1890s gave rise to piles of dinosaur bones that could not be linked together, and, as the piles grew higher, thousands were used to build a small cabin. The cabin became a museum and soon had a petrol station erected nearby. As business diminished the property was sold off.

rivers, soft hills and verdant valleys, inlets and creeks, lakes and natural meadows, where prehistoric people had made an easy living by gathering fruits, hunting, trapping and fishing. But the great plains are desolate. There is no water, little vegetation, and the scrub extends for hundreds of miles. From time to time a pool or a small lake appears to offer relief, but this is deceptive, for the water is often poisonous. These small lakes seem fringed with what appears to be white sand, but that 'sand' is comprised of chemical crystals. These lakes are concentrated solutions of soda – sodium carbonate, Na_2CO_3 – which crystallizes out as a natron, a white powder. The hopes of adventurers, seeking a supply of fresh water to drink as they passed through the western plains, were quickly dashed as these corrosive, alkaline pools of poison were all they found. Wagons passing through had to haul everything the explorers needed – dried food, fresh water – or they were doomed to die. Standing where they stood, I have experienced a deep sense of isolation and abandonment. These plains are not for the fainthearted. Life was easily lost in this wilderness.

Sedimentary rocks underlie these barren lands and date back to the late Jurassic period (156 to 147 million years ago). They are exposed along the front of the Rocky Mountains and extend out to the west as reddish-brown and yellow sandstones, siltstones and shales. These strata extend through 12 states, and they come to the surface mostly in Wyoming and Colorado, with outcrops in Idaho, Montana, North and South Dakota, Nebraska, Kansas, the panhandles of Oklahoma and Texas, New Mexico, Arizona and Utah. These huge deposits cover 600,000 square miles (1.5 million square km), though only a small proportion can be seen at the surface. They formed from material eroded from ancient mountain ranges far to the west, and deposited as sandy layers in great expanses of shallow lakes and rivers. Geologists categorize them as fluvial, floodplain, shallow lacustrine and even playa deposits – all terms that connote shallow expanses of water. Geologists believe that part of this vast prehistoric area was high in salt and soda and formed a vast stretch of water now known as Lake T'oo'dichi' and covering much of the eastern part of the Colorado Plateau – it would have been the largest-known prehistoric alkaline, saline lake.[8]

Those distinctive sandy rocky strata were first described from exposures at Morrison, Colorado. This is a small, arty, cheerful town at the entry to one of the canyons cutting into the Rockies just west of the mile-high city of Denver. The town was named after George Morrison, the most prominent businessman in the area in the 1870s. Not only are the rocks readily recognizable, but they revealed an extraordinary range of dinosaur fossils. European explorers found bones protruding from the ground, emerging into the daylight after millions of years as rainwater weathered away the rocks. As we have seen, the fossils had been known about for thousands of years to the Native American population, who told legends of monsters like monstrous bulls and of mythical beasts like the gigantic bison named Ahke (the Cheyenne word *ahk* means petrified) that surely originated with fossils of *Triceratops*. As palæontologists were soon to find, there were abundant examples of these petrified dinosaurs. The entire area had once teemed with dinosaurs; their fossilized remains were not hard to find.

The strata of the Morrison Formation had been overlaid by younger sedimentary rocks in the Cretaceous period that began 147 million years ago. By this time the climate must have been becoming wetter; and, in contrast to the sandy-hued strata of the Morrison Formation, these early Cretaceous rocks are mostly grey, sometimes with a greenish tint. Above these fluvial and floodplain deposits is a vast layer of beach sand, the Dakota Formation, which is about 100 million years old, and which marks the transition from freshwater to the marine deposits that were laid down in a great seaway extending from the Arctic right down to the Gulf of Mexico. Geologists agreed that this colossal seaway had existed, though its origin remained a mystery for decades. It now seems that it formed when the cooling slab of the Farallon Plate beneath the Pacific Ocean sank down and was subducted beneath the western fringe of North America. In parts, this enclosed sea was deep; its seabed probably plunged down some 2,300 feet (700 metres) or more. Large shoals of prehistoric fish – some of them of monstrous size – congregated across much of its extent, and scattered fossils of the plesiosaurs which fed on them have been commonly found. This seaway reached its maximum extent about 75 million years ago, and it disappeared shortly after the end of the Cretaceous, about 64 million years ago. In the extensive shallow areas of this huge sea, colossal dinosaurs met and multiplied. Their bones lay tumbled across each other in the coastal floodplain's mudstone, siltstone and sandstone – sediments that are reminders of the shallow waterways and lakes that once covered the area. The sediments from these shallow areas of sea yielded a treasure trove of dinosaur fossils.

The era of discovery was heralded by that most spectacular of astronomical events – the Sun disappeared from the sky. On August 7, 1869, a total solar eclipse swept across America. Its black shadow moved down across Montana and the Dakotas, touching Nebraska and on through Iowa. To the indigenous Native Americans it heralded disaster; and they were right. This was the time when their lands began to slip from their hands into those of the conquering Europeans. The eclipse heralded the end of their supremacy. It also marked the dawn of the great age of dinosaur discovery in the U.S., when an enthusiastic fossil hunter, Elias Root Beadle, a missionary from Philadelphia, came across the skull of a small dinosaur. Although he didn't know it, this was going to make history. He sent the details to Marsh at Yale, but Marsh was too preoccupied to reply.

It was Beadle's discovery that triggered the greatest wave of revelation that palæontology would ever know. When Beadle didn't hear back from Marsh, after a decent interval waiting for a response, he tried writing instead to Cope. Spurred on by Marsh's lack of interest, Cope had the fossil skull transported to him instead, and in 1870 he described it as a new genus, *Lystrosaurus*, from the Greek ιstroυ (*listron*, shovel) on account of its spade-shaped head. Marsh responded by setting out on his own voyage of discovery into the state on behalf of Yale, but nothing more was found. Exploration continued, and in 1877 a chance discovery triggered an explosion of interest. The Union Pacific Railroad Company had been constructing their segment of the Transcontinental Railroad and had met with the Central Pacific Railroad Company's tracks at Promontory Point, Utah, in 1869. There were repeated raids by the Native American tribes and guards were posted along the new stretches of the railway line. One of the armed Union Pacific foremen, William H. Reed, liked to go hunting bison and deer in his spare time, and in March 1877 he set off on a raised line of hills north of the rail track a few miles east of Fort Fred Steele (near the modern town of Medicine Bow). Suddenly he came across some gigantic fossilized long-bones and vertebrae protruding from the ground. A lake nearby had been named Como (after Lake Como in Italy, which has a similar outline) and the blunt hill overlooking the vale became known as Como Bluff. Wherever he looked, Reed could see dinosaur remains sticking out of the rocks. Reed privately told his friend William E. Carlin, an agent for the railroad company, and for several weeks they forgot about hunting and started collecting fossils. They worked away for four months before news of their discoveries reached Marsh – and his interest was immediately aroused.

Meanwhile, Arthur Lakes, a part-time professor at what became the Colorado School of Mines, had found a huge fossilized vertebra between Golden and Morrison, just west of Denver. He sent details of his discoveries to Marsh. At the same time a school teacher and amateur fossil collector named Oramel Lucas, the first school superintendent of Fremont County, was exploring the rocky strata around Cañon City, Colorado, and soon started to unearth dinosaur fossils. He hauled five wagon-loads of massive rocks into the town and the fossils were put on display in a curiosity shop owned by Eugene Weston. Lucas contacted both Marsh and Cope, in the hope of being engaged to find more. Marsh ignored his letter, but Cope promptly hired both Oramel Lucas and his brother Ira and soon they discovered a damaged vertebra dug from strata at Garden Park, near Cañon City. It was one of the biggest such bones ever discovered. Only part of the vertebra, including a portion of the body, the neural arch and the spine, were found, and they showed how fragile the bones would have been. This dinosaur would have dated from some 150 million years ago, and Cope named it *Amphicœlias fragillimus*. The genus of this new dinosaur was named for the concavity of the vertebra, αμφι (*amphi*, both sides) and κοιλος (*koilos*, concave), and the species name is from the Latin *fragillimus* (most fragile). The entire vertebra was calculated to have measured 8 feet 10 inches (2.7 metres) tall and 4 feet 10 inches (1.5 metres) broad, making *Amphicœlias* not only the largest dinosaur yet found at the time, but the largest ever discovered to this day. Cope formally published his findings in 1877. [9] That bone is now lost.

Some later calculations suggest that this monstrous beast could have been 190 feet (58 metres) long and weighed up to 120 tons, which makes it larger than the recently discovered titanosaurs. However, since the dimensions were based on relatively few bones – all those figures are conjectural.

It was now 1878, and by this time white people were everywhere across the Central United States, when another total eclipse of the Sun tracked its way across the dinosaur beds. On July 29, 1878, the Sun was again blotted out in a great swathe across Montana, Wyoming and Colorado, right across the regions where the greatest dinosaur discoveries in history were about to emerge. This time, with the Native American population largely subdued, it was widely observed by scientists. Thomas Edison, then aged 31, took a vacation in Wyoming specifically to observe and record the event. It was his way of proving his abilities, not just as an inventor, but as a serious scientist. As Edison was becoming the most celebrated inventor of his age, Maria Mitchell had become America's most famous female scientist and was teaching astronomy at the all-women Vassar College in Poughkeepsie, Arkansas. Mitchell was not invited on any of the male-dominated scientific expeditions, so she resolved to set up her own and launched the world's first-ever female team of female scientific observers.

An eclipse in that vast wilderness is an overwhelming spectacle – and it was one I also experienced in those latitudes almost one and a half centuries later when this book was being compiled. In August 2017, I drove 400 miles (640 km) across Colorado and Wyoming and stopped to observe the eclipse. The eclipse itself was a stunning and breathtaking experience. I have been present at two previous total solar eclipses; but this was the first where I was able to view it all in a completely clear sky. We all watched it in the middle of endless prairie that stretches for hundreds of miles, just scrub and occasional herds of bison – the buffalo of legend. Totality approached with the Sun now looking like the newest of new moons, and the lunar shadow could be seen to strike distant mountains on the western horizon, racing towards us at about 1,600 mph (2,500 km/h). As the last scrap of sunlight was shut off like a switch, the brilliant day was transformed to deepest night. All around, the distant sky at the horizon was red like twilight. Above, where the Sun once stood, was a perfect black disk from which hung gossamer wraiths like a bridal veil lacing into the sky. I took the chance to observe this soft silver crown through high-powered binoculars; you could only consider the strange thought that it's always there anyway, this magical corona, but the brightness of the Sun prevents us ever seeing it. This reminds us of the power of the endless radiation from the Sun, that nuclear reactor hovering in our sky. The stars shone in the middle of the day. The air cooled dramatically. The ring of ghostly light that replaced the Sun hovered above us like a halo. Just 2 minutes and 28 seconds later the totality ended – those distant mountains were suddenly shining in bright sunlight and the curtain of bright light raced towards us across

Dinosaur hunters in the American midwest went fully armed against attacks by indigenous tribes. Othniel Marsh (middle of back row) was photographed with his fully armed team in 1872, though normally he left excavating to his agents.

the prairie. Then a dazzling crescent of bright beads, like an eternity ring, suddenly emerged as the moon began to move away and sunlight flooded through the lunar valleys. They are known to astronomers as Bailey's Beads (after Francis Bailey, the British astronomer who first described them in 1836) so – down with the binoculars. When you think of the way a hand lens starts a fire by focusing the sun's rays on a newspaper, it reminds you of the danger of focusing sunlight on the retina of your eyes. Dramatically, everyone's shadows reappeared. The day was a sunny summer day once again, and we were left speechless with the memory of that pearly ring hovering high in the sky where the Sun had been, just seconds before. I was following the tracks of the explorers who had found the first American dinosaurs, and the majesty of that spectacular event made one feel a tenuous contact with those pioneers and the Native American tribes who had once lived and thrived right across the continent.

While Cope had been working on the fragmentary remains of his giant sauropod *Amphicœlias*, Marsh was presented with a far more rewarding find in Como Bluff, Wyoming, in 1879. His team had found the skeleton of a 40-foot (12-metre) sauropod that was almost complete. He named it *Apatosaurus*. The name derives from the Greek ἀπατηλός (*apatēlos*, deceptive) and σαῦρος (*sauros*, lizard). He used that irreverent term 'deceptive' because the chevron bones on the underside of the tail are like those of a mosasaur but are different from those of their distant cousins, the giant dinosaurs. This fine skeleton was found in the Rocky Mountains in Gunnison County, Colorado. Two years later, the virtually complete (but headless) skeleton of a still larger individual was excavated at Como Bluff, and Marsh – who was convinced that this was a different dinosaur – decided to name it *Brontosaurus excelsus*. A geologist named Walter Granger found a further example at Medicine Bow, Wyoming, in 1898; but there was a problem. Although it was nearly complete, the feet and the skull were nowhere to be found. Marsh surreptitiously added some sauropod feet that had been found nearby and topped off the skeleton with a made-up skull based on the 'biggest, thickest, strongest skull bones, lower jaws and tooth crowns from three different quarries.' Although it gave Marsh the satisfaction of seeing a completed mount, he knew it was not faithful to the original. The skull was conjecture. He had made a mistake as serious as Cope had done with his own elasmosaur skeleton: Cope had placed the skull at the wrong end, Marsh had the correct end, but the wrong skull! Controversy continued to surround the identity of these fossils, and many skeletons that were labelled *Brontosaurus* were subsequently classified as *Apatosaurus* instead. Because of the confusion, palæontologists decided that there was no such genus that could reliably be identified as *Brontosaurus*, and this genus remained invalid (to palæontologists) throughout the twentieth century. Thus, one of the best-known dinosaurs in the minds of the public was a fiction to science.[10]

Palæontologists from the Carnegie Museum subsequently discovered two more of the brontosaur skeletons in Dinosaur National Monument in Utah, and the matter was reconsidered. A paper published in 1975 showed that the skull Marsh had found actually came from *Camarasaurus*, a dinosaur that had an altogether larger head.[11]

The Reverend Henry Neville Hutchinson published *Extinct Monsters* in 1897. Plate IV is a vivid portrayal of 'A gigantic dinosaur, Brontosaurus excelsus' by Dutch artist Joseph Smit. Other artists were to use his picture as a reference source.

Because there was always uncertainty about the skull Marsh had used, some drawings of a brontosaur show that the artist was careful to indicate its outline by dotted lines. John S. McIntosh of Wesleyan University and David S. Berman of the Carnegie Museum were the palæontologists who eventually proved that the skeleton had the wrong skull. Between 1915 and 1937, the cast of the brontosaurus at the Carnegie had been correctly displayed, without any skull at all. The director of the museum, W.J. Holland, wanted to give it the skull of a diplodocus, just to make it look complete, but as soon as the president of the American Museum of Natural History, Henry Fairfield Osborn, heard about the idea he objected strongly, and the idea was abandoned. In 1937 the museum finally decided to make the skeleton look whole, so they added a cast of the *Camarasaurus* skull. 'Marsh needed a head, so he guessed,' said McIntosh later, at the public unveiling of the corrected skeleton in October 1981. 'Most of his guesses were remarkably good, but this one was not.' The *New York Times* reported that Karl Waage, director of the Peabody Museum, conceded that it would be 'Nice to get the right head on.' [12] Outside the museum in Yale, T-shirts were on sale bearing a picture of the new skeleton with the caption: 'I lost my head at the Yale Peabody Museum.' It had taken almost a century for this mistake to be corrected. The Field Museum, Chicago, subsequently swapped their incorrect skull for a plastic replica, though the William H. Reed Museum of the University of Wyoming in Laramie said they would leave theirs untouched.

Meanwhile, eager to match his rival's success, in 1870 Cope had decided to explore in Kansas. His mentor Leidy had made some important discoveries there, and Marsh had already been out exploring the rocky strata and inquiring in quarries. Benjamin Franklin Mudge was one of the local palæontologists who had helped Marsh, and Cope quickly signed him up to help his own efforts. By this time, he was showing signs of becoming manic, ignoring the need for food and water and experiencing nightmares in which he saw dinosaurs attacking him from every side. So long as there was daylight, Cope was at work with his team, marching, exploring, digging. Eventually he collapsed with exhaustion and was bedridden for weeks. Out west, meanwhile, work proceeded on the great Transcontinental Railroad, and the navvies were now turning up fossils as they excavated great cuttings through the mountains. Although these fossils were regarded as curiosities, there was no palæontologist around to evince interest, and so they were doomed to be discarded or destroyed. But word of the discoveries percolated towards the east coast, where the palæontologists began to wonder whether they should become involved. In those days, it was not so easy to travel westwards on a whim. The sites were not only some 2,000 miles (3,200 km) away from the universities across barren prairie, but were a mile (1.6 km) high. Cope managed to secure an honorary position with the U.S. Geological Survey under the direction of Ferdinand Hayden. Although it offered no personal salary, the post did involve travel to the sites of the new excavations where the dinosaurs had reportedly been discovered, and most of his expenses were paid. The fact that Cope wrote with flair and rapidity was perfect for Hayden, because he was keen to make a good impression with a series of vivid official reports. It worked for Cope, too, for this gave him the ideal outlet to publish his discoveries.

Cope launched his first expedition for Hayden in June 1872, heading out to survey the Eocene strata in Wyoming in which Joseph Leidy had already made several discoveries. This caused a break with Leidy, whom Cope had always admired so much. Hayden had always referred any new discoveries to Leidy, but the new situation meant that Cope now took charge. Leidy wrote immediately to Hayden to express his annoyance at being summarily replaced, but was told that there was little anyone could do. Cope's new job description was to carry out geological surveys, and faithfully to report what he found; he could not be prevented from doing so. Cope moved his family out to Denver, so that they would be closer to where he was surveying, while Hayden sought a way to keep him from prospecting in the areas where Leidy had already been working. Both men were eager for pointers as to where the next discoveries might be made, and one day a geologist named Fielding Bradford Meek sent a telegraph message to say that bones had been found where the railroad was being excavated near Black Buttes Station. When Cope arrived on site, he found part of the pelvis, some associated vertebræ and some ribs, so he decided it was a new dinosaur and named it *Agathaumas sylvestris*. This was a late Cretaceous dinosaur measuring 30 feet (9 metres) long and weighing about 6 tons. Cope was then expected to travel 100 miles (160 km) west to Fort Bridger, and quickly brought together a team with a cook and a local guide, together with three amateur palæontologists from Illinois. He didn't find out at first, but two of his teamsters were privately in the pay of another palæontologist – Marsh was working nearby. These workmen had turned to Cope because Marsh was slow to pay, and they did not care for his lofty and suspicious nature. When Marsh learned what had happened he turned on the hapless teamsters, who assured him that he remained their main priority, and added that they had gone to work for Cope only to lead him away from the best specimens.

One day in 1872, a fragmentary fossil was brought in for Edward Drinker Cope to examine. It had been dug out of the Lance Formation, Wyoming, and consisted of several ribs and the pelvis of a heavy dinosaur. As usual, there was no skull. Ever keen to announce a new discovery, Cope named it *Agathaumas sylvestris* and decided it was a duck-billed hadrosaur. Not until 1887 was any more material found, when a geologist, George Cannon, and his team from the U.S. Geological Survey found some fragments of a fossil, and Marsh was given a piece of skull bearing two huge horns that had been unearthed near Denver. In typical haste, Marsh decided that – because bison roamed in that area – this must be a prehistoric version. He concluded that it was a Pliocene creature, which he named *Bison alticornis*. During the following months, Marsh was sent a few more samples of horned skulls, and he eventually decided that these belonged to dinosaurs. He named the 'new' genus *Ceratops*. The following year, Marsh obtained another specimen from strata next to those in which Cope's first example had been discovered. This, he was now certain, was more *Ceratops* material. He quietly moved his earlier specimen over from *Bison* to *Ceratops*, transforming his mammal into a dinosaur in one stroke. Within a few months, he had changed that name too. All the fossils were now named *Triceratops*, the name we know today. After the first discoveries in Colorado and Wyoming, further examples were excavated in Montana and South Dakota in the U.S., and in the Canadian provinces of Alberta and Saskatchewan. The original species was named *Triceratops horridus* by Marsh. The species name does not translate as 'horrid', as most people think; the Latin *horridus* means bristly. In time, many other species were named: one group was comprised of similar dinosaurs, *T. horridus*, *T. prorsus* and *T. brevicornus*, another group contained two species, *T. elatus* and *T. calicornis*, and there were two additional species, *T. serratus* and *T. flabellatus*. Others later added *T. hatcheri*, *T. eurycephalus* and *T. obtusus* – but most were based on fragmentary skeletons, some of which were probably younger or older specimens of the same species, and it was soon clear that the classification had been undertaken with undue haste. Recently, all have been reclassified by Catherine Forster, a taxonomist at the George Washington University, and they are currently placed in just two species, *Triceratops horridus* and *T. prorsus*. It also emerged that these two species were excavated from strata at differing levels, so these two dinosaurs never co-existed.[13]

We now know *Triceratops* was a fearsome beast: its huge horns spanned over a yard (about a metre) and it weighed about 5 tons, so it was clearly able to travel on dry land. It is now one of the most easily recognized terrestrial dinosaurs, yet – though investigated first by Cope – the name we know today was eventually decided by Marsh. Such was the nature of their dinosaur rivalry. Cope's team found

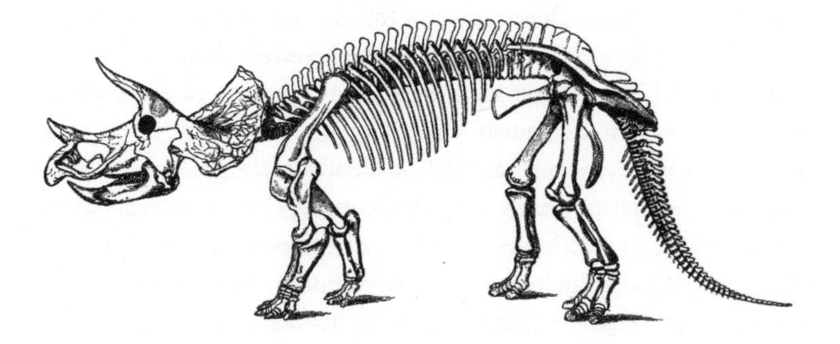

One of the most startling discoveries made by Marsh was *Triceratops prorsus*, which was first thought to be a buffalo-like mammal. Marsh published this detailed figure as Plate XV in a paper for the *American Journal of Science* in 1891.

many other dinosaur fossils that were entirely new to science. On one occasion, a crate of new finds that were meant for Marsh was sent, instead, to Cope. With commendable honesty, he forwarded them to the correct address, though Marsh remained obdurate. Throughout the 1870s, the three men – Cope, Marsh and Leidy – were all prospecting in roughly the same area. Priority became a problem, for they were often discovering the same species at the same time. When Marsh organized one of his prospecting trips from Yale in 1873, in addition to 13 students he had to travel with a platoon of soldiers as a show of force to the Sioux tribe. The students had to pay their own expenses, because previous trips had gone heavily over budget. In the future, Marsh decided to engage local enthusiasts to provide him with new fossils, while he worked through the vast amount of material he had already collected back at his office at Yale. Cope was also finding it increasingly hard to fund his expeditions, so he took a job with the Army Corps of Engineers. This gave him less freedom than he had previously enjoyed, but at least it kept him on site. In 1874, gold was discovered in the Black Hills of the Dakota territory. Marsh seized upon the extensive digging to see what fossils might be found, but tensions were running high between the American prospectors and the indigenous Sioux inhabitants. The solution was ingenious: Marsh would pay the Sioux people for any fossils they found, and, in exchange for their loyalty, he assured the chief that he would represent their interests at the Department of the Interior in Washington D.C.

For a decade Cope followed a rigorous schedule. Every summer he was out with his workers exploring for new fossils, and during the winter months he was busily writing up his discoveries for publication. He was in a perpetual battle against the academically minded Marsh and was determined not to be beaten. In 1874 Cope was signed up as geologist by the Wheeler Survey, a project led by George Montague Wheeler to map areas of the Western United States. In New Mexico, Cope discovered that neither Leidy nor Marsh had preceded him, so he was able to make a range of new discoveries. As a member of the survey team, Cope discovered that he had an added advantage – his findings were to be routinely printed, so he no longer had to pay for publishing. The rate at which he was publishing new discoveries was sometimes one every week, and he is quoted as having published more than 70 papers in 1880 alone, reporting the fossil finds that his teams had made in Colorado, Kansas, Oregon, New Mexico, Texas, Utah and Wyoming. There was a problem in this, of course; his hasty writing and description meant that accounts were often inaccurate, his conclusions sometimes unreliable, and some of the scientific names Cope proposed for newly discovered dinosaurs were either changed or, in some cases, had to be withdrawn.

In 1875, Cope was greatly saddened by the death of his father Alfred, though his feelings were mollified when he discovered that he had been bequeathed a fortune amounting to $250,000 (today this would be around £8 million, or over $10 million). This gave him a release from the search for funding, and he was at last able to concentrate on his first published book celebrating his newly discovered dinosaurs.[14]

This prodigious report ran to 622 pages, including 57 engraved plates that are crammed with illustrations. Still his new discoveries came tumbling out. In 1877, he purchased a half interest in the *American Naturalist* so that he could control the rapid publication of the new discoveries that he was continually announcing. Cope wrote every report himself. Marsh, by contrast, was working in a formal academic framework and he had teams of underlings to do his writing for him. His publications were often drafted by associates and he had editors, referees, and all the academic checks and balances to make the task as pain-free and accurate as possible. He repeatedly suggested that Cope could not conceivably be personally producing papers with the dates he claimed. To him, it simply did not seem possible.

Amateur enthusiasts were now following the reports of these dinosaur discoveries in the press, and it was in 1877 that Arthur Lakes set out hiking near the township of Morrison, Colorado. Suddenly, he turned a bend in the trail and chanced upon some dinosaur fossils at the surface of the track. These were of types nobody had seen before, and Lakes was wise enough to know that. As a teacher, he knew that it was sensible to ask your peers for advice if you are working in an unfamiliar field, and so he sent a letter to Marsh, describing a vertebra and humerus of a colossal 'saurian' skeleton that he had discovered.

Returning to the site, he dug out some more huge bones, none of which he recognized. Marsh, as was so often the case, took time to respond, so Lakes – concluding that Marsh wasn't going to be interested – sent a consignment of his new finds to Cope. Cope read the letter with interest, and when the news came out, Marsh immediately wired over $100 with the strict instruction that the matter was to be kept a secret. Lakes wrote back to say that he had also written to Cope, whereupon Marsh immediately dispatched an agent to Colorado to make sure that his claim was protected and to offer money for any new finds. Meanwhile, Cope had received a letter from O.W. Lucas, a botanist and amateur geologist, with news of yet another dinosaur site at Cañon City, Colorado, and was quick to ensure that Marsh could not interfere. Needless to say, Marsh was trying to claim that for himself too.

This continuing tit-for-tat continued every time something new was unearthed. It was the greatest personal rivalry ever seen in palæontology. The news soon spread that the two palæontologists were at war, and that there was money to be made, so the attitude towards these strange new fossils dramatically changed. Railroad workers now began to keep their eyes peeled for anything that might be of interest to the palæontologists. The *Laramie Daily Sentinel* was eagerly following the story, so when local workers told reporters of their fossil finds, they exaggerated the prices that Yale had been paying for the fossils. Cope was quick to react, and sent out an agent of his own, but it was not a success. The deal struck with Marsh had made the workers greedy, and Cope could not pay as much as they now wanted. Undeterred, he had his own agents recruit a new team of workers to go to Como Bluff in Wyoming and carry out excavations of their own. This historic site became the hub of the 'bone wars' when some of the team working for Marsh defected and began excavations for Cope.

There were now two nearby teams, both assiduously searching for fossils and each trying to sabotage the work of the other. When workers on the Union Pacific Railroad reached Como Bluff, they found strata that were rich in astonishing fossils. Without disclosing their real identities, and using false names, two of the workers sent word to

Marsh that they had found huge reptile fossils. They did not disclose precisely where these were, but asked if Yale University would be interested, and how much money they might offer. The letter added that, if Marsh felt unable to agree generous terms, then the men would write to Edward Cope. This was too good a chance to miss, so Marsh again dispatched his agent to agree a contract for the rights to any exciting discoveries and to purchase the rocky strata outright. To this day, it is the name of Marsh that is attached to these startling new revelations – even though it was not Marsh who made the discoveries.

The reports from America had meanwhile reignited the search for fossils in Britain. Great Quarry in Swindon, Wiltshire, had been used for centuries as a source of clay and building materials. Quarrymen used to sell the fossil shells they found as a sideline, and many of them had personal collections of curious shells. The strata at Swindon represent the bed of an ancient sea dating back 147–142 million years, late in the Jurassic period. Layers of sediment slowly built up over thousands of years, showing phases when the sea was replaced by a freshwater lake, and other times when it was transformed into an inland salt lake like the Dead Sea with high concentrations of dissolved sodium chloride (NaCl). The pebbles at the bottom of the water were eroded and stratified, and the strata were littered with shells. From time to time, bones from a fossil were excavated. And so, on the morning of May 23, 1874, workers reported to their manager that they had found some remarkable fossils that seemed too important to ignore. James Shopland, the manager of the Swindon Brick and Tyle Company, sent word to Richard Owen that his workers had unearthed some curiously large bones. Owen's palæontologist William Davies arrived on the scene to find bones protruding from a lump of hardened clay some 8 feet (2.5 metres) tall. They tried to winch them out of the ground, whereupon the mass broke into several fragments. They were all labelled and packed in crates that weighed a total of 3 tons. Owen put his most experienced excavator onto the task, a stone-mason named Caleb Barlow. At the middle of the mass was the pelvis of a large dinosaur with 6 posterior dorsal vertebræ, a full set of the sacral and 8 anterior caudal vertebræ, with several other bones scattered through the clay. The huge right femur (hip bone) was present, and so was the left forelimb. Barlow also revealed a damaged piece of the tibia and fibula of the lower leg with bones from the foot, a bony plate (from the right-hand side of the neck) and a strange spike.[15]

Owen described the creature in 1875 and originally named it *Omosaurus armatus*, though the generic name was subsequently changed to *Dacentrurus* because *Omosaurus* was already being used to describe a type of fossilized crocodile. This new discovery proved to be a stegosaur, the first such dinosaur ever to be discovered. The skeleton was reasonably well understood, though Owen wrongly concluded that the spike was part of the shoulder. We now know that it actually came from the right-hand side of the tail.

We have seen that the horn which Gideon Mantell had placed on the nose of the iguanodon was also in the wrong place, and it was a remarkable discovery in a Belgian coal mine in 1878 that was to prove the point. I first went to Bernissart in Belgium as a schoolboy; it is a former coal-mining town near the border with France. It remains largely unknown to the world at large, though everyone there knows about *Iguanodon*. Carboniferous plant fossils were regularly discovered in the seams, as they always are in coal mines. Miners used to find fossil ferns, or the impressions of cycad leaves, all beautifully preserved. On February 28, 1878, two miners, Alphonse Blanchard and Jules Créteur, were working 1,056 feet (322 metres) below ground level when Créteur called out: 'I have struck gold!' They had cut into a seam which contained what they took to be a golden tree trunk. On closer inspection it looked more like a large dinosaur bone. It was encrusted with fool's gold (iron pyrites, FeS_2) and it glistened in the light of their lamps like burnished brass. The supervisor of the mine, Alphonse Briart, recommended that a geologist should be consulted.

It did not seem possible that the fossil was a dinosaur bone: the coal measures were laid down during the Carboniferous period between 359 and 299 million years ago (remember that the name Carboniferous is derived from the era of coal-building), whereas dinosaurs existed between 243 and 66 million years ago, so they were on the scene 56 million years too late to be found in any coal seam. A

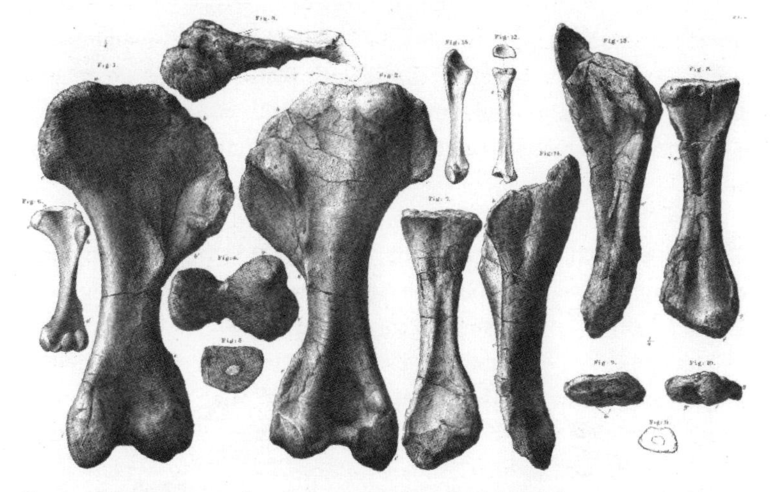

Bones of *Dacentrurus*, a large stegosaur originally dubbed *Omosaurus*, were formally described by Sir Richard Owen in 1875, as Plate 70 in his book *A History of British Fossil Reptiles*. This 8-metre (26-foot) long dinosaur weighed about 6 tons.

telegram was sent to the Royal Belgian Museum of Natural History in Brussels, and the specialist brought in to examine the site was a palæontologist, Louis de Pauw. In May 1878 he began to extract the fossils from the mine and they were packed into 600 heavy crates and shipped to Brussels. What he found at Bernissart was phenomenal, and we have rarely seen their equal in the history of science. A large number of fully grown iguanodon skeletons were discovered, all tightly crowded together. When the bones were separated out, they were moved to the chapel at the Palace of Charles of Lorraine near Brussels to be reassembled – it was the only building tall enough to accommodate the height of the dinosaurs. The task of reporting on the remains and reconstructing the skeletons fell to Joseph Dollo, born in Lille in 1857. With valuable input from correspondence with an Austrian colleague, Othenio Abel, Dollo was the first to recognize that fossils should be considered as part of an entire ecosystem, and he first laid down the principles of a new science: palæobiology.

Dollo's work began in 1882 and the remains of more than 30 iguanodons were eventually retrieved. Later that year the first of the skeletons was cleaned and any fragmented bones were stuck together

More than 30 *Iguanodon* skeletons were found crowded together in a Belgian coal mine in 1878. They were so complete that the correct position of the horn (which Mantell had assumed belonged on the snout) could finally be revealed.

by animal glue, before being assembled in an upright position, like a bear hugging a tree. In July 1883, people were admitted for the first time to view the fossils. The skeletons caused a sensation. They were later moved to the Royal Belgian Museum of Natural Sciences in 1891, where they remain on display to this day. The new species from the mine was found to be taller and more robust than the original *Iguanodon mantelli* (since renamed *Mantellisaurus atherfeildensis*, incidentally); Mantell's iguanodon was less than 16 feet (about 5 metres) tall, whereas the newly discovered skeletons from the mine all measured more than 20 feet (6 metres), and the tallest of all reached 24 feet (7.3 metres) from nose to tail. The new species was announced as *Iguanodon bernissartensis*. Nine of them are displayed in glass cases in the Brussels museum, all standing tall as originally constructed, and there are 19 more in the basement. Casts were made of the skeletons, and a replica of one stands in the Natural History Museum; another is near the entrance lobby of the Sedgwick Museum in Cambridge, and there is also one in the Oxford University Museum of Natural History. They are remarkable skeletons, and most are complete. The fact that so many of the skeletons were virtually entire and were still in their original configuration proved to be of crucial importance to the palæontologists, enabling Dollo to report on the strange spike that Mantell had put onto the snout of his iguanodon. The Belgian fossils confirmed that it really belonged on the forelimb, rather like a pointed thumb. Dollo mounted all these skeletons as if about to climb a tree, and they are so fragile that they remain in that position even though later work showed that the tail would not have been sufficiently flexible. Present-day reconstructions of *Iguanodon* have the spine parallel to the ground, not at right angles to it.

King Leopold II arranged for a plaster cast of the finest of the Bernissart iguanodons to be donated to the Sedgwick Museum of Cambridge University. It greets visitors to this day, still in its original upright pose, near the entrance.

The way that these iguanodons were crowded together initially led investigators to conclude that they were herd animals that had been drowned in a flood or some other disaster, but later it appeared that they had become trapped, one by one, in a fissure that had opened up into a pre-existing coal seam. The existence of pyrites in the fossils soon caused problems. As it oxidized in contact with fresh air it started to crumble, so in the 1930s the skeletons in Brussels were all impregnated with shellac and arsenic, which bound them together in a gluey mass. Needless to say, this didn't last, so in 2003 the bones were cleaned of the glue and shellac, and impregnated instead with cyanoacrylate and epoxy resin. It is hoped that this will suffice for a much longer time.

While that activity was creating news in Europe, another extraordinary discovery was making headlines in America. In 1877, in the mountains of Colorado, Arthur Lakes (a name you rarely find in accounts of dinosaurs) coined the name *Stegosaurus*. In the early spring, just as the snows were melting, he was looking at exposed rocks in the quarries of the Morrison Formation in Colorado – rocks of precisely the same vintage as the pebbly clay in Swindon's Great Quarry – and came across a strange skeleton with those huge protective plates clustered around its spine. He couldn't believe his eyes. Massive skeletons he had already found, but this was something different. Here was a dinosaur with massive plates down its back, like a reinforced rhinoceros. As soon as the news was out, the representatives from Yale descended on the site, paid handsomely for the fossil-bearing rocks and packed them securely in straw-lined crates for shipment back to Marsh. Once the crates were opened, the bones were quickly exposed to reveal a huge creature adorned with bony plates along its back and a spiked tail. In 1878, Marsh announced the new discovery. Following the discovery of the *Dacentrurus* in England, this was the second genus of the stegosaur family to be known to science. Since then, the existence of these monsters has been associated with Colorado and the name of Othniel Marsh, while the true story of their earlier discovery in England has been forgotten. The reconstruction that Marsh proposed was wrong. Because of the watery environment indicated by the sediments in which the skeletons had been found, he assumed that these creatures were some kind of turtle. He named this dinosaur as he did because he completely misunderstood how the dinosaur had looked. He believed that the plates lay flat on the back, like the shingles or slates on a roof, so that the creature was like a huge tortoise. The name *Stegosaurus* reflected this belief in a domed protective carapace: it came from the Greek στέγη (*stego*, roof) and, of course, σαῦρος (*sauros*, lizard).

Meanwhile, Marshall P. Felch had been prospecting for dinosaurs near Cañon City. Felch was a shoemaker from New England who became fascinated by dinosaurs and excavated fossils in a quarry to the west of Pueblo, Colorado. In February 1883 Marsh had written to him, asking if he could agree to undertake excavations exclusively for Yale. He would be well paid for his work. This provided Felch with the financial security he needed, and proved to give Marsh the opportunity to introduce a range of new dinosaurs to science. Some 65 dinosaurs were discovered in the Marsh-Felch excavations. Most of the remains were scattered bones, many of which were unearthed where there had been a particularly deep area of water, but among them

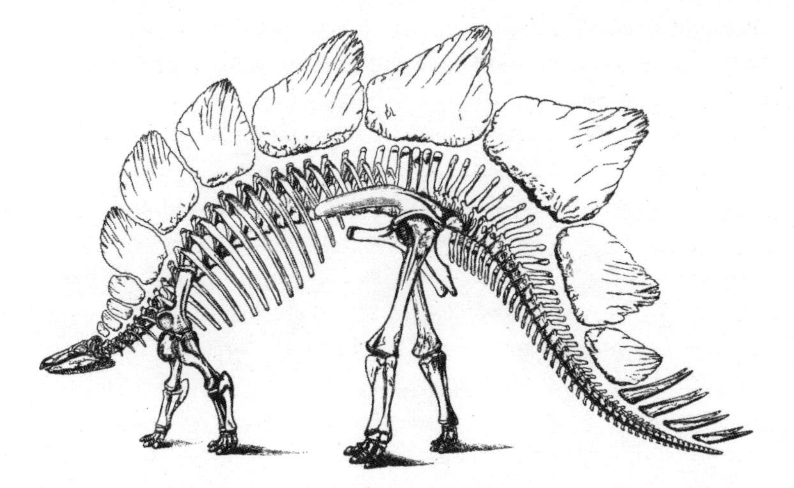

A complete skeleton of *Stegosaurus ungulatus* was published by Othniel Marsh in 1891. He showed the 8 tail spikes correctly and adorned the spine with 12 plates. Later studies showed that there must have been a double row of dorsal plates.

were several near-complete skeletons. In 1886 Felch dug up a beautifully preserved specimen of *Stegosaurus*. It was lying on its side, as though squashed flat, and it has been nicknamed the 'road kill' dinosaur. This skeleton confirmed how the bony plates lay along the spine, like a ridge, and were not laid flat on the back like a stony shell. The specimen that Felch discovered is the type specimen for *Stegosaurus stenops*, and more than 50 partial skeletons, including several with skulls, have since been found across Colorado, Utah and Wyoming. As our knowledge increased with the new finds, one particular problem began to present itself: the bony plates, if you lined them up along the spine, were too numerous to fit into the available space. There must have been two rows of plates running side by side along the animal's back. Even now, our concept of the stegosaurs' anatomy may not be correct.[16]

The best-known dinosaur skeleton in the world is a *Diplodocus* fossil, copies of which feature in capital cities around the world (like 'Dippy', which was on display at the Natural History Museum in London for almost a century). However, this was not the first *Diplodocus* to be unearthed. That discovery was made in 1877 by Benjamin Franklin Mudge, the lawyer-turned-geologist. He was a self-taught palæontologist and had also worked as an industrial

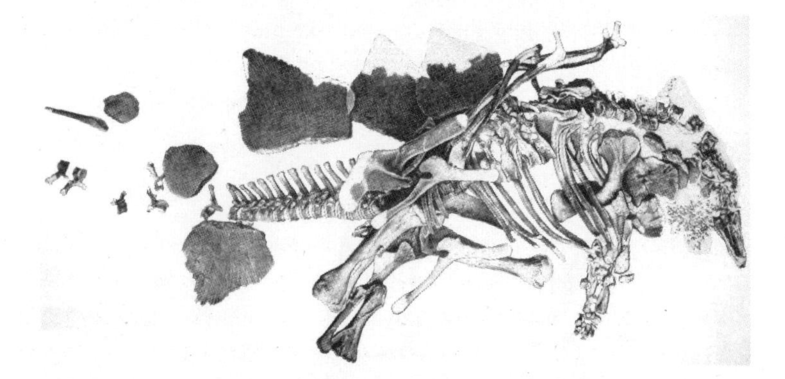

A remarkable skeleton of *Stegosaurus stenops* had been excavated in 1886 by Marshall Felch. Because of its condition, it was nicknamed the 'road kill' specimen. This engraving was published as Plate 2 in the *Bulletin of the U.S. National Museum* in 1914.

chemist before becoming a much-respected fossil collector. Although he is now forgotten, when he was active in the 1870s he was regarded as one of America's most noted palæontologists and has over 80 new species of fossilized animals and plants to his credit, including three that were named after him. Mudge first recorded and collected dinosaur footprints near Junction City, Kansas, in 1865, and in the following year he discovered fossil invertebrates and the outlines of Cretaceous leaves near Ellsworth. In 1870, he organized an expedition near Fort Wallace where he took part in excavations of fossilized saurodontid fish and plesiosaurs, and in 1871 added the 6-foot (almost 2-metre) toothed bird *Hesperornis* to his discoveries. So, when Marsh became interested in excavating new dinosaur fossils, Mudge was quickly signed up for the Yale research and he took a new recruit, Samuel Wendell Williston, under his wing. Williston was learning palæontology and was an excellent draughtsman, recording new finds with meticulous accuracy. It was in 1877 at Cañon City that they dug up their first *Diplodocus*. Marsh gave it the name *Diplodocus longus* when he formally described it the following year, the name deriving from the Greek διπλός (*diplos*, double) and δοκός (*dokos*, beam), an allusion to the double chevron bones in the tail. A later skeleton of *Diplodocus* was named *D. carnegii* after Andrew Carnegie, and this remains the best known, because of the copies exhibited at major museums around the world, but the key species for the genus remains the first one found by Mudge and Williston: *Diplodocus longus*. Curiously, even though many near-complete skeletons have since been found, there is not a single example anywhere with a skull. There was one example that was discovered by William H. Utterback in 1902 near Sheridan, Wyoming, and when it was described in 1924 it was named *D. hayi*. This was half of a skeleton, complete with skull; however, further research showed that it wasn't a member of the genus *Diplodocus* after all, and so it was renamed *Galeamopus* instead. Thus, the situation remains … no *Diplodocus* fossil anywhere in the world has ever had its head. That is extraordinary.

Nobody could work out how *Diplodocus* had moved about. In 1899 Henry Fairfield Osborn had concluded that it must have been able to swim from one piece of dry land to another. He inspected the

After the discovery of *Diplodocus* in 1877, Andrew Carnegie sent out teams of workers to excavate for dinosaurs in the 1880s. Rather than overalls, the team wore conventional clothing. Buttoned jackets, shirts, neckties and hats were normal.

elongated tail and speculated that the caudal vertebræ must have supported a broad tail fin. He reasoned that, although the bones were well preserved, the soft tissues of a fin would not have survived fossilization and so there could have been a tail fin even if nothing was visible in the rocks. To him, the tail would be 'of immense service as a propeller in enabling it to swim rapidly through the water.' On land, he imagined the tail providing a very different purpose. It would then act like a third limb, so that the dinosaur could rear up on its hindlegs, using the tail as a rear-end support – much like the way the iguanodons had been reconstructed in Belgium. This, Osborn concluded, would give this quadrupedal dinosaur the same advantage that a bipedal iguanodon would have. It would have firm and steady support on a tripod, and would be well able to reach up to browse for high leaves on tall trees. Not only that, he reasoned, but the weight of the tail would act as a counterbalance to the elongated neck. Trying to imagine how *Diplodocus* moved about on land was proving difficult. The first depictions showed it as a gigantic lizard, with limbs spread out from either side. Oliver Hay reconstructed it as a vast, sprawling creature, and his view was supported by a leading European reptile expert, Gustav Tornier, from Dąbrowa Chełmińska, in Poland. Herpetologists (who specialize in the study of lizards) liked this new theory, as it showed that the vast dinosaurs were little more than monstrous lizards with lofty aspirations.[17]

But William Holland would have none of it. He could see *Diplodocus* as standing tall on firm, columnar limbs; and he showed that, if it sprawled, as Hay was claiming, it would have needed a deep groove cut into the earth to accommodate its belly. This was not just an amplified lizard, he believed – it was a proud, monstrous dinosaur, something radically different from anything now alive.[18]

The size of these creatures continued to tempt palæontologists into making extraordinary decisions about how they lived their monstrous lives. Waddling dinosaurs, and tails that bore fins for swimming, were only the beginning. The enormously elongated necks of the sauropods were equally impossible to explain, so in 1897 Edward Drinker Cope introduced his own proposal. He postulated that this was the breathing apparatus for a submerged dinosaur. According to Cope's new

The way *Diplodocus* moved was studied by American palæontologist Oliver Hay, who realized it was too heavy to walk. He thought it must have been a sprawling lizard and in 1910 he instructed an artist, Mary Mason Mitchell, to prepare this sketch.

theory, the massive dinosaur lived deep under water, walking on the bed of a lake and invisible from above. Its neck served as a snorkel – it could poke its head above the water and breathe at will.[19]

This was first depicted by Charles Knight (p. 287) and in 1941 a reconstruction was painted Zdeněk Michael František Burian, a Czech illustrator. In 1951 the fact that water pressure would prevent them inhaling was explained by Kenneth Kermack, a palæontologist at University College, London.[20] Even more obvious to me is the fact that they would have floated. Flesh is approximately as dense as water, and a scuba diver – even equipped with heavy tanks – needs lead weights to sink. Those sauropods could never have stayed submerged.

One of Charles Knight's most memorable paintings dates from 1897 and shows a group of brontosaurs: one is trudging across the landscape, while others are resting in swamps. Such ideas were soon

Edward Cope was also puzzled by the difficulty large dinosaurs would have walking on land. He sketched a group of *Amphicœlias* on the bed of a lake, using their necks as snorkels. Water pressure on the chest would make this impossible. These dinosaurs would have floated at the surface.

Aged 25, Charles Robert Knight is seen making a stegosaur model from clay. Short-sighted and irascible, he was the first American artist to use extinct creatures as his inspiration, and visitors have seen his work in many American museums.

cast aside, as dinosaurs were determined by palæontologists to be terrestrial creatures. Indeed, although Charles Knight's image of the resting brontosaur is published on Wikipedia, it is accompanied by this cautionary note in case anybody should risk speculating that this might have been an aquatic creature:

> This historical image is not a factually accurate dinosaur restoration. The idea that Brontosaurus was wholly or mostly aquatic is now considered outdated. Tail dragging and Camarasaurus-like skull are also inaccurate … Note that this image may be appropriate to illustrate obsolete paleontological views.

Images published in *Life* magazine in 1953 were stirring studies by Antonio Petrucelli that showed a brontosaur resting in swamps, as though taking the weight off its feet. Only that one dinosaur in his picture was allowed the luxury of relaxation; the others – a range of the well-known types from *Triceratops* to *Stegosaurus* – are all as the palæontologists preferred, firmly on dry land. The terrestrial dinosaur won the day.

Although gigantic dinosaurs like *Diplodocus* and *Brontosaurus* were attracting all the attention, there had been other monstrous finds in earlier decades. The very first find of a fossil from an *Allosaurus* skeleton had been made in 1865 by the American physician and palæontologist, Ferdinand Vandeveer Hayden, near Granby, Colorado. Isolated vertebræ used to turn up in the nearby rocks from time to time, and they had long been known to the local people who believed them to be petrified horses' hoofs, but when Hayden sent a newly discovered fossil to Joseph Leidy, he sent it with the opinion that it was a dinosaur vertebra. Leidy agreed. He named it *Hadrosaurus*, and at first thought the new bone belonged to the theropod genus *Poekilopleuron*, but later changed his mind and gave it the name of a new genus: *Antrodemus*. In their rivalry for fame and fortune, neither Cope nor Marsh paid much attention to Leidy's research. Cope gave the same dinosaur the name *Epanterias*, while Marsh, in his haste, gave the fossils two new names, *Creosaurus* and *Labrosaurus*. Mudge and Williston had already found better fossils of *Allosaurus fragilis* and *Ceratosaurus nasicornis* in their quarry, and their research led to the conclusion that *Allosaurus* was a large theropod measuring over 30 feet (9.5 metres) in length and weighing at least 5 tons. *Allosaurus* is the name that science eventually settled upon, though it could well have been *Antrodemus*. That was, after all, the first accepted name. In due course, Marsh decided to resume excavations at Garden Park, and in 1883 Felch discovered an almost complete *Allosaurus* along with the remains of several partial skeletons. *Allosaurus* has become a well-known dinosaur, while the rarer *Ceratosaurus nasicornus* is named for the curious horns it had on its head. It was derived from the Greek κερατος (*keratos*, horn) and the conventional σαυρος (*sauros*, lizard) and was in many ways similar to the allosaurs, though with a characteristic knife-like horn on the snout, and two smaller horns over the eye sockets. It would have measured some 18 feet (5.5 metres) in length and weighed over a ton.

By this time, these enthusiasts were discovering new types of dinosaur almost every month. Many of the earliest discoveries are now on public display together at the Smithsonian Institution in Washington, D.C. Curiously, this remarkable museum was set up by an English benefactor, James Smithson, who never visited America. At his death in 1829, Smithson bequeathed 105 cloth bags of gold coins, at the time amounting to almost 2 per cent of the entire federal budget. The legacy was formally accepted by Congress on July 1, 1836, though administrative discussions dragged on for a decade before the U.S. Senate passed the act which formally established the Smithsonian Institution. It was eventually signed into law on August 10, 1846, by President James K. Polk, and the building opened in 1855. It now houses 150 million items in its collections, including the embalmed body of Smithson, which was disinterred from his grave in Italy and was brought to Washington by Alexander Graham Bell in 1904. This legacy from a distant Englishman gave the U.S. one of the world's finest museums, and served greatly to stimulate further research into palæontology.

Although Cope and Marsh successfully contrived to dominate the world of palæontology, many key discoveries were made by palæontologists who escaped from beneath their umbrellas. We have already observed some of the brilliance of John Bell Hatcher, and he still had a crucial contribution to make. Like many palæontologists of his era, he came from a farming family. In his vacations, he had a student job as a miner in a coal pit and quickly realized that there were fossils concealed in the seams. Hatcher had matriculated from Grinnell College, Iowa, in 1880 and transferred to Sheffield Scientific School at Yale in 1884, where he showed his collection of Carboniferous fossils to his tutors. He was assigned to work under Marsh. Although Hatcher remained one of the technical team until 1893, and was an excellent field palæontologist, he became deeply unhappy at working for such a tyrannical head of department. Marsh insisted on naming and publishing all the discoveries made by his group, and Hatcher deeply resented the way his own success was hidden from view. In 1889 at Lusk, Wyoming, Hatcher had excavated two fossil skulls of a strange new dinosaur – similar in many respects to *Triceratops* – and both had been transferred to Yale for Marsh to examine. This was a new find: the skulls had massive bony frills which were perforated at intervals, so Marsh dubbed the new dinosaur *Torosaurus* in 1891. The name is derived, not from the Spanish 'toro' meaning bull (alluding to the horns) as most people believe, but from the Greek verb τορέω (*toreo*, to perforate), which describes the perforations or fenestræ in the extended frill. In *Triceratops* the frill was solid; only in *Torosaurus* was it perforated. Hatcher eventually transferred to Princeton University as Curator of Vertebrate Palæontology and an assistant in Geology. In 1891 Hatcher discovered a dinosaur now known as *Edmontosaurus* in late Maastrichtian Upper Cretaceous strata of the Lance Formation in Wyoming. The skull showed this was a curious, duck-billed dinosaur, and the fossils were dutifully sent to Marsh at Yale for scientific appraisal. They proved to be the remains of a large plant-eating dinosaur some 44 feet (13 metres) long and weighing at least 10 tons. Marsh named it *Claosaurus*.[21]

So complete were the skeletal remains that the dinosaurs were reconstructed in the workshops. When the work was completed in 1901, it was the first time any fossil dinosaur skeleton had stood on its own feet anywhere in the United States. This project acted as a dry run for the rebuilding of *Diplodocus*. In the final years of the nineteenth century Hatcher led many fossil-finding expeditions across South America, and it was here that he noted the remarkable similarity between the plant fossils he found and those reported from Australia. This was an essential component of his belief that the two continents had once been joined.

Marsh kept his teams of the best palæontologists working assiduously; they discovered, excavated and collected the fossils, while he spent most of his time with clean hands and shiny shoes, analyzing the specimens as they came in and publishing the results as quickly as he could. Marsh continued to take the credit for everything. Cope, on the other hand, spent his time out in the field. He was a hands-on palæontologist, and although his lack of institutional backing meant that he was slower to find new specimens, his rapid rate of writing and his outlet through the *American Naturalist* gave him the edge in terms of speed. Cope could ensure speedy publication. A new report would take only a day to hand-set in metal type, and a day to check and place on the galley, ready for printing, so he could record a new find within a few days. Unlike Marsh, Cope was generous in acknowledging others. Much to Marsh's annoyance, although he ran the larger operation – and was describing new dinosaurs quicker than Cope – the discoveries being made by his rival amateur were increasingly admired by the world of science, and it was Cope who was invited back to England in 1878 to attend the annual meeting of the British Association for the Advancement of Science. Marsh had continually attempted to destroy Cope's academic reputation, but as Cope travelled on to England and then to France he was greeted with warmth and respect. Whenever he attended a conference, he exhibited stunning plates of the reconstructions of the latest dinosaur discoveries that he had announced, which had been beautifully produced for him in Philadelphia by zoologist and embryologist John A. Ryder. Cope met up with many leading palæontologists, including Owen.

Meanwhile, Marsh had managed to have himself appointed Head of the Consolidated Government Survey. Both men would take fossils from under each other's nose if they could. If ever an opportunity arose, they arranged for the destruction of diggings and the damaging of derricks that had been erected to shift rocky strata. They sent out scouts equipped with binoculars to keep watch on each other's progress, bribed team followers for secret information, and even stole specimens from each other. Marsh developed a novel method of recovering fossils, which is still in use today. Rather than simply carving the bones out of the surrounding rock, with the risk of incurring damage, he devised a system in which the fossil was buried in protective plaster of Paris, which was strengthened with layers of sackcloth or burlap, sometimes with poles or sticks of wood to guard against damage. This allowed Marsh to collect his specimens still embedded in strata and transport them back to his laboratory with greater facility, and by 1885 he was clearly in the lead. He had such generous funding from the estate of his millionaire uncle George Peabody that he could lay out more for digging crews and teams of surveyors. Cope could not keep up – and meanwhile, teams from other east coast universities were starting to join the rush to discover new dinosaurs. The advantage Cope had in rapid publication of his findings could not be matched by Marsh, so Marsh kept a careful eye out for mistakes in Cope's results – which were inevitable – and then trumpeted each error whenever he could.

Cope became increasingly preoccupied by the superior attitude of Marsh, and sought an opportunity to turn the tables. As luck would have it, Marsh had irritated many of the team at the U.S. Geological Survey, and in 1884 Congress announced an investigation into the conduct of the operation. This gave Cope the chance he had hoped to find: he arranged for many of Marsh's workers to testify against his activities, and then released the contents of all the problems in his personal notebooks to the press. For twenty years, he had kept a detailed record of every slight Marsh had committed, each mistake, and all the underhand activities in which he had indulged, and he sent the lot to a staff journalist at the *New York Herald*. On Sunday, January 21, 1890 they ran a full-page article headed 'Scientists Wage Bitter Warfare', illustrated with engravings of the two rivals. This brought the affair to the attention of the public, and then they had to publish fierce responses from Marsh in which he, in turn, listed every misdemeanour he could recall against Cope. In the end, Marsh was forced to resign from the U.S. Geological Survey.

Cope, meanwhile, was appointed head of the National Association for the Advancement of Science. He was developing his continuing interest in the mechanism of evolution, and formed the view that mind could profoundly influence matter. Indeed, to Cope, this was the thrust behind evolution. Animals changed because they had a deep-seated wish to do so, and it was the power of an animal's mind that directed its evolutionary progress. His book, *Theology of Evolution*, is a transcription of a lecture, and runs to fewer than 40 pages; but the informal friendliness of Cope's approach gives an insight into his personality and the way he communicated his ideas.[22]

Neither of these more philosophical concerns interfered with the active state of warfare between Cope and Marsh, which continued unabated until 1892, by which time the dinosaurs so familiar to us as children had been given the names that we all recognize: *Triceratops*, *Allosaurus*, *Diplodocus*, *Stegosaurus* and *Camarasaurus*, together with scores of others. Of the two men, the academic Marsh scored the greatest number of discoveries, with 80 new dinosaurs, compared with 56 logged by Cope. The self-taught Cope, however, had the moral victory, and remained personally the more popular of the two.

Marsh devoted his last years to writing a definitive description of the North American dinosaurs, extending to more than 300 pages, which was published by the U.S. Geological Survey in 1896. In spite of its title, the book also included brief accounts of the known European dinosaurs.[23]

The book is rich in resonances of Marsh's self-centred personality: he is quick to describe a specific dinosaur as being 'first named' or 'first described' by him, without reference to the person who actually discovered the fossils in the first place. His descriptions are confident, if fanciful; for example, he devotes a lengthy section of text to describing in detail the skull of *Diplodocus*, neatly avoiding the fact that no skull had ever been discovered. When writing about *Stegosaurus*, he omits any reference to his erroneous original description of the bony plates – as huge scales lying on the back – merely adding that their position was initially in doubt. In the section on *Triceratops*, he makes no mention of his original diagnosis of the animal as a prehistoric bison. On page 105 of this volume, he has a section on the European dinosaurs, before moving on to anatomical descriptions of each type of dinosaur. These are detailed and concise accounts and show full mastery of anatomy. And then there are the plates: 85 of them, mostly of groups of bones shown in graphic detail, interspersed with entire skeletons simply engraved and posed as he imagined them in life. These are pictures to capture the imagination of the non-palæontologist, and his descriptions are vivid. Dinosaurs, he writes, included 'the largest land animals known', whereas some were 'very diminutive'. He sums up the dinosaurs as 'quadrupeds which usually walked on their hind feet, yet sometimes put their fore feet to the ground.' Dinosaurs were certainly becoming clearer in the public mind, and in 1864 a French science writer, Camille Flammarion, wrote a story about prehistoric monsters turning up in the modern era. The book, published in Paris, included an image of a huge dinosaur reaching up to the fifth floor of a city building. Not only did it have the stance of an *Iguanodon* that had been popularized decades earlier, but it still had a horn on its snout – the feature that Mantell had originally ascribed to this dinosaur, before it was realized that it belonged on a limb.[24]

Some accounts say that Mantell's error was quickly corrected, but it isn't so. As we shall see, there was a horn pictured on the snout of *Iguanodon* as late as 1963, when a paperback edition of Edgar Rice Burroughs' *The Land that Time Forgot* was reprinted in New York. The artwork showed a sabretooth cat with a pterosaur swooping overhead and a pair of dinosaurs wrestling in the foreground. The monster on the right was an iguanodon – and on its snout it still bore that horn, some 85 years after its existence had been disproved.

Dinosaurs were becoming part of popular culture. Everybody knew what they were. The time was ripe for vivid re-creations of dinosaurs – images that were not cartoons, like Flammarion's gigantic Godzilla-like reptile, but which captured the essence of what dinosaurs were like. The artist who first succeeded in painting popular pictures of these colossal creatures was a young New Yorker who claimed that he was virtually blind – one eye had been damaged by a flying rock, and the other suffered severe astigmatism. He was Charles Knight. He was born in 1874 and by the age of 12 he was studying at the Metropolitan Museum of Art in Manhattan. By the time he was 16, he had been offered commissions to work on designs for stained-glass windows, and was soon providing artwork as a freelancer for *McClure's Magazine*. This was a stimulating time; he became acquainted with writers including Sir Arthur Conan Doyle and Rudyard Kipling, and he became fascinated by capturing anatomically correct animals in his paintings. He was a persnickety and difficult person with whom to work, and he always took criticism badly. One day in 1897, Jacob L. Wortman stopped by to introduce himself. He had been a palæontologist under Edward Cope, and was now working at the American Museum of Natural History. He had met Knight when the young artist had been making drawings of stuffed animals at the Museum. Would he be interested in painting an impression of one of the animals that Wortman had been working on? This was a fossilized skeleton of a prehistoric mammal then named *Elotherium* – the name has since been changed to *Entelodon* from Ancient Greek ἐντελής (*entelēs*, complete) and ὀδών (*odōn*, tooth) – a hog-like animal from about 35 million years ago. There was a complete skeleton at the Museum, and with guidance from Wortman, Knight felt confident he could re-create its appearance. It was his first commission for painting a prehistoric animal, and Knight was inspired. He went on to paint dinosaurs of every kind with stunning vividness. Some academics decried his work, pointing out that these were lurid pictures of creatures nobody had ever seen; but Knight retorted that, although they were works of his imagination, they were based on skeletons that he had studied as closely as anyone. It was the start of a lifetime that would give the public their first published ideas of how dinosaurs might have looked in life. The addition of a certain artistic licence was a small price to pay, and many of Knight's startling images stand up well today.

At the American Museum, Knight's reconstruction of an extinct mammal caught everyone's attention and he was soon introduced to the director, Henry Fairfield Osborn. Osborn had recently opened his Department of Vertebrate Paleontology, and could see a great future for the young artist. Knight was given a commission to paint more prehistoric monsters, and Osborn took him to meet Cope. Knight spent two weeks studying with Cope, discussing dinosaurs, looking through Cope's extensive sketchbooks, bombarding him with questions, and asking for advice on how dinosaurs should be portrayed. This was when Knight had painted the submerged dinosaurs breathing through their snorkels, all accomplished under Cope's direction. It was a formative time for the young Knight, but Cope passed away within a month, in April 1897. Cope had spent all his money on the quest for dinosaurs, though his dying wish was not for his collections, but for his brain. He died insisting that his skull be examined and measured after his death, because he was convinced that his brain was bigger than Marsh's. Two years later, Marsh died, also in financial difficulties after years of over-expenditure in his bitter rivalry with Cope. Marsh declined to take up the challenge left by Cope, and was buried with his head intact, while Cope's detached skull remains in storage at the University of Pennsylvania. The saga attracted the attention of one of America's greatest novelists, Michael Crichton, author of *Jurassic Park*. Crichton wrote the draft of a novel telling the story of William Johnson, a student at Yale whom he follows as he works, at different times, for both Cope and Marsh. The novel was found among

Crichton's papers after his death in 2008 and it was published to critical acclaim in 2017.[25]

By the end of the nineteenth century, dinosaurs were good copy; newspapers knew that they could always sell extra copies if there was a good dinosaur headline on the front page. And it worked both ways: explorers searching for dinosaur fossils, and anxious for fame and fortune, knew that they could always rely on eager newspaper reporters to inflate the humblest discovery into an eye-catching sensation. The story of William Harlow Reed, born in Connecticut in 1848, reveals how it worked. He was not a palæontologist, and had no academic training, but became one of the most significant figures in the unfolding story of dinosaur discovery. As we have already seen, William Reed (he was always known as Bill) became the station foreman for the Union Pacific Railroad at the Como Station in Wyoming. He and the station agent, William Edwards Carlin, had become adept at spotting dinosaur bones in the jumble of excavated rocks. Although Marsh signed Reed up as a prospector, Reed found Marsh to be difficult, vain, unreliable and suspicious, and at Easter 1883, he quit. Managing the farmlands was becoming profitable, so Reed joined forces with another disaffected member of the Marsh team, George Bird Grinnell, to go into sheep ranching. It didn't work. Within a year he was seeking other sources of employment and started working as an independent fossil collector. Here he excelled. Reed could pick out the remains of some gigantic creature in a wilderness of scattered rock and he soon came to the attention of Wilbur Knight, Professor of Mining and Geology at the University of Wyoming. Knight was on the lookout for specimens. He needed to make a name for himself, because he had just been given an exciting new job – as the curator of the University Museum – so in 1896 he put Reed on contract as his official collector of fossils.[26]

The following year, Reed found a single fragment of a dinosaur femur in the strata of the nearby Freezeout Hills. Returning to the site from time to time, he could find nothing more; but he was anxious for progress and reported what he had found to newspaper journalists. Reed was keen to please his professor, while Knight was keen for the museum to be at the forefront of discovery, so there was no reason to hold back on the potential of this discovery. The result could hardly have pleased them more: a report in an issue of the *New York Journal*, published in December 1898, containing a full-page feature on their expeditions. It was decorated with a lurid artist's impression of a massive monster poking its nose into the eleventh storey of a city skyscraper, under the headline: 'The most colossal Animal ever on Earth just found out West.' There was a photograph of footprints, heavily retouched for dramatic effect, a reconstruction of the skeleton with a man in the picture reaching just above the monster's ankles, and a vivid description of a terrifying dinosaur captioned: 'How the Brontosaurus giganteus would look' … truly, their discovery had hit the headlines.[27]

Public interest in dinosaurs was regularly nurtured by sensational stories in the press. In 1912 the *Washington Herald* newspaper featured an exaggerated impression of *Gigantosaurus* demolishing domestic dwellings. Explorers at the time were hoping to discover living dinosaurs in Africa. Another report was of an 80-foot (24-metre) dinosaur and six pictures dominated the front page of *The Republic* newspaper, published in St Louis in October 1903. Gigantic dinosaurs are not recent news.

36 feet of NECK on SHOULDERS 20 feet HIGH!

What Newest Gigantosaurus That's Found Has and Now the Scientists Are Going to Search for a Live One in Africa's Swamps.

ABOUT IN AMERICA

MILLION YEARS AGO

50 Feet Long Whose Bones are in a Museum.

SYMPHONY SEASON SOON TO BEGIN.

And then came the consequences. In New York, the news was read eagerly by Andrew Carnegie, then the wealthiest person in the world. Born into poverty in a weaver's cottage in Scotland, he had emigrated to the U.S. with his parents and became enormously successful in building America's infrastructure – roads, railroads, bridges etc. – soon amassing a fortune, the equivalent of around £80 billion ($100 billion) today. He gave more than 90 per cent to charitable causes, including the public understanding of science, and during his lifetime he opened 3,000 public libraries, the first at Dunfermline, Scotland, in 1883. Carnegie was fascinated by palæontology and he decided he wanted the newly discovered dinosaur for himself. He had recently opened his Pittsburgh Museum, so he scribbled in the margin of the newspaper report: 'Buy this for Pittsburgh?' and he mailed the copy over to the curator of the museum, William Holland, with a payment of $10,000. Further inquiries disclosed the truth: 'Brontosaurus giganteus' did not exist. There was no skeleton, only a fragmented femur. Reed was undeterred and explained that the size of the bone indicated that there had once been a massive dinosaur at that site, and so he offered to return and look for more, just as soon as the winter snows had melted. All he needed was the money. Without delay, Carnegie ordered that Reed should be signed up – so he immediately resigned from the Wyoming Museum, and became a fossil collector for Carnegie and the Pittsburgh Museum on a 12-month exclusive contract.

It is tempting to assume that, by the close of the nineteenth century, the 'bone wars' had ended, since Cope and Marsh had died. Yet there were still rival operations in Wyoming, each still trying to outpace the others. The old rivalry soon revived, and a new bone rush was under way. The University of Wyoming, having been deserted by Bill Reed, obtained funding from the Union Pacific Railroad and launched its Fossil Fields Expedition to search for more dinosaurs. The railway company provided free passage to their station at Laramie, where Wilbur Knight from the university met the teams and gave guided tours of the most propitious sites. Carnegie now commissioned Bill Reed to find the rest of his huge new brontosaur, and Reed set out across the wilderness, reporting back to William Holland. Meanwhile, the American Museum of Natural History sent out a rival team from the Department of Vertebrate Paleontology where Barnum Brown still ruled. One member of the team was Arthur S. Coggeshall, only 25 years old but already their top preparator and conservator; another was Jacob L. Wortman, their fossil curator. They soon met up with the Carnegie team, their hottest rivals, and were photographed together with a plate camera in an informal group picture. But that particular battle soon fizzled out. Carnegie could offer far more money than the museum could provide, so Coggeshall and Wortman – much to Barnum Brown's annoyance – handed in their notice and went to work for the Pittsburgh team instead.

Travelling to the Freezeout Hills was a difficult and hazardous trip. They had to take tons of supplies and excavation equipment on horse-drawn wagons that were exposed to the elements in the thin air of the high plains, and it all had to be transported manually when crossing tumbling rivers or the precipitous ravines. Reed was relieved when they all reached the site where his huge dinosaur femur had been found, but – after two solid months of digging and prospecting – they found little more. Word reached them that they might have more luck at Sheep Creek Basin in the Morrison Formation, which was emerging as a rich source of dinosaur finds, so they packed up their camp and trudged across with their wagons, arriving on Monday July 3, 1899. Next morning, being the Fourth of July and a public holiday, Reed went out with his gun, scouting across the massive tumbled rocks nearby, when suddenly a fossilized bone caught his eye in the sunlight. He summoned Arthur Coggeshall and Jacob Wortman, and they immediately set to work excavating. This was extraordinary – as they cleared that first bone, another appeared; within hours it was obvious that they had found something unprecedented. Most of the skeletons that palæontologists discover are broken up and fragmented, but here they had something like an entire dinosaur. Palæontologists from other teams came over to see, including Henry Fairfield Osborn, Coggeshall's and Wortman's previous employer, who had made the 10-mile (16-km) journey from his own dig to congratulate the Carnegie team. The new skeleton needed a name; Coggeshall proposed 'the star-spangled dinosaur' because it was found on the Fourth of July. But once it was obvious that this was a *Diplodocus*, it was soon nicknamed 'Dippy'.*

As the bones were meticulously catalogued by the workmen, they turned out to represent the largest dinosaur skeleton that had ever been discovered. The press eagerly reported the discovery as a vast creature measuring some 130 feet (40 metres) long and originally weighing 55 tons. The first bones were transferred to Pittsburgh by rail late in 1899. On his return from South America in 1900, Hatcher was appointed curator of Palæontology and Osteology at the Carnegie Museum of Natural History, and here he took on the personal responsibility for the scientific investigation and exhibition mounting of the famous *Diplodocus*. News reached him of the death of Marsh as this work was under way, so Hatcher felt it diplomatic to name it 'Diplodocus (Marsh)' in the title of his formal description. This became the first paper on the first page of the first volume of the journal of the Carnegie Museum. It was truly a landmark discovery.[28]

Hatcher described these dinosaurs as being 60 to 100 feet long (roughly 20–30 metres) and as 'the largest land animals known to science'. He also took pains to remind the reader of the palæontologists who actually worked on the remains; Jacob Wortman in 1899 and O.A. Peterson in 1900. It is interesting to see just how many people Hatcher mentions in his book; he is assiduous in giving credit to those who discovered the dinosaurs. This is a meticulous account and often he devotes a page or more to describing a single bone. He is quick to point out that 'unfortunately, there is no skull of *Diplodocus* in our collections' and he correctly reports that, for instance, 'the skull of *Diplodocus* reported in Science, Nov 9, 1900, p 718, when freed from the matrix, proved to belong to *Stegosaurus*.'[29]

It took more than two months for the entire skeleton of the newly discovered dinosaur to be excavated, packed in crates and shipped back to Pittsburgh. This new fossil posed unusual problems. Coggeshall introduced sandblasting to separate the fossils and small

* The original notebooks of many of these collectors have been digitized and are available online through the research division of the American Museum of Natural History at: http://research.amnh.org/paleontology/notebooks/

pneumatic hammers to clean off extraneous rock, a task involving immense diligence. These are still used today; I have learned how to use them and they are remarkably efficient. So many of the bones had been found that he now had to find ways to support an entire skeleton. Coggeshall relied on the methods developed for the iguanodons at the Royal Belgian Museum of Natural History in Brussels, manufacturing curved steel rods on which the vertebræ could be supported, with custom-built bands of metal to support the ribs and limbs. Meanwhile, Hatcher drew and measured them. He soon reviewed his provisional name and decided that it was diplomatic to call it *Diplodocus carnegii* (sometimes spelled *carnegiei*) out of respect for his benefactor. To complete his description of the *Diplodocus* skeleton, Hatcher simply cited the details already published by Marsh, adding his published words in quotes. Hatcher knew there was no real skull, and managed to complete his description while reminding readers that Marsh was his primary source of information. Everyone wanted to know more about this astonishing fossil, because dinosaurs were becoming established as creatures to be seen and appreciated by the public at large.

When Hatcher finally examined the assembled skeleton, he found it to be somewhat less massive than Marsh had led the press to believe – even so, he calculated that it measured 84 feet (25 metres) long from nose to tail, and would have stood 15 feet (4.5 metres) tall at the hips. Hatcher's detailed scale drawings were framed and hung on the wall of Carnegie's study at his castle in Skibo, Scotland, overlooking the Dornoch Firth; the castle is now a luxury private country club. When

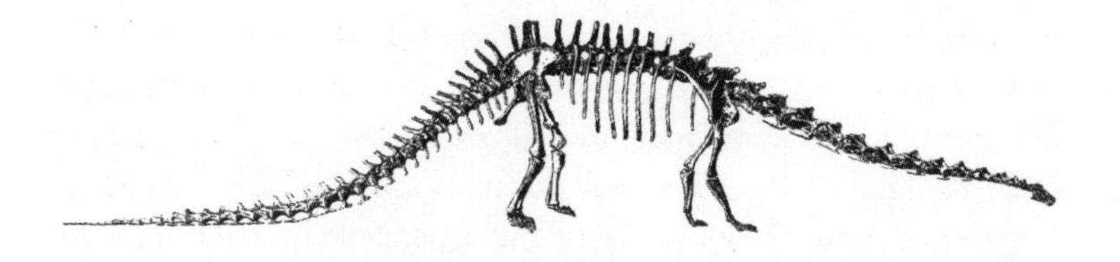

John Bell Hatcher published this skeleton of a dinosaur he named *Diplodocus (Marsh)* in 1901. He described this species as 'the largest land animal known to science' and calculated that it was up to 100 feet (30 metres) long from head to tail.

Carnegie was on vacation and residing at his castle, King Edward VII called to visit and inquired about the drawings on the wall. The king wanted a dinosaur for London. Carnegie explained that the skeleton could not be donated but he agreed to have moulds prepared, so that casts of the fossil bones could be made and put on display. It was arranged that the first set would be commissioned for the king, and would go on display in the galleries of the British Museum (Natural History) in South Kensington, London.

Four men worked on the project for 18 months. When the casts of the skeleton were ready, they were temporarily assembled at the Western Pennsylvania Exposition Society in Pittsburgh before being packed and shipped to London. On May 12, 1905, the skeleton was unveiled to an astonished audience. The replica bones had been mounted on a low platform in the middle of the entrance hall of the museum, the graceful neck of the dinosaur arcing over the assembled audience and its elongated tail snaking across the wooden floorboards. The museum's director, E. Ray Lankester, rose to speak. He addressed the audience confidently, emphasizing the pre-eminence of Britain in the field of dinosaur discovery, even implying that the

When the Natural History Museum opened in May 1905, the centerpiece was the 105-foot (32-metre) cast of the *Diplodocus* skeleton sent to London by Andrew Carnegie. E. Ray Lankester, the director, gave a speech that minimized its significance.

The diplodocus replica, nicknamed 'Dippy', dominated the Museum's entrance hall from 1979 until 2017 when it was taken down for a tour of regional British museums. Since 1993, its tail was repositioned, and raised high on metal supports.

museum's existing collections of skeletons almost rendered this American gift superfluous. So crowded were the specimens in their hall of palæontology, he explained, that there was no room for this new arrival. In acknowledging Carnegie's success, he reminded everyone that Carnegie had been British by birth, adding that the enterprise shown in America was merely an offshoot of the mother country.

Carnegie's gracious address was much more magnanimous. He explained that it had been a request from the king that had led to his gift; Carnegie reminded his audience that Edward VII was active in so many areas 'advancing the interests of his country in every department of national life, from the peace of nations to the acquisitions of his museum.' (The king had formerly been a member of the museum's board of trustees.) Carnegie added that, in his view, Britain and the U.S. should never have become separated, and might one day be reunited. For Carnegie, this was an important gesture – a huge dinosaur skeleton given by one of the youngest museums to one already well established. This was not only a palæontological wonder, but could have political implications. When John Lubbock, 1st Baron Avebury, rose to respond, he could not resist the temptation to bring the glory back to Britain, announcing in a dismissive tone that 'the size of the animal does not indeed add much to the interest.' None of those British speakers could be commended for their civility.

It was the young American from Pittsburgh, William Holland, who restored a certain dignity to the proceedings. He reminded everyone that it had been Marsh who had coined the name *Diplodocus* after it had been discovered in Colorado, and Carnegie who had led the investigation of the skeleton. He also paid generous tribute to John Bell Hatcher, whom many had expected to attend the ceremony. This promising palæontologist had died of typhoid the previous year at the age of 41 and Holland spoke of 'the labours of Professor J.B. Hatcher, my learned colleague, who for fully 18 months devoted himself to superintending the restoration of the object before us, but who unfortunately was stricken down in the midst of his activities by the hand of death.' It was a fitting tribute to a prolific young palæontologist.

5

DRIFTING CONTINENTS

John Bell Hatcher had studied more than palæontology. By the time he was 30 he had formulated a radical new theory: he believed that South America, Antarctica and Australia had once been part of a single landmass. Hatcher was a pioneer of the theory of continental drift, though he never lived long enough to progress the idea. This – the idea of what we now call plate tectonics – owed much to the study of palæontology and illustrates how disparate areas of scientific understanding can originate from a single discipline. However, the idea of continental drift is not as new as we think, and most of the pioneers have since been forgotten. Let me trace the many philosophers who reached the same conclusion. There are more than you might imagine.

First in the field had been the Flemish philosopher Abraham Ortelius, who in 1596 speculated on the similarity in outline between the eastern coast of the Americas compared with the west coast of Europe and Africa that was emerging from the early world maps, and he wrote that the Americas seemed to have been 'torn away from Europe and Africa … by earthquakes and floods.' Ortelius went on to say: 'The remains of the rupture reveal themselves, if someone were to bring forth a map of the world and consider carefully the coasts of the three continents.' There you have the idea of plate tectonics dating back more than four centuries.[1]

Ortelius may have been the first to record the similarity between the outlines of the continents, but he was soon followed by Francis

Bacon, who commented on the similarity of the continents of the Old World (Africa) and the New World (America). Bacon remarked that the continents had similar shape, somewhat like pieces of a present-day jigsaw, so that the Gulf of Guinea seemed to fit with eastern Brazil, though he looked no further for an explanation.[2]

Further evidence came from the realm of palæontology, when a French philosopher (sometimes erroneously described as American) named Antonio Snider-Pellegrini compared the types of fossils found on opposite sides of the Atlantic. In his book, he argued that the evidence suggested the continents must surely have once been conjoined and had subsequently moved apart.[3]

He was followed by the brilliant Charles Lyell in 1872, who concluded: 'Continents, although permanent for whole geological epochs, shift their positions entirely in the course of ages.'[4]

Robert Mantovani, an Italian geologist, who resided for a time on the Indian Ocean island of Réunion, also theorized that the continents had once been a single primordial landmass.[5]

Surprising as it seems, by the 1890s the notion of continental drift was becoming widely accepted. In 1891 Alfred Russel Wallace, the

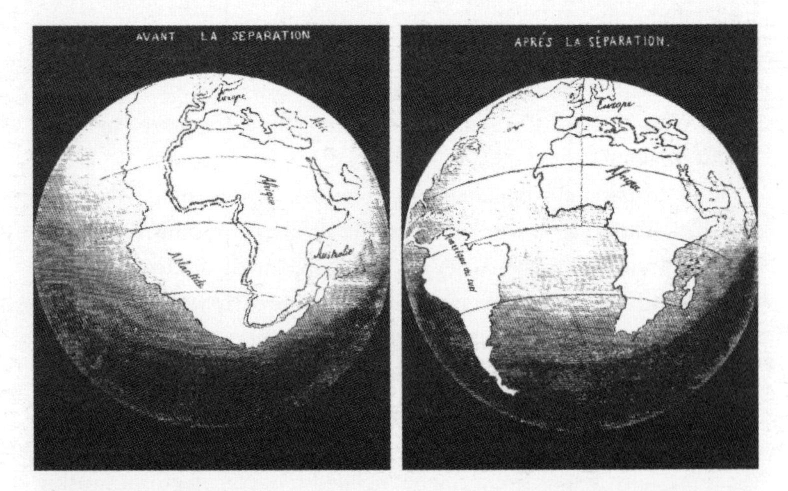

Continental drift was pictured by Antonio Snider-Pellegrini after a detailed comparison of fossils found in America and Africa. In 1858 he published his ideas in *La Création et ses mystères dévoilés* (Creation and its Mysteries Unveiled).

pioneer of evolution by natural selection, wrote that the concept was commonplace among geologists: 'It was formerly a very general belief, even amongst geologists, that the great features of the Earth's surface, no less than the smaller ones, were subject to continual mutations, and that during the course of known geological time the continents and great oceans had again and again changed places with each other.' You see? The idea was common currency even in the nineteenth century.[6]

Early in the twentieth century, William Henry Pickering, a prominent American scientist, speculated how the moon had once been part of the Earth and the scar from its breaking away from Earth had become the Pacific Ocean. Pickering theorized that America, Asia, Africa and Europe had originally been a single continent, which fragmented only when the Moon broke away.[7]

Two years later, Viennese geologist Eduard Suess published a four-volume book on the face of the world in which he proposed the concept of a single primeval continent which he named Gondwanaland. He also postulated a single ocean which he named Tethys;[8] even though Suess has been forgotten, both of the terms he proposed are in use today.

The idea of continental drift was proposed at the same time by a self-taught American geologist, Frank Bursley Taylor. He enrolled at Harvard but found the university ambeance restrictive, so he left without graduating. (You see? Another one.) Taylor embarked instead upon an independent career in geology, financed by his wealthy father. He proposed his idea of 'continental creep' at a meeting of the Geological Society of America on December 29, 1908, arguing that the continents had slowly shifted about on the Earth's surface and that the shallower zone in the middle of the Atlantic showed where Africa and South America had separated. He concluded that the collisions of continents had caused the uplift of mountain ranges. His ideas were based on his studies on mountains including the Andes, Rockies, Himalayas and Alps. Only the collision of moving continents, he argued, could produce sufficient force. Thus, the Himalayas had arisen because of an ongoing collision between Indian and Asian tectonic plates. This was a timely and correct suggestion, and he was closely in line with our present-day theories. Most geologists at the time ignored his views, and those who referred to the theory rejected it. Taylor was ridiculed wherever he went, but he was right.[9]

The name that we most readily associate with the theory is Alfred Wegener, a German meteorologist and an expert on the polar regions. Wegener hypothesized that Africa and South America had once been connected. He proposed a single, primeval landmass (as Suess had done some years before), which he named Pangea. Yet, for all the growing mass of evidence, Wegener's proposal was rejected by the scientific establishment and he too was dismissed as a crank. The theory was assailed by everyone, and Wegener – as a meteorologist, rather than a geologist – was dismissed as an ignorant outsider.[10]

Many geologists roundly attacked Wegener, including Thomas Chrowder Chamberlin who founded the *Journal of Geology*. The drift theory 'takes considerable liberties with our globe,' he wrote, adding that it ignored 'awkward, ugly facts' and 'plays a game in which there are few restrictive rules.' He quoted this clarion call: 'If we are to believe Wegener's hypothesis we must forget everything which has been learned in the last 70 years and start all over again.' At a meeting in London of the Royal Geographical Society, speakers berated the theory as nonsense and someone rose to express relief at seeing the theory 'blown to bits', thanking the absent Professor Wegener 'for offering himself for the explosion'. In the U.S., the president of the prestigious American Philosophical Society described the theory as: 'Utter damned rot!' – and this in an era when such expletives were severe. To this day, Americans will bleep out the word damned from song lyrics, or replace it with a euphemism (like 'darned'), so at that time, prior to World War I, the condemnation must have had the vernacular force of the F-word today. Few scientists wanted to accept continental drift, and American geologists were devotees of the theories published by James D. Dana, a mineralogist and volcanologist at Yale. He had argued powerfully for the permanence of the continents and the ocean basins in his *Manual of Geology*.[11]

Dana's ideas permeated much geological thinking and militated against continental drift being accepted in America. Indeed, the American Association of Petroleum Geologists organized a symposium, which Alfred Wegener and Frank Bursley Taylor both attended, specifically to humiliate the proponents and to reject this 'ridiculous' theory.[12]

The concept of continental drift began to gain further acceptance in Britain after research by Arthur Holmes a British geologist from Gateshead, in the north of England. He was particularly interested in dating strata using the measurement of radioactive decay in rocks, and he soon became a firm proponent of Wegener's theory.[13]

I was taught about the principles of continental drift in junior school in north London during 1949, some 20 years before the idea was accepted in America. Indeed, there was a poll taken at the close of a meeting of the American Association of Petroleum Geologists in 1968 that overwhelmingly rejected the concept of plate tectonics and sea-floor spreading. It was still not acceptable in America until, in a startling reversal, a poll taken at the annual meeting of the same Association in 1972 found the theory was almost universally accepted. That isn't long ago …

Wegener's idea had actually been born centuries earlier, and there were at least 10 scientists who wrote on continental drift before his time. Once again, the convenient orthodoxies of present-day science are shown to be misleading.

The young John Bell Hatcher would have carried on research into this theory, though he never lived to tell the tale, and it is his great success with the *Diplodocus* skeleton that is a fitting testimony to the labours of his short life. During his lifetime, palæontology had emerged, not just as a backward-looking study remote from the real world, but as the key to understanding more about how our world works. Once the replica skeleton of Dippy had been unveiled in London, the original skeleton was displayed at the new dinosaur hall of the Carnegie Museum of Natural History in Pittsburgh, and in the following two years other copies went on display in Berlin, then Paris, Vienna and Bologna. St Petersburg obtained their plaster skeleton in June 1910 (it is now on display in Moscow), La Plata in Argentina had theirs in 1912, and in 1914 one was erected in the National Museum, Madrid.

These are all the museums to receive one of the replica skeletons:

- British Museum (Natural History), London (given in May 1905)
- Humboldt Museum, Berlin (April 1908)
- Muséum National d'Histoire Naturelle, Paris (June 1908)
- Museum of Natural History, Vienna (September 1909)
- Aldrovandi Museum, University of Bologna, Italy (October 1909)
- Paleontological Institute, St. Petersburg (July 1910), later moved to Moscow
- Museum of Natural Sciences, La Plata, Argentina (September 1912)
- National Museum, Madrid (November 1913)
- National Museum of Natural Science, Mexico City (April 1930)

One further cast was exchanged for European fossils with the Bayerische Staatssammlung für Paläontologie, Munich, in 1934, but the 33 sealed crates were never unpacked and were not rediscovered until 1977. Another was sent to Utah Field House of Natural History in Vernal, Utah, in 1957 and was eventually erected outside on the lawn. It was removed a few years ago and used to make new moulds, from which a new version of the skeleton was cast and subsequently erected inside the museum. The London skeleton remained a prominent tourist attraction, and for decades it stood towering above the visitors crowding into the Hinzte Hall as they entered the museum, its tapering tail snaking across its wooden stand and a magnet for souvenir hunters.

The death of Hatcher of typhoid fever was formally announced in 1904 in Pittsburgh, Pennsylvania. He had died as he was completing a paper on *Ceratopsia* that had been partly prepared by Marsh. The monograph was not completed until 1907, when it was published by Richard Swann Lull. Hatcher's body was buried in the Homewood Cemetery in Pittsburgh and it lay unmarked for more than 90 years. In 1995, at the annual meeting in Pittsburgh of the Society of Vertebrate Paleontology, members agreed to fund the erection of a headstone. In addition to his name, and his dates, the headstone has a silhouette of the new genus of dinosaur that he had discovered: *Torosaurus*. This remains one of the most remarkable of all the dinosaurs. Fossils of *Torosaurus* have now been found across North America in a band stretching from Saskatchewan down to Texas. They reveal that it was an herbivorous dinosaur that flourished during the late Maastrichtian stage of the Cretaceous some 67 million years ago. It had one of the largest and most impressive skulls of all, measuring more than 9 feet (2.8 metres) long. The adult dinosaur measured 30 feet (9 metres) long and weighed up to 10 tons. It differs from *Triceratops* in having an elongated frill that is perforated by those characteristic fenestræ, and it has a number of small horns at the rear of the frill, in contrast to the single nose horn that is a feature of *Triceratops prorsus*. Otherwise, the two genera have elements in common. In 2010, at the Museum of the Rockies at Bozeman, Montana, a revolutionary new idea was proposed. Perhaps *Torosaurus* is simply an older *Triceratops* – the dinosaur's frill changing throughout life, eventually losing its main horn and acquiring several smaller versions, while developing holes in the frill as an inevitable consequence of ageing.[14]

Other palæontologists have pointed out other differences, so many remain convinced that these are two separate genera. Only time will tell if the Bozeman conclusions are correct ... if so, then the genus *Torosaurus* may suffer an untimely demise, much as its discoverer did in 1901.

By the turn of the twentieth century, the discoveries through the 'bone wars' era had revolutionized palæontology and had given science unimaginable new revelations of 136 of the strangest creatures anybody could imagine. Although the bitter rivalry in America was regarded with disdain by the European palæontologists, it did confer two important benefits. First, it gave an enormous impetus to palæontology and, in particular, to the popularity of dinosaurs. Second it fuelled the fight for discovery – without the burning sense of competition, the rate of new revelations would never have been so rapid. Othniel Charles Marsh and Edward Drinker Cope created a worldwide eagerness to know more about dinosaurs, and many of the monsters with which we became familiar as children were originally announced by these two men. The skeletons everybody sees in museums around the world largely owe their popularity to those two rivals, and we must not forget that it was their enmity and mutual hatred that can teach us lessons in the way scientific competition operates. We can also learn from their fallibility; remember, both are known for making the same mistake – mounting the wrong skull in the wrong place.

There are more modern examples of resonances from the 'bone wars' of the Victorian era. One intriguing proposal emerged in September 2012 when HBO television announced that Steve Carell and James Gandolfini would star in a movie to be made about the dinosaur controversy. Plans for a film on the 'bone wars' story proposed that Carell would play Cope and Gandolfini would portray Marsh. The provisional title of the movie was 'The Bone Men', and Gandolfini was reported as saying it would be a 'very funny' movie. However, on June 19, 2013, Gandolfini died unexpectedly, at the age of 51. He was visiting Italy with his family en route to collecting an award at the Taormina Film Festival, when he was found unconscious in his hotel after a strenuous day sightseeing. He died of a cardiac arrest. So far, there is no information as to a plan for the movie to progress. Meanwhile, there is an Android app entitled *Dino Hunters*, which invites players to 'journey to a hidden, untouched Jurassic island and kill the most ferocious animals in history. Encounter Jurassic beasts long thought extinct, from the docile stegosaurus to the terrifying T. rex.' Similarly, Digital Dreams has a game entitled *Carnivores, Dinosaur Hunter Reborn*, in which peaceful dinosaurs, as playful as puppies, are relentlessly stalked across an arid landscape by a game-player with a high-calibre weapon.

While the American palæontologists were struggling for supremacy in their new bone rush, a British palæontologist was attempting to tackle the way they should be classified. The fused sacral vertebræ proposed as a diagnostic feature by Richard Owen had been recognized as unreliable. Other palæontologists tried to classify dinosaurs from their teeth, or – according to the German palæontologist Christian Erich Hermann von Meyer – their limbs, but none proved reliable. The new approach to the classification problem was proposed by a Londoner, Harry Seeley, who had looked more diligently at dinosaur skeletons and was confident he could find an answer. Seeley was born in London in 1839 and studied at the Royal School of Mines before being appointed an assistant to Adam Sedgwick at the Woodwardian Museum in Cambridge. Seeley turned down positions with the British Geological Survey and the British Museum, only later becoming Professor of Geology at King's College, Cambridge. He published his paper in 1887, proposing to divide all dinosaurs into two main groups, depending upon their bone structure: ornithischian, or 'bird-hipped', and saurischian, or 'lizard-hipped', dinosaurs.[15]

The idea increased in popularity for over a century, and Seeley was better at classifying dinosaurs than naming new discoveries. He came up with a range of strange new names, including *Aristosuchus*, *Craterosaurus* and *Thecospondylus*, none of which caught on. The essence of his classification remains widely used today, even though recent research has shown that today's birds are distantly derived from the saurischian dinosaurs, rather than the ornithischian types. That is curious, since the ornithischian types were named after birds.

So many strange and colossal creatures had been dug out of the ground, reassembled and scientifically named, that dinosaurs were now universally well known and there was a sense that the great era of discovery was winding down. Little did people realize that some of the greatest discoveries were just about to dawn. First to emerge was a huge skeleton that was discovered in 1900 – once again, on the Fourth of July – during exploration of the Morrison Formation by Elmer S. Riggs and a team from the World's Columbian Exposition in Chicago, which was later to become the Field Museum of Natural History. The team had gone to the area after Riggs had corresponded with Stanton M. Bradbury, a dentist and amateur fossil collector living in Grand Junction. Bradbury wrote to say that dinosaur bones had been regularly collected as curios since 1885. One of the team, H.W. Menke, was digging in what is now known as Riggs Quarry No. 13 when he came across some enormous bones. A preliminary report, published in 1901, did not name the specimen but claimed that it must be the largest dinosaur ever discovered.[16]

Riggs measured the huge humerus compared with the length of the shorter femur and described the enormous size of both. He recognized that this new dinosaur had proportions like those of a giraffe. Later, in 1903, he named the new discovery *Brachiosaurus altithorax*, deriving the genus from the Greek βραχιων (*brachion*, arm) and σαυρος (*sauros*, lizard), and the species from the Latin *altus* (deep) and the Greek θώραξ (*thorax*, chest). Thus, a huge new dinosaur entered the lexicon – though these were not in fact the first *Brachiosaurus* bones to be discovered. A skull of this dinosaur had already been unwittingly unearthed in 1883 at Garden Park, Colorado, at Felch Quarry No. 1, and had been sent to Marsh. This, we now know, was the skull that Marsh foolishly had added to the top of his *Diplodocus* skeleton before naming his hybrid *Brontosaurus*. Each time one of the new bones was cleaned by Riggs' technicians and was fit for display, it was sent to Hall 35 in the Fine Arts Palace of the World's Columbian Exposition and set up in a glass case for the public to see. By 1908 all the bones were on view, though only 20 per cent of the skeleton had been retrieved, so it was not possible to attempt a full reconstruction. Even so, it was clear that this would be the biggest dinosaur skeleton ever put on public display. More recently, casts have been made of the missing bones by using skeletons of brachiosaurs in other museums, so a complete copy has been assembled. In 1999 it was erected in the B Concourse of Terminal 1 at Chicago's O'Hare International Airport. A second cast replica stands outside the Field Museum on its northwest terrace, where it overlooks the teeming traffic on Lakeshore Drive. The few original fossils of *Brachiosaurus* currently on show are a humerus and two other bones in the Mesozoic Hall of the Evolving Planet exhibition in Chicago's impressive Field Museum. This is where Sue, their restored *T. rex* skeleton, will be permanently on display. In 2018 it was decided she was too small for their main hall, so Sue (in the language American scientists love to use) was 'de-installed' and replaced with a plastic cast of *Patagotitan mayorum* like the one in the American Museum of Natural History (A.M.N.H.), New York.

The new century had begun with the announcement of a bigger dinosaur than had ever been seen before, and it was soon followed by the fiercest. This new discovery became famous around the world as the legendary *Tyrannosaurus rex*. Or was it so new? The first fragmentary remains had been a few isolated teeth excavated in 1874 in

Colorado, followed by portions of a skull that were discovered in Wyoming in 1889. Marsh decided to name it *Ornithomimus grandis*. A few more bones were discovered by explorers working for Edward Drinker Cope in South Dakota in 1892, and Cope decided to name it *Manospondylus gigas*. In 1902, the assistant curator of the American Museum of Natural History set out to find more. He was Barnum Brown, the fossil collector (named after P.T. Barnum, the celebrated circus owner and entertainment impresario), and the fossil skeletons he unearthed were far more complete. The world was about to greet the biggest and most ferocious meat-eating dinosaur yet discovered, and the names proposed by Marsh and Cope were ignored when the president of the museum, Henry Fairfield Osborn, decided in 1905 to give this dinosaur the name everybody now knows: *Tyrannosaurus rex*. This is the only species of the genus *Tyrannosaurus*, although similar meat-eating dinosaurs in the same family have since been discovered and given different names, including *Albertosaurus*, *Gorgosaurus* and *Appalachiosaurus*. Most have been found in North America, though some more recent discoveries have been made in Asia. Because the new discovery, *T. rex*, was a huge carnivorous dinosaur, the name has always been considered to be singularly appropriate. It comes from the Greek τύραννος (*tyrannos*, tyrant) and the familiar σαῦρος (*sauros*, lizard). The *rex*, of course, means 'king'. Here was a dinosaur that had populated the whole of the western U.S. and Canada around 67 million years ago, living on a subcontinent that geologists now call Laramidia. It was reconstructed as standing on its massive muscly hindlegs, with ridiculously reduced forelimbs, and with a formidable skull seemingly counterbalanced by a long, muscular tail. An adult measured about 40 feet (12 metres) long, 12 feet (3.7 metres) tall, and could have weighed 20 tons. Osborn portrayed it with a massive tail dragging along the ground, just as Winsor McCay had drawn Gertie. This dinosaur has been categorized as a terrifying hunter that tore its prey to shreds, and it is claimed that it could run surprisingly fast, like the *T. rex* in the *Jurassic Park* movie. Since those early discoveries, more than 50 specimens of *T. rex* have been discovered, many of them virtually complete. The presence of proteins and the remains of soft tissues have even been claimed, though this remains a controversial subject. Even so, *T. rex* has become the best-documented dinosaur in the world.

It was Sue Hendrickson of the Black Hills Institute (a private company that searches for dinosaur fossils and sells what it finds) who discovered the biggest and most complete *Tyrannosaurus* skeleton in the Hell Creek Formation, South Dakota, in August 1990. Over 90 per cent of the bones were retrieved. It immediately became embroiled in a legal dispute over ownership; in May 1992 a dawn raid was mounted by the Federal Bureau of Investigation (FBI), and their agents crated up all 10 tons of the skeleton. The FBI confiscated research material and photographs, company accounts and business records, adding up to some 50,000 items. It took seven years of legal argument before it was finally ruled that the skeleton was legally the property of the occupier of the site, Maurice Williams. The fossil remains were auctioned by Sotheby's, and the Field Museum of Natural History in Chicago – with the help of funding from the Walt Disney Company, McDonald's and California State University – paid a gross price of $8.4 million, making it the costliest dinosaur skeleton in history. For over two years, the Field devoted more than 25,000 man-hours to cleaning the bones, and the entire skeleton was then sent to Phil Fraley Productions, a fine art foundry near Trenton, New Jersey, where the steel mount to hold the skeleton was constructed. On May 17, 2000, the skeleton, named Sue after its discoverer, was back in Chicago and was unveiled in the Stanley Field Hall, the vast main concourse of the Field Museum. The palæontologists concluded that the skeleton was female, and that this *T. rex* had died aged 28. She showed many signs of an eventful life – the left leg had been broken, but had healed; and there were further healed breakages in the caudal vertebræ and in a rib. There was an old injury to the jaw, and damage to the neck where an embedded tooth was found. Although the back of the skull was crushed, the experts did not conclude that this was the fatal injury (they claim the damage was probably due to trampling postmortem), though there were signs of abscesses in her bones that I have photographed. It seems to me that Sue may well have suffered chronic infections, perhaps being infested by parasites. There are also holes in the front of the skull, which were quickly interpreted by

Sue the *T. rex*, which stood in the Stanley Field Hall of the Field Museum in Chicago, was 'de-installed' in February 2018. The skeleton re-appears in 2019 as part of the Evolving Planet exhibition. The skeleton is over 40 feet (12 metres) long.

palæontologists as tooth marks inflicted by an attacker, but which must surely be erosion resulting from abscesses.

More *T. rex* skeletons have since been found in the same area. One was named Stan, to recognize the amateur palæontologist, Stan Sacrison, who discovered it in the Hell Creek Formation near Buffalo, South Dakota, in 1987. This skeleton was given a world tour in 1995–1996 and is now on display at the Black Hills Institute of Geological Research in South Dakota. Casts of Stan have been exported to other museums, and you can buy one for $100,000 (about £80,000). The replica skull alone will cost you $9,500 (about £7,000). Stan is the most complete of the presumably male *T. rex* skeletons yet found.[17]

Like Sue, Stan shows signs of a fearsome life, including ribs and vertebræ that had been broken (and then healed), and there is a hole in the rear of the skull which has been interpreted as the result of an attack by another carnivorous dinosaur. Subsequently, five more *T. rex* skeletons were discovered at Fort Peck Reservoir, Montana.

Tyrannosaurus rex has become an iconic dinosaur known all around the world. The rock group formed by Marc Bolan in 1967 was named *Tyrannosaurus rex* from the start, though the name was shortened to *T. rex* in 1970. Their convention was precisely in line with correct scientific terminology. Other popular versions you may encounter in the press, such as *T Rex* and T-rex, are incorrect. The full name correctly gives a capital letter to the genus (Tyrannosaurus) and a lower-case initial to the species name (rex); both are then printed in italics. When the name of a genus is repeated in a section of text, scientific convention has the generic name reduced simply to its initial (so that *Tyrannosaurus* becomes *T.*), though the specific epithet 'rex' should never have a capital letter, so T Rex is also incorrect. It is curious that the singer and musician Marc Bolan always had it right, while the media around the world – who should know better – so often have it wrong. You would never have written President-tRump; that's how bad T-Rex looks to a scientific eye.*

Other new types of dinosaur were still appearing. Charles Hazelius Sternberg was an amateur palæontologist from Texas who had learned his trade in the 1870s from Benjamin Franklin Mudge. Sternberg used

* When describing types of dinosaurs colloquially, the name is used without a capital letter. You could write generally about stegosaurs, for example, or tyrannosaurs, or plesiosaurs. However, when formally denoting a genus or a species, then the names are printed in italics – and the genus always has an upper-case initial. Then, instead of stegosaur, you would write *Stegosaurus* (the species would be printed as *Stegosaurus ungulatus*, the name given by Marsh to the specimen he described in 1879). For over a century there was a tradition where the initial letter of a species would also be printed in upper-case where it derived from the name of a person or place. Thus, the dragonfly named after Sir David Attenborough would always have appeared as *Acisoma Attenboroughi*, while the bacterium used to make bread in California was always known as *Lactobacillus Sanfranciscensis*. In the new millennium that convention has been dropped in the interests of typographical consistency, and those species are now *A. attenboroughi* and *L. sanfranciscensis* with a lower-case species initial. Similarly, the dinosaur named *Megalosaurus Bucklandii* by Gideon Mantell to commemorate William Buckland in 1827 is now *Megalosaurus bucklandii*. In an era of digital typesetting, this consistency makes life easier for the compositor. Incidentally, it is always 'species'. Sometimes people affect to say 'specie' as though it were the singular, but it is not; *specie* means money and has nothing to do with taxonomic nomenclature.

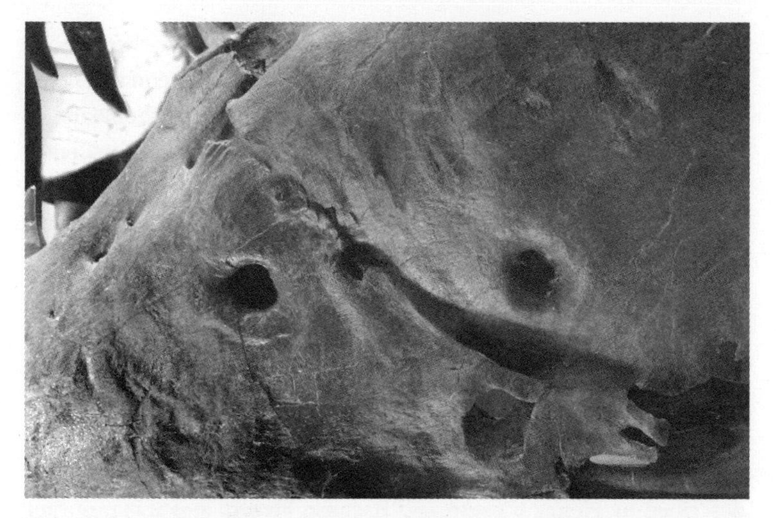

Holes in the *T. rex* skull have often been taken as teeth marks. However, the spacing is not correct, and a row of teeth would crack a skull (rather than perforate the bone). These are certainly lesions left by chronic infection.

to prospect for fossils with his three sons. He had tried unsuccessfully to obtain a contract from both Marsh and Cope, and eventually secured sponsorship from the British Museum (Natural History), which agreed to pay generously for any suitable finds. In 1908, in the Lance Formation strata near Lusk, Wyoming, Sternberg and his team made one of the most astonishing discoveries in the history of palæontology. They managed to excavate an entire skeleton of a duck-billed dinosaur, and in addition to its bones, they could even see the perfectly petrified impression of its scaly skin. This wonderful discovery was originally named *Trachodon annectens* but was later accepted as a member of the genus *Edmontosaurus*. The impression of the leathery skin in the sedimentary rock was perfectly preserved, looking like a gigantic designer handbag and lustrous in its detail.

Once the news of this unprecedented discovery had spread, everyone wanted to acquire the specimen. As soon as Henry Fairfield Osborn heard about it, he resolved to acquire it for the American Museum of Natural History (A.M.N.H.), in New York and he immediately dispatched one of the museum's palæontologists, Albert Thomson, to secure it for the museum's collections. Osborn's message was simple: although Sternberg had a contract with the British Museum, his patriotism should override commercial concerns, and surely he would prefer the fossil to remain in America. Thomson was told that the price would be $2,000, though the specimen was already protected by plaster and sacking, ready for transportation, so he could not closely inspect it. Within a few hours, William Jacob Holland, who was director of the Carnegie Museum of Natural History, also arrived on the scene with his cheque book in hand. Thomson became concerned that they would lose possession, so he wired Osborn for instructions; and, by the end of the day, negotiations were complete and the A.M.N.H. had acquired the fossil. It was a close call. Once back at base, the specimen was methodically cleaned and the skin was conserved by their technician, Otto Falkenbach. It was formally described by Henry Osborn and Barnum Brown, and was mounted – just as it was found – in a glass case without further restoration. It stands in the museum to this day, and remains one of the most remarkable specimens in the field of dinosaur discovery.[18] Another of Sternberg's seemingly mummified edmontosaur skeletons is on display, as we shall see later, in the Senckenberg Museum in Frankfurt, Germany. Sternberg went on to discover 17 new species of dinosaur, and eventually a movie was made about his life. It was entitled *Charlie*, the name by which he was always known, and was produced by the National Film Board of Canada.

Barnum Brown continued to lead dinosaur-hunting expeditions in the early years of the twentieth century, and some of the finds made by his team were spectacular. In part this was due to his novel way of prospecting. He realized that travel over water was so much less problematic than hauling supplies on land, and so in 1910 he decided to use a specially designed barge on which he and his team would navigate down a river, stopping off at strategic sites and venturing ashore to search for bones. Their finds could then be packed aboard and shipped back to base. In 1911 they unearthed an extraordinary fossil of a hadrosaur that (like Sternberg's *Edmontosaurus*) did not merely reveal the skeleton and the high vaulted skull but also had details of the dinosaur's scaly skin preserved. It was discovered near the Red River in Alberta, Canada, and looked more like a mummified corpse.

This specimen was dubbed *Corythosaurus*, from the Greek κόρυς (*korus*, corn), and the skeleton proved to be almost complete, lacking only a few phalanges from the front limbs and part of the tail. It was one of the most remarkable fossil finds to date. There were patches of skin still visible on its surface which formed as faceted scales that fell into folds near the nape of the neck, just as it would have been configured in life. The outline of some of the soft tissues of its body could be traced. The entire animal was lying on its side, and had not been disturbed since it died. It would have lived about 76 million years ago and measured 30 feet (9 metres) long, with a weight between 8 and 10 tons. This fossil truly gave an impression of how the dinosaur would have appeared when alive. It is on display, along with other similar, more recent, discoveries, at the A.M.N.H.[19]

Brown described two new hadrosaur skeletons in 1914, noting their curious duck bills and the high-domed skull. Because he realized this specimen was somewhat smaller than *T. rex*, he wittily decided to name it *Hypacrosaurus* from the Greek υπο (*hypo*, under) and ακρος (*akros*, tall).

As Brown was continuing to explore for fossil dinosaurs in America, the largest fossil expedition ever mounted set out from

Mounted in the Senckenberg Museum in Frankfurt am Main is this cast of *Stegosaurus stenops*. Note that it possesses 20 plates along the spine, in two rows. The original skeleton is in the American Museum of Natural History in New York.

Germany. In 1907, some intriguing finds had been reported in Tendaguru, a settlement that lay four days' march inland from the important coastal port of Lindi in German East Africa. Before World War I, a large area of East Africa had been under German control. This territory covered much of what is now Tanzania, Burundi and Rwanda. Germans had arrived in 1885 and signed agreements with local chieftains in Zanzibar, so – when the Sultan of Zanzibar protested – the Germans turned up with a flotilla of five heavily armed warships and simply took over much of the African Great Lakes territory. Germany lost these overseas African territories as part of the Versailles agreement after World War I. The fossil site had originally been discovered by a mineral prospector named W.B. Sattler from the Lindi Prospecting Company. Sattler uncovered the site while searching for valuable mineral resources. He reported finding huge dinosaur bones protruding from the rocky scrub. As luck would have it, Eberhard Fraas, a professor of palæontology, happened to be visiting the colony at the time, and he reported back that the Tendaguru site could be one of great importance to science. He brought back a few fossil specimens and showed them to his peers. The director of the Berlin Museum, Wilhelm von Branca, was ecstatic and insisted that an expedition should be sent out to search for more. Within a year, funding had been raised from the German government and from academic sponsors, and in 1910 a team was sent out to Africa under the leadership of Werner Janensch, the curator of Fossil Reptiles at the Berlin Museum, and his assistant Edwin Hennig. This was the biggest single palæontological expedition that the world had ever seen. Within a year, they had 500 native workers on the project, and with the German palæontologists and ancillary staff, the whole expedition employed almost 1,000 people, which posed continual difficulties in finding food and water for everyone. The workers excavated deep trenches across a site about 2 miles (roughly 3 km) across, and discovered large numbers of bones, many of them fractured. The fossils were wrapped and plastered (in the manner that Marsh had perfected) and then carried on slings, or balanced on their heads, by local workers for the four-day trudge to the coast. About 5,000 separate consignments of fossils were eventually shipped back to Berlin, weighing a total of over 250 tons. The process of separating the bones went on for years, and as each new skeleton was uncovered it was assembled and put on display in the Humboldt Museum in Berlin.*

Their dinosaurs remain on public display in the same building. Among the specimens discovered in Africa is a magnificent brachiosaur which was the most perfect and complete dinosaur skeleton in the world. None of the bones is a replica; almost every bone in its body comes from a skeleton that was complete, apart from a few tail-bones, and those were substituted by identical vertebræ from another brachiosaur that was excavated nearby (in 1988 it was renamed *Giraffatitan* by Gregory S. Paul). Not only was this the most complete skeleton ever discovered, but when erected it was also said to be the most massive in the world. The museum authorities have displayed their copy of the *Diplodocus* fossil alongside, and they say that, although the American *Diplodocus* is longer (at 90 feet, or 27 metres) than their brachiosaur (73 feet, or 22 metres), theirs is taller and far more bulky. Recent discoveries of titanosaurs show that they were much bigger again, but (as we have discovered) dinosaur specialists are given to deep-seated rivalry and blatant overstatement. The brachiosaur in Berlin is still described as the world's tallest mounted dinosaur skeleton, and alongside is proudly displayed a framed certificate from *Guinness World Records*, stating that 'The tallest mounted dinosaur is the *Brachiosaurus brancai* which measured 13.27 m (43 ft 6 in) high [it probably means *tall*] on 1 June 2007.' During the following decade there were many larger finds, though the museum makes

* The title 'Humboldt Museum' is ambiguous. The Natural History museum, where the dinosaurs may be seen, is officially the *Museum für Naturkunde, Leibniz-Institut für Evolutions und Biodiversitätsforschung* and is situated on the Invalidenstraße in Berlin. It is popularly known as the *Naturkundemuseum*. The actual Humboldt Museum is completely unrelated – it is housed in the Tegel Palace, and commemorates the life and work of the Prussian brothers Friedrich and Alexander von Humboldt, who were active in exploration and politics in the nineteenth century. Alexander was Germany's foremost naturalist, while Friedrich founded the Humboldt University, which for decades housed their museum, but the two institutions separated in 2009. Neither is connected with the Humboldt Museum in Nevada; that is another matter altogether.

In the atrium of the Museum für Naturkunde, Berlin, stands this magnificent skeleton of *Brachiosaurus brancai* (now *Giraffatitan*), found in 1910 in Tendaguru Hill, in what became Tanzania. It is mounted erect, reaching to the glass ceiling.

no mention of any bigger specimen being subsequently discovered.[20] There is a great deal of rivalry, hidden animosity, misrepresentation and plain dishonesty, in the world of science.

So popular had dinosaurs become that in 1894 *Punch* magazine in London began to feature a cartoon series drawn by Edward Tennyson Reed and entitled 'Mr Punch's Prehistoric Peeps', in which cave dwellers were portrayed living alongside playful dinosaurs. In 1905 a 4-minute film was made entitled *Prehistoric Peeps*, based on the *Punch* cartoons, in which dinosaurs were represented by actors in costume. It was the world's first dinosaur film, directed by Lewin Fitzhamon and starring Sebastian Smith as Prof. Chump. Clearly, there was money to be made from a popular novel that would bring these fasci-

When the American Museum of Natural History assembled a plastic reconstruction of *Argentinosaurus*, it was too large for their biggest gallery. Part of the shoulder, some ribs, 15 vertebrae, pelvis and thigh bones had been found. This dinosaur was far larger than any brachiosaur, and even bigger 'ultrasaurs' may yet be discovered.

GUINNESS WORLD RECORDS

CERTIFICATE

The tallest mounted dinosaur skeleton is the
Brachiosaurus brancai
which measured
13.27 m (43 ft 6 in) high
on 1 June 2007.
The 150-million-year-old dinosaur skeleton
is displayed at the
Museum für Naturkunde der
Humboldt Universität zu Berlin, Germany

GUINNESS WORLD RECORDS LTD

Dinosaur palæontologists are intense rivals. *Guinness World Records* (for which the author was the scientific editor) claimed their brachiosaur as the 'world's tallest' dinosaur, after mounting it with its neck vertical. Other skeletons are larger.

nating monsters to a wide readership. The project fell to Arthur Conan Doyle, a practising physician who had first become famous at the age of 29 when he published his novel *A Study in Scarlet* in 1887. The heroes of that novel became two of the best-known names in world fiction: Mr. Sherlock Holmes and Dr. John Watson. Conan Doyle was a curious man. He qualified in medicine at the University of Edinburgh Medical School, the first of that school's alumni in this book actually to complete his degree. Then he was a doctor in the English Midlands, a surgeon on a whaling ship, and eventually went into private practice in Portsmouth and started to study spiritualism. His first academic publication was a letter to *The British Medical Journal* in 1879 on the toxic plant *Gelsemium* and was signed A.C.D. In it he described his taking progressively larger amounts of extracts from this poisonous plant, until he was able to tolerate doses larger than those thought to be fatal.[21]

Resonances of this inquiring attitude appear throughout his Sherlock Holmes novels, though he also ventured into other realms of

In 1899 the London humorous magazine *Punch* published this drawing by Edward Tennyson Reed, the leading cartoonist of his day. His fanciful portrayal of cave dwellers wining and dining with dinosaurs inspired the first dinosaur film in history, 'Prehistoric Peeps', which was released in 1905.

On February 8, 1925, the silent movie of Sir Arthur Conan Doyle's novel *The Lost World* opened at the Astor Theatre in New York. Produced by First National Pictures, the film starred Wallace Beery as Professor George Challenger.

fiction. Like so many *cognoscenti* of the time, Conan Doyle was captivated by dinosaurs, and in 1912 he published the first popular novel to feature them in the plot. It was entitled *The Lost World* and, like many other novels of the time (those of Charles Dickens among them), it first appeared as a serialized story in a magazine. As a book, it remains in print from a variety of publishers.[22]

His story told of a brave young newspaper reporter, Edward Malone, who asks for a dangerous assignment in order to impress his new girlfriend. Malone's editor tells him that a notorious academic, Professor Challenger, has returned from travelling in South America and has brought back incredible stories of a world where prehistoric creatures still survive. Eventually, a team sets off to explore these hidden places, and after a trip rich in adventure they arrive at a high plateau that extends up above the jungle and is virtually inaccessible. When they eventually reach the top, they are assailed by several iguanodons and swooped upon by aggressive pterodactyls. The group later encounters *Stegosaurus*, *Allosaurus*, *Megalosaurus* and a

The Lost World was a success and released in other countries – in Paris as *Le Monde Perdu*. The original version was soon edited down from 90 minutes to 1 hour and the full length film was believed lost, until in 1992 when a full print was found in the Filmovy Archiv of the Czech Republic.

Brontosaurus, while ichthyosaurs and plesiosaurs cruise menacingly through the lakes. Doyle's book was a huge commercial success and has been the inspiration for at least five movies. After his novel appeared in print, he quickly sold the film rights to the British film producer, J.G. Wainwright for £500 ($650), a sum that would be worth about £15,000 ($18,000) in today's money. The film proved too much for Wainwright to handle, so he sold the rights on to an American producer, Watterson Rothacker, who managed to raise over $1 million (today about $30 million, or £25 million) to make the movie.

Other film-makers were quick to jump on the bandwagon. After *Prehistoric Peeps*, the first ever movie to show live-action model dinosaurs was *Brute Force, Primitive Man*, directed by D.W. Griffith and released in April 1914, just two months after Gertie – as a cartoon character – had debuted in Chicago. Then came *The Dinosaur and the Missing Link: A Prehistoric Tragedy*, produced as a humorous cartoon by Willis O'Brien in 1915 and released by the Thomas Edison company, Conquest Pictures, to be followed in 1918 by *The Ghost of Slumber Mountain*, a 15-minute film in which Uncle Jack uses a magical telescope to look back in time, and watches living dinosaurs at play. The special effects on this movie were produced by Willis O'Brien, although he is not credited in the film. In 1920, a 10-minute short film *Along the Moonbeam Trail* was released, telling the tale of two children who fly to the Moon and witness battles between prehistoric monsters. It is often said that the film has been lost, though the Academy of Motion Pictures in Los Angeles has recently located two separate and slightly different versions in long-forgotten archives. It has still not been digitized.

So successful were the effects of stop-frame animation in *Slumber Mountain* that O'Brien was brought in to make Conan Doyle's *The Lost World*. The first time anyone saw some of the results was at a dinner of the Society of American Magicians hosted by the world-famous escapologist Harry Houdini (the stage name of Erik Weisz) in June 1922. Some short sequences of the dinosaurs were screened for the delighted audience without explanation, many of those attending

Conan Doyle's story was resurrected in a film released on November 3, 1954 as *Gojira* (ゴジラ) in Japan. Eizo Kaimai, the 85-year-old chairman of Kaimai Productions, says that their 'dinosaur' was made from bamboo, cloth, paper and wire. The Japanese *Gojira* was later changed to *Godzilla*.

imagining that the dinosaurs were filmed in real life, and the story made headlines next day in the press. 'Dinosaurs Cavort in Film for Doyle', said a long single-column article in the *New York Times*. Were these simply a 'joke on them magicians,' asked the writer? The reporter emphasized that Conan Doyle refused to be questioned about the film:

> Sir Arthur said the [movie images] were "psychic" and also that they were "imaginative", and announced in a firm tone, before they were shown, that he would submit to no questions on the subject of their origin. His monsters of the ancient world or of the new world which he has discovered in the ether, were extraordinarily lifelike. If fakes, they were masterpieces.[23]

There was a backstory to this. At Christmas 1920 in *The Strand Magazine* in London, Arthur Conan Doyle had enthusiastically reported the existence of fairies, and published photographs to support his ridiculous beliefs. Five prints showing fairies had been produced by the 16-year-old Elsie Wright of Cottingley in Yorkshire, starting in 1917, and Conan Doyle was convinced that they were genuine. These images became known as the Cottingley Fairies, after the town near Bradford where they were taken, and they clearly show fairy images. Not until 1983 did Elsie and her young cousin Frances admit how they faked them, using pictures cut from a magazine and supported by hat pins. Both women died in the 1980s, and prints from their negatives were sold at auction in 1998 for £22,000 (roughly $30,000). The original negatives were later sold for the modest sum of £6,000 (roughly $8,000) and have since disappeared. Nobody knows who bought them.

When first released, *The Lost World* was promoted as a more serious movie by far. It was described on posters as 'One Kind of Photo play you Never saw Before! Astounding Amazing Adventures!' and when the film was eventually released in 1925, it was the most expensive and technically complicated movie yet made. It was repeatedly shown thereafter, though in 1929 it was cut down by 45 minutes in a version marketed by Kodascope (this is the version most archives possess today). It was eventually followed by a Cinemascope widescreen remake in colour that was directed by Irwin Allen of Saratoga Productions Inc. in 1960. Their chief screenplay writer was Charles Bennett, whom I knew well in Beverly Hills; when Bennett was staying with me at home in Cambridgeshire in his 90s he told me that he was embarrassed by the crude special effects, which were based on using iguanas and small crocodiles, sometimes with appendages attached. The same story was made into a movie in 1992, in a film directed by Timothy Bond, in which the location was switched from South America to Africa; again in 1998, with Ben Keen as director; and most recently as *King of the Lost World*, directed by Leigh Scott in 2005. There has also been a range of radio plays based on Conan Doyle's story, and several serializations on television.

The idea of animating clay models of dinosaurs was taken up in a Buster Keaton movie *The Three Ages*, which was released in 1923. This was a parody of the D.W. Griffiths film, *Intolerance*, and in the first of the trilogy – the Stone Age – we see Keaton standing on what appears to be a huge rock, but which turns out to be the back of his domesticated dinosaur. The dinosaurs were made of clay, each model about 2–3 feet long (less than 1 metre). Of the innumerable movies that followed, one in particular should be mentioned: *Monsters of the Past*, produced in 1925 in the Pathé Review series. It shows *T. rex* and *Triceratops* battling it out in a forest clearing (*T. rex* loses the fight). One peculiarity is that the tyrannosaur is shown standing erect like the iguanodon skeletons on display at the time in museums, its large tail resting on the ground like that of a kangaroo – and, when it becomes agitated, it hops like a kangaroo by jumping with hind feet together. That was a first, and hasn't been seen since. The movie redeems itself with an ingenious sequence that re-creates a skeleton from fossilized bones: they emerge from the sandy ground and methodically re-assemble themselves. A separate sequence reveals how the models were made, and records how their sculptor, Virginia May, handled the clay to create the dinosaur models.

Conan Doyle's *The Lost World* was an immediate success, and it was quickly plagiarized. In 1915 Russian scientist Vladimir Obruchev wrote his own version, *Plutonia*, where the intrepid explorers encounter their dinosaurs inside a hollow Earth. In 1918 Edgar Rice Burroughs produced a series of magazine articles under the title *The Land That Time Forgot*, where the dinosaurs were found in an isolated part of the Antarctic. It was released as a novel in 1924 with a startling and vivid cover image by J. Allen St. John, a Chicago artist who is regarded by many as the godfather of present-day fantasy art. His cover illustration showed a highly stylized iguanodon with that curious horn still mistakenly placed on its snout (yes, the same anatomical error was still there in 1963 when the book was reprinted).[24] Although Burroughs' *Tarzan* novels were released in a torrent of movies, starting in 1918, *The Land That Time Forgot* had to wait until 1974 for its production as a film.

Movies were not the only way in which the public encountered visions of dinosaurs. There were also some stunning paintings. Back in 1889, with great entrepreneurial flare, Henry Osborn of the American Museum of Natural History enlisted the support of a patron, J. Pierpoint Morgan, a wealthy and highly influential industrialist. Morgan agreed to fund an art project that Osborn had in mind for Charles Knight. After his success in creating riveting images of prehistoric life, Knight was commissioned to produce works of art that re-created the era of the great dinosaurs. Altogether, 20 dinosaurs appeared before an astonished audience; most were paintings, some were sculptures. Copies of the pictures were printed off and donated to schools, while the vivid originals were hung in the museum. And what images they were – for the first time, the public had an impression of huge, juicy dinosaurs in their natural environment. Yes, there are small adjustments we might make nowadays. In his imagined *T. rex*, the eyes are in the wrong position and the forelimbs end in three fingers, whereas they should end in two, but these are details – the images themselves remain as landmarks in the story of dinosaurs. Nothing as realistic had ever been seen before. Knight created artwork for each edition of the *Century Illustrated Monthly Magazine*, he also undertook independent commissions (two of his life-size heads of elephants are still on display at the Bronx Zoo in New York), and he then set to work on a huge project: vast murals to hang in the new Hall of Man in the American

Museum of Natural History. These kept him occupied until 1923, when the Natural History Museum of Los Angeles commissioned him to paint a huge mural of prehistoric life from the tar pits of La Brea – a key site of interest for any geologist, and one of the most unappreciated sites in that great city of Los Angeles. Visit La Brea if you can. The significance of this fascinating fossil site is widely ignored, even by those who live nearby. There are just five areas in the world where asphalt bubbles to the surface from oil-rich strata that lie deep beneath. The others are at Lake Guanoco in Venezuela, Pitch Lake in Trinidad, McKittrick in Kern County, California, and Carpinteria, 12 miles (20 km) south of Santa Barbara, California. At La Brea, the tar was used by native American tribespeople to caulk their canoes. The fact was first recorded by Europeans in 1769, when a Spanish adventurer named Gaspar de Portolá took a group of explorers to investigate the coast. Historian John R. Kielbasa writes that the cleric in the team, Father Juan Crespí, recorded:

> While crossing the basin the scouts reported having seen some geysers of tar issuing from the ground like springs; it boils up molten, and the water runs to one side and the tar to the other. The scouts reported that they had come across many of these springs and had seen large swamps of them, enough, they said, to caulk many vessels. We were not so lucky ourselves as to see these tar geysers, much though we wished it; as it was some distance out of the way we were to take, the Governor did not want us to go past them. We christened them Los Volcanes de Brea.[25]

The tar pits lie at the heart of the Hancock Park area of Los Angeles, and the name, La Brea, means *the pitch* (or tar) so the normal name (La Brea Tar Pits) has a hint of tautology. La Brea is enough.

A later account came from the Jedediah Smith expedition to California in 1826, telling of his team's exploration of the Mojave Desert, the Great Basin and the Sierra Nevada. A diary of the expedition was kept by Harrison G. Rogers, who wrote that the local people were also using the asphalt to seal the roofs of their houses against the rain.[26]

The area was settled as the Rancho La Brea, and from time to time bones emerged from the tar pits on the Hancock Ranch. They attracted little interest – the ranch had horses and cattle, dogs and cats, even camels, and once in a while they would become trapped by the sticky tar and were unable to escape. Bones were not unexpected. However, in 1901 a petroleum prospector from the Union Oil Company, William Warren Orcutt, took a closer look and he realized that the bones actually came not from present-day farm animals at all; they represented extinct species. One of the first to be discovered was the La Brea coyote, and it was named *Canis latrans orcutti* in honour of Orcutt's ground-breaking insights. From 1915 onwards, there were scientific excavations on the site, and in 1932 Professor John C. Merriam and his postgraduate student Chester Stock described one of the most iconic fossils of all, *Smilodon californicus*, the sabre-toothed cat. Generations learned of this creature as the 'sabre-toothed tiger' but pedants have recently insisted on 'cat' instead. Until 10,000 years ago, this great carnivore ranged across the Americas; now it is known only as fossils. Many thousands of *Smilodon* fossils have been found, represented by hundreds of thousands of bones. This impressive beast is now the official State Fossil of California. That state has other trophies, including the official State Grass (*Nassella pulchra*), even an official State Fabric (denim). In March 2017 state assembly member Richard Bloom introduced a proposal that *Augustynolophus morrisi* should be made California's official state dinosaur. Governor Jerry Brown signed a bill putting the official seal of approval on the proposal in September 2017. This herbivore was a hadrosaur that inhabited California some 66 million years ago, and two specimens are known, one excavated in Fresno County in 1939 and the other found in 1941 in nearby San Benito County. These dinosaurs were about 26 feet (8 metres) long and weighed up to 10 tons. *Augustynolophus morrisi* is a very modern dinosaur, even having its own Twitter account: @augustynolophus.[27]

Once the excitement of discovery dwindled at the La Brea site, the diggings were deserted and eventually caved in. Not until 1969 did work recommence, and in 1976 a covered visitor centre was constructed to allow the public to watch the excavations as they continued. This work ceased in 1980, but it was resumed in 1984 for a two-week season to entertain visitors to the Los Angeles Olympic Games. During 1985 it was opened again for a summer season, and the digging proved so productive that the investigators currently return to work on the site each summer. Some of the fossils are dramatic, including a massive mammoth with perfectly preserved tusks. These fossils are younger than the era of the dinosaurs. The most ancient La Brea bones are 38,000 years old, while the mammoth dates back a mere 10,000 years. This is an endlessly fascinating site, and is testimony to the birth of the oil industry of Los Angeles. Crude oil still oozes from the ground, and the manicured lawns that surround the excavation site and the museum are sometimes marked by sticky pools of tar that well up from beneath. You would imagine that these would be flagged as highlights of your visit, but no; instead of having signs pointing excitedly to these symbols of the past, they are simply marked by traffic cones wallowing in the morass, warning people to keep away because of the risk of ruining their clothes. This is rare and graphic evidence of something that underpins American prosperity, but today the oozing oil is dismissed as a nuisance.

The animals excavated from the tar pits feature in a series of vivid Charles Knight paintings that still hang in their original positions, dominating the displays. Knight's work was soon attracting increasing attention, and the success of these dramatic studies in Los Angeles caught the attention of other museums. Murals were becoming fashionable internationally as an artform, and between 1926 and 1930 Knight worked away on designs for 28 huge panels for the Field Museum in Chicago. They illustrated life from its beginnings, through the ages of primitive animals, on through dinosaurs until culminating in the dawn of humanity. We should note, however, that Knight himself did not personally paint these huge murals. Rather, he produced the artwork in the form of small detailed canvases which were then measured, enlarged and copied onto the vast display panels by teams of commercial artists. Knight came along at the end of the process, to make small adjustments and touch up the details. Many people have seen his paintings without realizing their significance, or who painted them. They gained further renown when they were reproduced in full colour in a special edition of the *National Geographic* magazine published in 1942. Although the haunting pictures remain, the vision of their painter – bent low over his easel, squinting in the light and balancing spectacles on his nose – has been largely forgotten by the public. Knight's vivid pictures gave rise to others; in September 1963 *Life* magazine produced a special edition that looked at prehistoric life. The artists were Rudolph F. Zallinger, James Lewicki and Antonio Petruccelli, and it was Zallinger's portrayals of dominant dinosaurs in their lush landscape that quickly caught the public attention. These images clearly owed much to Knight's inspiration, though the images created for *Life* were not as convincing as those Knight had produced 30 years earlier. Charles Knight's pictures were also featured in a movie from 1952 entitled *The Beast from 20,000 Fathoms*. In that film, scientists try to identify their strange monster by referring to some of the paintings by Knight. A year later, aged 79, he was dying, pleading with his daughter to make sure that nothing untoward would happen to his art. Altogether he left nearly 800 drawings and more than 150 oil paintings. He would be pleased to know that they are admired to this day.

Clay models remained the most popular method of re-creating dinosaurs for the cinema audience. This giant sauropod was animated by Ray Harryhausen and Willis O'Brien in a documentary entitled *The Animal World*, which premiered in 1965.

So often we remember the images, but not the person who created them. Equally unheralded are the tireless palæontologists and anatomists who worked on the fossils back at the laboratory, sifting and cleaning, identifying, repairing and assembling the bones. The eagerness in the public mind for news of dinosaurs had many of the pioneering palæontologists jostling for cult status, and the names remembered in the standard reference books are largely those of individuals who assiduously promoted themselves, while the people who made the original discoveries are generally overlooked. Few people have heard of Peter Kaisen, a palæontologist and surveyor who joined an American Museum of Natural History expedition to find new dinosaur fossils. The year was 1906, and the expedition leader was Barnum Brown. In the Hell Creek Formation, Montana, Kaisen discovered a strange fossil and excavated most of the skull and teeth, part of the shoulder, some ribs and the backbone, together with more than 30 of the bony tile-like scutes. Brown published the details of this armour-plated dinosaur in 1908, creating the name *Ankylosaurus* from the Greek αγκυλος (*ankulos*, crooked) and σαυρος (*sauros*, lizard). The species name *A. magniventris* came from the Latin *magnus* (great) and *venter* (belly) in allusion to the breadth of the creature's abdomen. Because of the portion of the skeleton that was missing, the first reconstructions of this dinosaur were inaccurate, and it was not until 1910 that a better fossil was found, when a further A.M.N.H. expedition (again with Brown in charge) discovered an *Ankylosaurus* skeleton near the Red Deer River in Alberta, Canada. This was an improvement; most of the skeleton was retrieved. It was a crucial discovery.

Another forgotten pioneer is Charles Whitney Gilmore, curator at the Department of Paleobiology at the Smithsonian Institution. Gilmore was the last of the museum specialists to work with the Victorian dinosaur hunters. Born in 1874, he had been taken to museums from an early age, and in 1880 his aunt took him to visit one of the suppliers of specimens to American museums, Ward's Natural Science Establishment in Rochester, New York. The young Gilmore was captivated and soon began to collect his own specimens of birds' eggs, neatly mounted insects, and fossils. He qualified to enter the University of Wyoming in Laramie in 1900, and there he met with John Bell Hatcher, who was visiting from the Carnegie Museum. So impressed was Hatcher that he hired Gilmore for a team searching for late Jurassic dinosaurs. The expedition went well, and in 1902 Gilmore was offered the job of skeleton preparator at the Carnegie Museum, where work was already under way on the *Diplodocus* skeleton. The death of Othniel Charles Marsh in 1899 had left a huge amount of fossil material awaiting assessment. During his time as Vertebrate Paleontologist to the U.S. Geological Survey, which had started in 1882, huge crates of fossil-bearing rocks had been regularly sent back by rail. It was now decided that the Marsh fossils should be taken to the U.S. National Museum in Washington D.C. They got more than they asked for. More than 80 tons of rocky strata and marl arrived, and the museum faced an insurmountable task in extracting the bones, identifying the remains and rebuilding the skeletons. In 1903 Charles Gilmore signed a contract to restore one of the skulls of *Triceratops* that Hatcher had discovered, for display at the museum. The job was so impressively performed that he was offered the full-time post of preparator within a year. Still saddened by the death of his mentor, Gilmore nicknamed the dinosaur specimen 'Hatcher' and he resolved to reconstruct the entire skeleton. Together with his life-long collaborator Norman Boss, he succeeded in mounting the skeleton with its huge skull, ready for public display, and then reconstructed the skeleton of the *Edmontosaurus* fossil. These were highlights of the displays at the old Arts and Industries Building, and construction soon began on the new premises for the U.S. National Museum. There was great excitement in 1911, when the Hall of Extinct Monsters was opened to the public with the two dinosaur skeletons greeting all who entered.

Gilmore was now recognized as the nation's leading restorer of fossil skeletons, and for the following decade he devoted long hours to mounting skeletons of dinosaurs including *Stegosaurus*, *Brachyceratops* and *Thescelosaurus*. As a senior specialist, in 1923 he was able to arrange an expedition with Boss to the Dinosaur National Monument in Utah. He had tired of indoor work and was glad to get out into the field. In a quarry at the site they identified some exposed bones as those of a gigantic dinosaur, and the next three years were devoted to exposing the skeleton, retrieving the fossil-bearing rocks, and separating the bones from the rocky matrix. It proved to be a gigantic adult *Diplodocus* measuring some 70 feet (22 metres) long, and it was not fully assembled until 1931. It instantly became a main attraction in the main hall of the Smithsonian, and is in precisely the same position some 90 years later. Gilmore was said by all who knew him to be a quiet, unassuming, kind and helpful individual. He held the position of curator until he died in 1945, bequeathing to posterity a range of impressively rebuilt dinosaur skeletons.

'Quiet' and 'unassuming' would not describe another of the most notable palæontologists of the early twentieth century, Roy Chapman Andrews. This remarkable man was a traveller, adventurer, explorer and avid anthropologist who lived a life of daring and extraordinary variety. He loved painting a vivid self-portrait as a risk-taking adventurer. Andrews once wrote:

> In fifteen years [of field work] I can remember just ten times when I had really narrow escapes from death. Two were from drowning in typhoons, one was when our boat was charged by a wounded whale, once my wife and I were nearly eaten by wild dogs, once we were in great danger from fanatical lama priests, two were close calls when I fell over cliffs, once was nearly caught by a huge python, and twice I might have been killed by bandits.[28]

Andrews is frequently cited as the original Indiana Jones, and it is claimed in innumerable sources that he was the inspiration for that character. When asked, however, neither George Lucas (who created the character) nor Steven Spielberg (the movie producer) has confirmed that he was. There were other dynamic characters on whom Indie could have been based, such as Lieutenant Colonel Percival Harrison Fawcett, a British Victorian artillery officer turned explorer, who spent decades in risky adventures in South America searching for the lost city of El Dorado. Or perhaps it was an American adventurer, W. Douglas Burden, staff scientist at the A.M.N.H. and an intrepid explorer who went out to Southeast Asia in search of the Komodo dragon.

In any event, Andrews was an unforgettable character. He was born in Beloit, Wisconsin, in 1884, and from his earliest years he was out playing in the local woods of his rural home. He described himself as being 'like a rabbit' and was only happy when running about out of doors. For his ninth birthday, he was given a shotgun and soon became skilled as a marksman. He didn't simply hunt for sport, but took the cadavers back home and taught himself how to stuff and mount them. In time, he became skilled at taxidermy and earned enough money from this hobby to pay for his college education. His adventures became more daring until, when he was 21, a trip ended in tragedy. He was out in a rowing boat hunting wildfowl on the Rock River in Wisconsin with a college friend, Montague White, when their boat was driven against rocks and both were tipped overboard.

Roy Chapman Andrews was a sailor, explorer and fossil hunter – an archetypal adventurer who some have said inspired the character of Indiana Jones. George Lucas (who created Jones) and Steven Spielberg (the producer) deny those rumours.

Andrews eventually managed to struggle ashore but White, even though he was the stronger swimmer of the two, was seized by cramps in the cold water and drowned. It was an event that Andrews said never left him, and it may have been instrumental in laying down the principles he followed when arranging his future trips.

Andrews became famous for meticulous planning and risk assessment before setting off – but, even so, his life was one of extraordinary adventure marked by a series of near-catastrophes. He had supreme personal self-confidence and was always determined to achieve an ambition once it had formed in his mind. His aim as a graduate was to work at the American Museum of Natural History in New York, so he immediately applied for a staff position. No job, he was crisply told, was available. Undeterred, he submitted an application for a job as a cleaner – anything to get under that famous roof. This application served him better, and he soon became the museum's odd-job man. He was so keen that he was soon given work closer to his interests; including tasks that nobody else wished to undertake. He was interested in mammalian anatomy (as he had been since his teenage interests in taxidermy), and, when a dead whale was washed up on Long Island, he and a colleague were sent down in freezing and stormy weather to secure the skeleton. It was thought to be a hopeless task, and nobody expected them to succeed, but the two young enthusiasts brought back the entire skeleton, perfectly cleaned; indeed, it is on display in the museum's Department of Mammalogy to this day.

Andrews soon gained the full support of the museum's director, Henry Fairfield Osborn, who agreed in 1922 that a team should go to explore the Gobi Desert. This was largely unknown territory to the Western world, and several of the team members were wealthy enthusiasts who helped fund the venture. Two were young members of the Colgate toothpaste family, S. Bayard Colgate and his nephew Gilbert Colgate, whose grandfather, Samuel Colgate II, was president of the board of Madison University (later to become Colgate University). The team's intention was to find archæological evidence that humans had first evolved in Asia, but they found no remains of humans. Instead, they came across several entire skeletons of dinosaurs and large sections of skeletons of some monstrous fossils. As they travelled, Andrews also collected mammal specimens and insects. Several of the animals were new species and the specimens ranked as the most extensive collections of new fossil species ever collected in Central Asia. The museum was ecstatic, though Andrews insisted that they were only scratching the surface. In 1923, the palæontologist on the expedition, Walter Granger, found a small skull preserved in Cretaceous sandstone among the dinosaur remains. It turned out to be one of the earliest mammals – an historic find for science. Even more exciting was a discovery made by George Olsen, an assistant palæontologist, who ran into the camp one morning to say that he had discovered some fossilized eggs. The team went out to see, and indeed there were three dinosaur eggs showing in the rocky strata, and 23 more emerged during careful excavation. Each measured about 9 inches (23 cm) long.

This was another historic development. Fossil eggs had occasionally been collected in previous centuries, though nobody had associated those with dinosaurs. One early dinosaur discovery, dating from 1859, was of fragments of fossilized eggshells discovered in the south of France by a Catholic priest and amateur naturalist, Jean-Jacques Pouech. He believed them simply to be the remains of a large bird's eggs. Ten years later in the same area, a geologist, Phillippe Matheron, found fragments of eggshells along with bones from a monster that he named *Hypselosaurus*. Matheron believed it to be a crocodile, though we now know it was a large dinosaur measuring 40 feet (15 metres) long. Matheron took all the specimens to the Muséum National d'Histoire Naturelle in Paris, where their palæontologist, François Louis Paul Gervaise, showed that the structure of the eggs was similar to those of a turtle. Because dinosaurs were reptiles, it had always been assumed that they must have laid eggs, though none had ever been formally described before.

Olsen's discovery of dinosaur eggs was a tremendous coup, and Henry Fairfield Osborn was delighted with the success of the expedition. Roy Chapman Andrews could not have imagined that their discovery would have made such an impact, but in October 1923 he was featured on the front cover of *Time* magazine, and public interest in dinosaurs was widespread. When he gave his first lecture after returning from Mongolia, there were more than 4,000 members of the public standing in line for the 1,400 seats in the auditorium. Osborn wanted to capitalize on these important discoveries, and so the American spirit of enterprise came to the fore. Osborn and Andrews decided that the most propitious way to capture the world's attention would be to offer a dinosaur egg – for sale. When the announcement was made, Andrews told reporters that, with 25 fossilized eggs in their collection, one would not be missed. He added that they needed the money more than the egg, to help cover the costs of their expeditions. There was so much interest in the forthcoming sale that the museum attracted more than $50,000 (now about $700,000, or over £500,000) in donations. At auction, the egg raised $5,000 (today worth $70,000, about £50,000). The egg was purchased by Colonel Austen B. Colgate, who generously donated it to the university in 1924. Their president's report noted that the egg had given Colgate University 'some distinction as well as publicity'.

The dinosaur egg gained more unexpected publicity in March 1957 when two students took it. They removed it from its glass case in Lathrop Hall. The police, the sheriff, even the FBI, were called in to investigate the prank, and the egg was quietly returned at night, being left on the doorstep of the university's priest. There was no note, and nobody realized it had been given back. The egg was accidentally kicked into the garden and lay there undetected until someone retrieved it from the shrubbery nearby. The alumnus organization certainly knows who did it, but they have never publicly revealed a name. I have discussed the affair with Daniel DeVries, the current Media Relations Director at Colgate, but to no avail. After its adventures, the fossilized dinosaur egg at Colgate University was eventually put on display in their Robert M. Linsley Geology Museum, named after their distinguished Professor of Paleontology who taught students from 1955 to 1992 with an infectious enthusiasm and boundless laughter. That egg in his department proved to have a propitious purpose, because Colgate University frequently returned to the subject and has since become a leading centre of research into the eggs of dinosaurs. The aim of bringing dinosaurs before the public gave rise to a spectacular display in 1933, when an exhibit was planned for the Chicago World's Fair. It was (inappropriately) entitled 'The World a Million Years Ago' and was installed close to the Field Museum. It presented a range of animatronic models, including *Apatosaurus* and *Triceratops* in threatening poses. The exhibits were manufactured by Messmore and Damon Company Inc., and versions of the exhibition toured the world until 1972.

Although the production of life-size animatronic dinosaurs is big business for some companies in America, it is China that produces most of them. In the Sichuan city of Zigong there is currently a zone under expansion that is acquiring a reputation as 'Dinosaur City.' More than 30 different companies are at work full-time, turning out full-size models of dinosaurs that growl and roar, pivot and move, and are exported to countries all around the world. They claim that 90 per cent of the model dinosaurs in the world come from Zigong. There's a reason for this: between 1970 and 1990 a leading Chinese palæontologist, Dong Zhiming, led a team of technicians who excavated a huge amount of unique fossil dinosaurs from the Dashanpu Formation some 4 miles (7 km) outside the city. Now that the enthusiasms of the local people had been ignited, it seemed only logical to start making new models once the fossil-bearing strata had been exhausted. Dinosaur City was founded on dinosaur discovery.

So many dinosaur fossils had been smuggled out of Mongolia a century ago that the authorities – realizing the value of the dinosaurs that were being discovered – passed a law in 1924 that prohibited the export of any more fossils. This was well timed; a joint Chinese/Swedish Palæontological Expedition returned to the Gobi Desert in 1927 and they turned up portions of a skeleton that were recognizable as a *Tyrannosaurus*. Nothing more was done until 1946, when the Paleontological Institute of the Soviet Academy of Sciences arranged with the Mongolian People's Republic to search for more fossils. They found excellent fossils all over the southeast part of the Gobi Desert, including most of the skull and many vertebræ from another gigantic tyrannosaur. These were the first significant finds of tyrannosaurs in Asia. Two years later the American palæontologists were back, and on May 9, 1948, a team member named James Eaglon suddenly discovered an almost complete tyrannosaur buried in sandstone. It measured 33 feet (10 metres) long and was lying in the Upper Nemegt Beds of the Gobi Desert. These dinosaurs remain uncommon in America, although one new specimen was unearthed in the Hell Creek Formation in Montana as recently as August 2016. The Gobi Desert, by contrast, seemed to be littered with them. The expeditions of 1948 and 1949 soon discovered 7 nearly complete skeletons, all of which were shipped back to Moscow for examination. Everywhere the explorers looked, they found pieces of skulls and partial skeletons. Nothing like it had ever been seen before. A leading Soviet palæontologist, Evgeny Maleev, named the new species from the Gobi Desert *Tyrannosaurus bataar* in 1955, though a decade later it was suggested that the genus should be changed to *Tarbosaurus*. Few people were interested in the name. Some of these finds were sent to London, and one of the skeletons was then exported on to the United States. In these cases, the Mongolian authorities were not informed. When the *Tyrannosaurus* was shipped on to America, it was done with faked export documents. For decades, the Mongolian authorities complained about the unlawful trade in what they felt to be their fossils, and matters came to a head in 2012, when the Texas-based company Heritage Auctions, Inc. offered a *Tyrannosaurus bataar* skel-

Dinosaur City is the name now given to Zigong in Sichuan Province, China. Thirty companies are making dinosaur sculptures that roar and move. They say that 90 per cent of the model dinosaurs in the world now come from Zigong.

eton for auction in New York City. It was a specimen that had been excavated in the western Gobi Desert between 1995 and 2005. This time, the Government of Mongolia asked for a Temporary Restraining Order that would prohibit the auction or movement of the fossil, and the President of Mongolia, Tsakhiagiin Elbegdorj, asked for assistance in preserving 'Mongolia's cultural heritage in this rare national treasure.'

America needs to be friendly with Mongolia. Both countries have signed an Investment Incentive Agreement, along with a Trade and Investment Framework Agreement and a Bilateral Investment Treaty. There are strong commercial pressures – a decade ago the Mongolian gross national product was about $1 billion, and within ten years it had topped $12 billion. Exports from the U.S. to Mongolia increased from $40 million in 2009 to more than $650 million in 2012, even though that level has since declined. American cars and trucks are major exports, while Mongolia supplied America with crucial commodities like iron and steel, a range of rare minerals, plus art, antiques and jewels. The minerals are particularly important. Whereas Mongolia was once a far-off country of little interest to the West, it is now recognized as a trading partner of importance – and this can only increase, as scarce minerals become rarer. After taking advice, Heritage Auctions went ahead with their sale and the skeleton sold for over $1 million – but the deal depended upon the outcome of the proceedings being taken by the Mongolian authorities. Shortly after the sale, the U.S. ordered Homeland Security Investigations to seize the skeleton. Because there was no formal documentation identifying the origin of the fossil, scientific reports were called for and they confirmed that the distinctive hue of the bones proved without doubt that it had been taken from the Gobi Desert. This was a crude examination. I would have recommended subjecting the bones and the rocks that surround them to analysis by X-ray fluorescence (XRF). That would give an exact indication of the ratios of the elements, and could safely substantiate a common origin for two similar samples.

A few weeks later, Eric Prokopi, a commercial dealer in fossils, who had originally imported the skeleton into America, felt the full force of the law. He was arrested for conspiring to smuggle illegal goods into the U.S., for the possession of stolen property, making false statements, and one additional count of interstate sale and receipt of stolen goods. In December 2012, he made a plea bargain with prosecutors and admitted illegally importing fossilized skeletons of numerous dinosaurs. He agreed to hand over the bataar fossil, a second almost complete *Tyrannosaurus bataar* skeleton, the skeleton of a duck-billed dinosaur, *Saurolophus angustirostris*, and ownership of a third *T. bataar* skeleton that was still in England. During the next year, other specimens were added to the sheet, including five oviraptor skeletons that had been secretly excavated and illegally exported.

In England they traced their path to Christopher Moore, who runs a fossil-dealing business at Charmouth (close to the Jurassic Coast where Mary Anning had once plied her trade) and who had allegedly been helping to smuggle dinosaur skeletons. He released the third tyrannosaur skeleton, along with four skeletons of a 26-foot (8-metre) ornithosaur named *Gallimimus*, skeletons of *Ankylosaurus* and *Protoceratops*, and a display of dinosaur eggs. All were shipped to the U.S. Attorney's Office for return to Mongolia, and it was agreed that Moore would face no further charges because he had given up the contraband fossils without delay.[29]

Eric Prokopi was not so lucky. In July 2014, he was sentenced to three months in jail – but it could easily have been more.[30]

Every country in the world now wants to hold on to its heritage, and in recent years thousands of artefacts and other relics have been returned to their countries of origin by the U.S. authorities. In February 2016, a Canadian dealer who had been indicted for selling dinosaur fossils illegally imported from China at Tucson Gem, Mineral, and Fossil Showcase was sentenced to five years' probation and fined $25,000 (about £19,000) following a probe by U.S. Immigration and Customs Enforcement's Homeland Security Investigations. Jun Yang, of Richmond, British Columbia, was indicted in Tucson for selling smuggled *Hadrosaurus* eggs. He sold them for $450 each (roughly £330) and was offering for sale a smuggled *Psittacosaurus* fossil for $15,000 (£11,000). The psittacosaurs were small relatives of the better-known *Triceratops*. Throughout the twentieth century, dinosaur fossils were traded internationally –

sometimes smuggled across borders without proper documentation – but nowadays this is easier to control. The authorities are on the lookout, and they regularly score successes.[31]

Palæontologists, meanwhile, had been continually discovering new sites to excavate, and in 1939 work began in earnest on the Cleveland-Lloyd Dinosaur Quarry, near Cleveland, Utah, which turned out to contain the most concentrated accumulation of Jurassic dinosaur fossils that has ever been discovered. Stone had been quarried at this site since the 1800s, and the quarrymen often noticed bones protruding from rocky strata. In 1927, Ferdinand F. Hintze, of the Department of Geology at the University of Utah, organized an expedition to see what was there, and they retrieved about 800 bones. Excavation work began in earnest in 1939 when a team from Princeton University arrived on the site under the direction of William Lee Stokes. They dug for three seasons, collecting over 1,200 dinosaur bones, and Stokes eventually had 8 new dinosaurs named after him. No more work was done on the site until 1960, when digging resumed, and in 1974 a new dinosaur was formally described by James H.

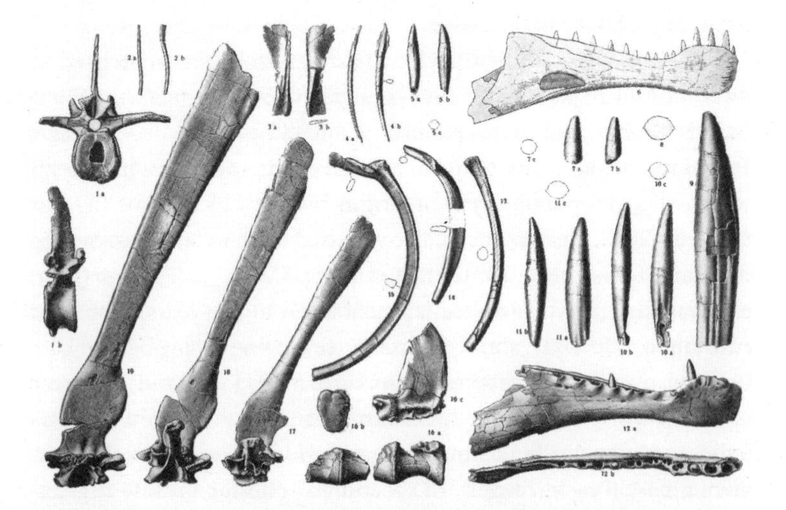

An expedition to Egypt was organized in 1912 by a prominent German palæontologist, Ernst Freiherr Stromer von Reichenbach. He retrieved several bones from a huge meat-eating dinosaur he later named *Spinosaurus ægypticus*.

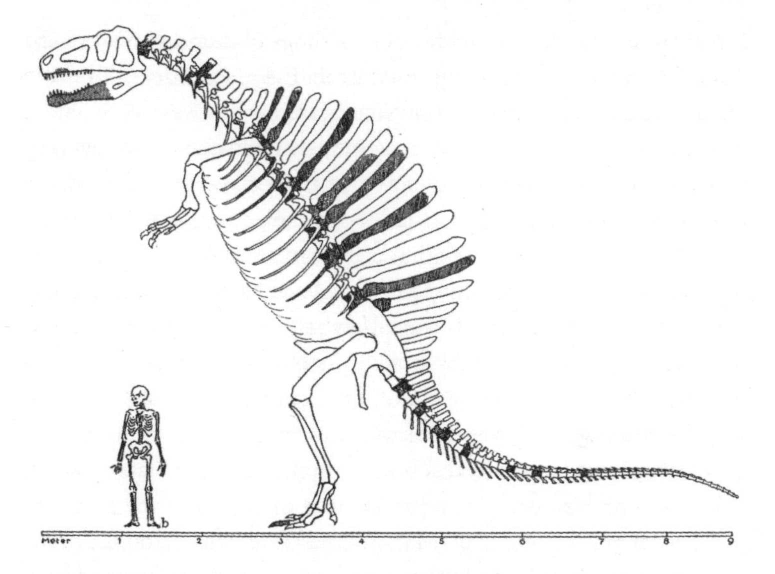

Stromer often speculated about the dinosaur whose few bones he had recovered, but not until 1936 did he publish a preliminary sketch of how the entire dinosaur might have appeared (the bones he had retrieved are shaded).

Madsen, Jr., who was the Assistant Research Professor of Geology and Geophysics at the University of Utah. He named it *Stokesosaurus clevelandi*. Two years later, another new dinosaur was identified, *Marshosaurus bicentesimus*, and in 1987, palæontologists from Brigham Young University discovered a fossilized dinosaur egg, which was at the time the most ancient yet discovered. Since then, some 16,000 bones have been discovered at this site. In October 1965 it was recognized as a National Natural Landmark, and most of the skeletons are now on public display at the Natural History Museum of Utah, including magnificent examples of *Stegosaurus*, *Allosaurus*, *Camptosaurus* and *Camarasaurus*. A substantial selection has been sent overseas, and dinosaur fossils from the Cleveland site are now exhibited in more than 65 museums around the world. Thus, it transpired that the casual reports of sheep-herders led to one of the greatest sites for dinosaur bones ever to be discovered.[32]

Egypt first began to emerge as a site for dinosaur discovery early in the twentieth century when the German explorer Ernst Freiherr

Stromer von Reichenbach unearthed a range of dinosaurs, including *Bahariasaurus*, *Carcharodontosaurus* and *Ægyptosaurus*. Then, in 1907, the American Museum of Natural History in New York, under the direction of Walter Granger, launched an expedition to the oasis town of Fayoum, some 60 miles (about 100 km) southwest of Cairo. His deputy was George Olsen, and they were accompanied by Henry F. Osborn. This was the first time an expedition of American palæontologists had ever been mounted outside of North America, and it proved to be a great success; when they sailed back to New York they were accompanied by 75 large crates containing fossils. In 1912, a second expedition was organized by Stromer. This gave them an unexpected prize – the remains of a vast meat-eating dinosaur that was just as terrifying as *T. rex*, but was apparently even larger. It was unearthed by Stromer's assistant Richard Markgraf in the Bahariya Formation of western Egypt. Markgraf had settled in Fayoum in 1906 to collect antiquities for Western museums. One of his first finds was the jaw of a fossilized primate later identified by Osborn as *Apidium phiomense*, the first discovery of a primate fossil in the whole of Egypt. That was a tiny mammal, but his new find was a monstrous meat-eating dinosaur. When Stromer published his official account of the remains in 1915, he gave it the name of *Spinosaurus ægypticus*. This was bigger than *T. rex* – it was the largest carnivorous dinosaur species ever discovered.

6

REPTILE DYSFUNCTION

Skeletons of familiar dinosaurs like *Tyrannosaurus* and *Stegosaurus* were becoming well known around the world, but in the field of palæontology attention was turning to the other signs that dinosaurs might have left behind. Dinosaur tracks had been known since ancient times, and they were among the first signs of dinosaurs to be investigated – even though all the early records show they were believed to be footprints made by gigantic birds. The first scientific account had been published in the U.S. a decade before dinosaurs were even recognized as a group.[1]

The greatest concentration of dinosaur tracks yet discovered had been unearthed in 1929 by Barnum Brown at Cameron, Arizona, where he recorded 300 dinosaur tracks that he considered to come from four different dinosaurs. Wrote Brown at the time: 'These will aid us greatly in determining posture and foot structure of early dinosaurs.' In the following year, Charles M. Sternberg announced the discovery of numerous footprints in Peace River Canyon in British Columbia. About 100 were later excavated, though thousands remain and are now flooded at the bottom of a reservoir created by damming the river. Also in 1930, a range of dinosaur footprints was unearthed in Denver, Colorado, in what we now know as the 'Dinosaur Freeway', and the following year saw more sauropod tracks come to light north of Kenton, Oklahoma, along with the skeleton of a 70-foot (22-metre) *Apatosaurus*. Dinosaur hunting had by this time become an attractive pastime for many people. Around 1908, Thomas and Grace Boylan set up home just below Como Bluff and later established a gas station to provide petrol for adventurous travellers who were venturing along the newly built dirt roads – this was the first transcontinental highway that America had seen. The Boylans became acquainted with many of the bone prospectors, and in 1915 Thomas Boylan started his own excavations on the hills behind his homestead in the hope of constructing an entire dinosaur skeleton. His excavations were not methodical, and he had soon collected thousands of assorted bones. Boylan asked visiting palæontologists from the University of Wyoming what he should do, and was told that the bones were just a random assortment; so he changed his plans. Instead of constructing a skeleton, he decided to create something unique in the world – a cabin made of dinosaur bones. Painstakingly, he used 5,796 dinosaur bones weighing 102,166 pounds (46,341 kg, almost 50 tons) and cemented them together producing a rectangular building. By 1932 it was ready to open to the public, on a site beside U.S. Route 30/287 some five miles (8 km) east of Medicine Bow. On public display were relics, photographs and newspaper cuttings. From 1935 he had postcards printed which emphasized the uniqueness of the site. It was promoted as 'the strangest building in the world' and 'the building that used to walk'. Newspapers reported its curious nature and it featured in the long-running series 'Ripley's Believe It or Not!'. The gas station and the curious museum were run by Thomas Boylan until his death in 1947, after which his wife Grace took over until 1960, when – with business falling off – the gas station closed, and the petrol pumps were removed. In 1974 the museum was no longer financially viable, and Grace Boylan finally sold the building and the rest of the homestead. The revamped museum struggled on until 1992 when it finally closed its doors, and it remains deserted to this day, though it is still listed in the National Register of Historic Places. This is a ghostly place; the signs have faded, though behind the old museum and surrounded by weeds I found the painting of a *Brontosaurus* that was still impressive, though faded with the years. Como Bluff looks down forlornly over the site. The excavations are now privately owned, and casual visits are no longer allowed.

Elsewhere in the 1930s, dinosaurs were successfully exploited. The Sinclair Oil & Refining Co. Inc. decided to launch a publicity campaign, with dinosaurs including *T. rex* and *Triceratops* featuring in their advertisements. Most popular was their creation 'Dino the Brontosaurus'. The idea caught on. To this day, the green Dino is the logo for Sinclair gasoline. The company soon became involved in sponsoring a huge excavation in Wyoming. Barnum Brown had spent 1931 and 1932 excavating on the Cashen Ranch in the Crow Indian Reservation and had found some bones from a small carnivorous theropod, but these were not impressive – so he simply put them in store to examine later. Eventually, these would lead to a revolution in the way dinosaurs were envisaged. More interesting was a quarry near the tiny settlement of Shell, Wyoming (population 50), that had come to the attention of the U.S. National Park Service. There were reports of dinosaur bones being discovered, and in 1934 Barnum Brown was funded by the Sinclair Company to carry out a survey of the site. The land belonged to a rancher, Barker Howe, and it proved to be an astonishing site; more than 4,000 fossilized sauropod bones were discovered packed into a rocky basin. Even so, the site was then left

During the 1930s, the Sinclair Oil & Refining Co. Inc. launched a new campaign, with dinosaurs including *T. rex* and *Triceratops* featuring in their advertisements. Their creation 'Dino the Brontosaurus' is still their logo today.

undisturbed until a Swiss palæontologist, Kirby Siber, leased the 'bone rights' from the landowner and began to search again in 1995. Siber owned and operated the Saurier Museum at Aathal in Switzerland, and during the following 15 years he carefully excavated a range of incredibly well-preserved skeletons of dinosaurs, many of which became world-famous. One was a complete skeleton of *Allosaurus* nicknamed Big Al, which became the subject of major television documentaries. Then came Baby Toni, the only known juvenile sauropod and estimated to have been six months old at death. There were three specimens of *Stegosaurus* which were nicknamed Sophia, Victoria and Moritz, an *Apatosaurus* they called Max, two *Camarasaurus* (E.T. and Paula) plus seven *Diplodocus* skeletons. This is the biggest selection of near-perfect dinosaur skeletons ever found. Sophia the *Stegosaurus*, for instance, was purchased by the Natural History Museum in London and put on public display in the Museum's Exhibition Road entrance as the most complete *Stegosaurus* skeleton known to science. They did not disclose the purchase price.

The entire Bighorn Basin in Wyoming already has 1,000 different fossil collection sites, and more are emerging. Brown concluded that the mass of dinosaurs indicated they had died of thirst, and this became a memorable scene in Walt Disney's cartoon film *Fantasia*, released in 1940. Dinosaur National Monument at Jensen, Utah, continues to make headlines. This is an extraordinary place. The dinosaur fossil beds were first described by Earl Douglass, a palæontologist surveying for the Carnegie Museum of Natural History in 1909. Douglass and his team unearthed thousands of fossils and crated them back to Pittsburgh, Pennsylvania, to be cleaned and analyzed. So impressive were the results that President Woodrow Wilson had the area designated as Dinosaur National Monument in 1915. Utah became so excited at these discoveries that *Allosaurus* was declared its state dinosaur. In time, people forgot; when Kenyon Roberts, a 10-year-old dinosaur enthusiast, asked State Senator Curt Bramble why Utah had *Allosaurus* as its official state fossil, Bramble replied: 'I didn't even know we had a state fossil!' Young Kenyon said that, although *Allosaurus* fossils had been excavated across Europe and Africa, it had never been found in Utah. He thought that

In the entrance to the Earth Hall of the Natural History Museum, London, is a *Stegosaurus* skeleton that is 85 per cent complete (the rest being made up of plastic casts). She was seen at a fossil fair in the United States and bought for an undisclosed sum.

Utahraptor would be more appropriate 'because it had Utah in its name.' There is just one species, *Utahraptor ostrommaysorum*, first discovered by collector Jim Jensen, in the Dalton Wells Quarry near Moab in 1975. The species was named after the palæontologist John Ostrom and Chris Mays, a one-time pilot who set up the short-lived Dinamation company in California importing and making moving models of dinosaurs. Kenyon Roberts asserted that fossils of *Utahraptor* had never been found anywhere but in Utah.

When Charles Whitney Gilmore had excavated his enormous *Diplodocus* skeleton near Jensen, Utah, with Norman Boss in 1923, he recognized that the area was ripe for further exploration and the original 80-acre (33-hectare) reserve was extended in 1938 to its current area of more than 200,000 acres (80,000 hectares). The valley that passes through the site was perfect for building dams, and in the early 1950s a scheme to construct 10 dams costing over $1 billion (£750 million) was announced. There were immediate objections, and a campaign was launched to keep the Monument site untouched. Against the protestors were powerful commercial and political groups arguing for the dams to be built. The hydro-electric power from the dams would be crucially important, they argued, and the impounded water would be a valuable resource. Eventually a compromise was agreed. The Colorado River Storage Project Act was passed in April 1956 with this proviso: 'No dam or reservoir constructed under the authorization of the Act shall be within any National Park or Monument.' The palæontological site was saved in perpetuity. And what a site it is: in a section of sloping cliff that leans back at 70 degrees and measures about 45 feet (roughly 15 metres) wide and 30 feet (10 metres) tall, the density of dinosaur bones is so great, and the fossils themselves so compacted together, that no attempt has been made to excavate them all. Instead, it has been roofed and is available for visitors to see as the 'Wall of Bones'. It is a breathtaking sight. Now scientists have discovered a similar dinosaur fossil site in Pu'an Township, Yunyang County in southwestern China. Since October 2016, some 5,000 dinosaur fossils have been excavated. These too form a wall of fossils measuring 500 feet (150 metres) wide and more than 25 feet (8 metres) tall – the largest fossil wall in the world. It may be even bigger, for the team believe that the dinosaur fossils may extend 65 feet (about 20 metres) below the level of their excavations.[2]

In recent years, some wonderfully well-preserved dinosaurs have been unearthed. One life-like specimen is *Nodosaurus*, an armoured dinosaur, which was discovered in Alberta in 2011 and taken to the Royal Tyrrell Museum where curators spent five years restoring it for public view. It is about 110 million years old and looks almost as though it has just emerged from the mud. It isn't flattened, as are most fossils, but is one of those that looks mummified, as if it had recently died. Conventional restorations of these beasts show them strutting about on land, and recent books by David West, Don Lessem, Kenneth Carpenter and a host of others all portray these dinosaurs as the powerful, terrestrial monsters that everyone believes them to be.[3]

Movies featuring dinosaurs have continued to appear. I have mentioned the 1940 release of *Fantasia*, in which the demise of the dinosaurs in a hostile habitat is spectacularly portrayed; then in 1954 the Japanese company Toho Films released a movie entitled *Gijora* (ゴジラ) that featured a colossal dinosaur. It was apparently inspired by the 1953 American movie *The Beast from 20,000 Fathoms*, in which an Arctic dinosaur was awakened by a nuclear test. The Japanese story

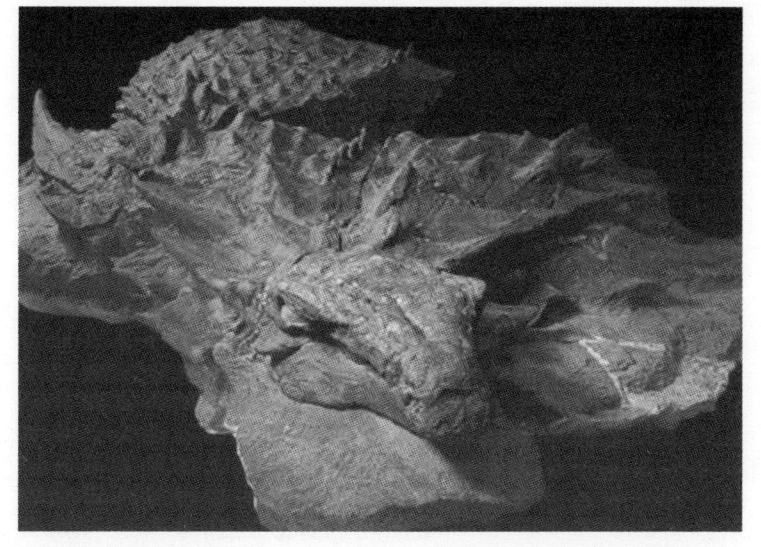

Armoured dinosaurs have been well preserved, and in 2011 a specimen of *Nodosaur* was excavated in Alberta, Canada. Technicians at the nearby Royal Tyrrell Museum reported that it looked more like a mummified body than a fossil.

told of ships that were being mysteriously destroyed by a sea monster, a huge dinosaur that was some 165 feet (50 metres) tall and had been released from deep within the Earth as a result of hydrogen bomb tests. Various methods were tried to control it in the movie, including a vast electric fence and a system that deprived it of oxygen. One of the characters in the movie complains that they once had to endure the nuclear bombing of Hiroshima by the Americans and now they must face this new threat. Towards the end of the movie it is said that, if nuclear tests continue, then there may be more monsters unleashed. The director, Ishirō Honda, said that the movie was intended to be an allegory about the hazards of the nuclear arms race. Many American Japanese saw the film, and it was so commercially successful that it was re-edited and released in the United States. This time it had the title everybody knows – *Godzilla*. In Japan, 27 sequels were produced by Toho, and in the U.S. five more movies were made. Warner Bros reissued them all in 2014, and they remain highly popular among fans of the genre.

While the public remained captivated by dinosaurs as objects that inspire awe and astonishment, dinosaur palæontology was losing momentum. There were a few expeditions organized by Polish, Chinese and Soviet palæontologists, to countries in North Africa, along with Argentina and Brazil, and when Ostrom discovered fossils of an 11-foot (3.4-metre) carnivorous dinosaur that he named *Deinonychus* in Montana in 1964, he revived the age-old notion that birds were descended from a common ancestor with dinosaurs. By that time, more than 500 dinosaur species had been identified. Fossilized dinosaurs had been found on all the continents apart from Antarctica, though that changed when the remains of an *Ankylosaurus* were discovered on Vega Island in the Antarctic in 1986. By and large, dinosaurs were - regarded as a wound-up area of investigation. Although graduate palæontologists still emerged from the universities, they were quickly absorbed by oil companies who were eager to use their knowledge of microscopic fossils to characterize the rocks through which the engineers were drilling. It was the tiniest fossils that were the centre of attention, while dinosaurs attracted little scientific interest. A search for published papers on the extinction of dinosaurs at the end of the Cretaceous period showed only about three or four appearing every year in the 1960s and 1970s; by comparison, since 1980 plenty of new papers are published every week.

Youngsters remained fascinated by the prospect of these great creatures, and – like many of my school friends – I used to collect fossils with eagerness. On the Isle of Wight I liked to scrape the multicoloured sands into test tubes in Alum Bay (you cannot do that today), then go hunting for ammonites in the Chines. Sometimes at Shanklin we would find small pieces of dinosaur skeleton in the sea-washed rocks. There have been so many dinosaur discoveries on this diminutive island, no more than 23 miles (37 kilometres) long and only 4 miles (about 7 kilometres) off the coast of southern England, that it has been dubbed 'dinosaur island'.[4]

Back on the mainland, I used to scurry along beneath the cliffs east of Lyme Regis, as shards of disintegrating rocks skittered down to the beach from time to time. Adults would warn all of us children about the dangers of a rock fall, but we took no notice. It was a real threat;

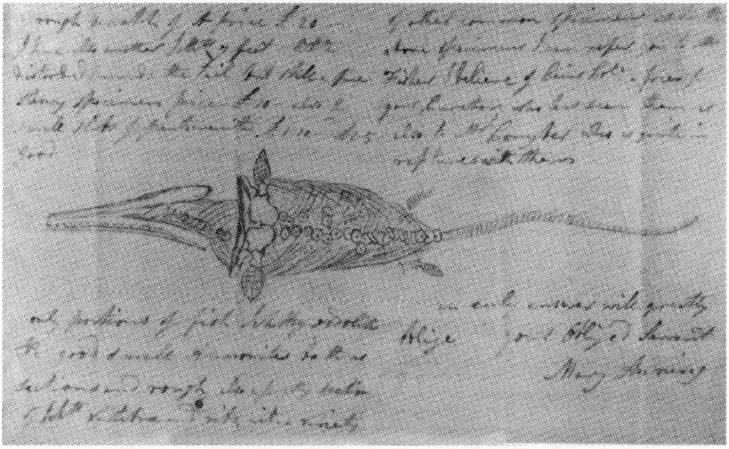

an early answer will greatly
oblige your obliged Servant
Mary Anning

On May 4, 1843, Mary Anning wrote this note to Adam Sedgwick in Cambridge, offering to sell him a newly excavated ichthyosaur. Her proposal is on display in the Sedgwick Museum to this day. Her sketch accurately outlines the details of the fossil.

hunks of ancient rock regularly crash down to the beach. In July 2012 there were several massive collapses, and one brought down 400 tons of strata, burying alive 22-year-old Charlotte Blackman, who was on holiday from Derbyshire. We could easily have been crushed, but as youngsters we were oblivious to the risk. This was an exciting place,

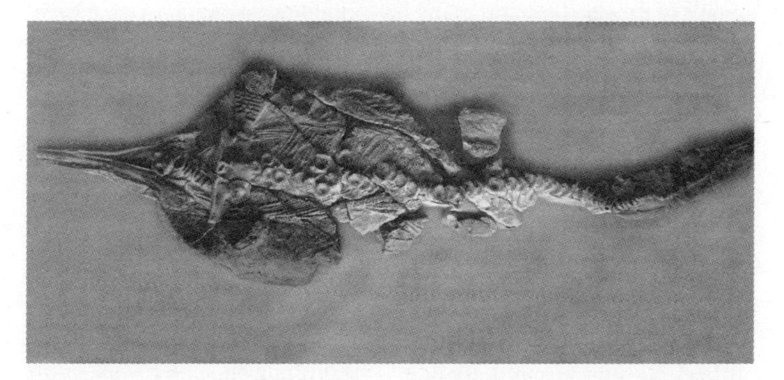

The fossilised specimen Anning sent to Cambridge can be compared with the careful drawing in her letter of May 1843. The vertebræ (known as 'verteberries' by the local quarrymen) are clear, and the icthyosaur's paddles are perfectly preserved.

where substantial signs of plesiosaurs and ichthyosaurs lurked among the strata. On holiday from school in North London, we used to head directly to the British Museum (Natural History) in Kensington, as it then was, to check out the stuffed animals and vivid wax models of plants, and then just stand in awe, looking at the huge dinosaur skeletons. I learned more about dinosaurs at boarding school in Hampshire, and as a teenager at the King's School in Cambridgeshire I spent many happy summer days picking through the clay beds at the nearby brickworks, collecting ammonites and gazing at their perfect preservation, glinting like flint spearheads in the sunlight. When we were in Cambridge I liked nothing better than to explore the Sedgwick Museum, where that proud cast of the Belgian iguanodon skeleton rears up imposingly above the human visitor, and original fossils from Mary Anning and Gideon Mantell lie perfectly preserved in the display cases. Later I used to take friends and relatives on visits to the National Museum of Wales in Cardiff. All along the South Wales coast were extensive exposures of the Blue Lias beds that also emerge on the Jurassic Coast, and numerous fossils have been donated to the museum over the years. On the weekends, I used to go down to Sully, where there is a small tidal islet half a mile (800 metres) long and half a mile offshore with its outcrops rich in fossil crinoids. As a student I mapped the little island, showing the various rare plants and some details of the geology, and many memorable summer days were spent at that beach or barbecuing on the island. With teenaged friends, I would often explore the smaller outcrop of Bendrick Rock or head west past Lavernock, where in the rocky strata we found the scattered remains of ichthyosaurs and occasional dinosaur footprints, some of which ended up in the National Museum of Wales. Occasionally, fossil hunters would pry out sections of rock and sell what they found to collectors, so in the 1980s the area was declared a Site of Special Scientific Interest (S.S.S.I.). Further along the coast, at Southerndown and at Llantwit Major, you could find fossilized invertebrates in unending numbers, and amateur palæontologists never went home unsatisfied.

Dinosaurs continued to fascinate me, just as they captivated the hearts of many youngsters, though my burning desires were increas-

Blue Lias strata represent thousands of fine layers of muddy sedimentary shale interspersed with limestone. They formed at the bottom of vast seas in the early Jurassic period. This photo is from Lyme Regis, but similar exposures occur in South Wales.

ingly directed towards microscopic living cells rather than the bones of gigantic dead reptiles. Everyone else was studying dinosaurs, and few scientists were studying how cells live their busy lives; and so palæontology remained for me just a casual, bystander's interest. I liked the idea of studying dinosaurs, but only when somebody else was doing it. There were many fossilized prehistoric reptiles on display at the National Museum of Wales. The director was Dilwyn Owen, with whom I had many interesting discussions about the museum's specimens. I used to examine the serried rows of cabinets filled with microscope slides (and later compiled a full catalogue of them all), and I adored delving into the collections in their departments of zoology and botany; but the dinosaurs remained a puzzle. Like others, I had found university to be confining and regimented, so I resolved to pursue science in a free-wheeling, all-embracing manner. Science seemed to me to be moving towards an enterprise based on opportunism, and academic funding seemed increasingly to support the tried-and-tested conventional conformity, rather than encouraging radical innovation. It was developing into a commercial concern geared to self-preservation, rather than openness and honesty. Indeed, I satirized it all in a book published in 1971. As a metaphor for the long-winded obscurity of this commercialized movement, I gave it the longest and most obscure title in publishing history. Though I adored scientific inquiry, the conformity and subservience of the orthodox academic establishment was not for me.[5]

My book was reported on television, was duly discussed in the press and was anthologized, while a Spanish translation was subsequently published in South America. The translation was meticulous and yes, they even rendered the full title, though they added 'How We Falsify Science' at the start. I should have done that with the original English edition, but I have never been very good at titles.[6]

Much of what the book prophesied came to pass, and I returned to the theme in a follow-up with a simpler title, *The Cult of the Expert*, in 1985. It showed how the simplest and most obvious of ideas can be dressed up to look impressive and emphasized that topical subjects could always be made to look more promising than they are, in order to gain prominence. In its German translation it was *Der Experten-Kult*, as the perceptive mind may already have realized, and gained a subtitle: *vom Maximalen Minimum*, meaning 'making the most of very little'. It all showed how the modern trend in science was to exaggerate everything and to ensure that the public (and the funding authorities) were kept strictly at arm's length, while the Experts retained their air of mystery and omniscience. I showed how easily the public and the press were being bamboozled by the use of obscure scientific language. Scientists do not use those long words primarily to communicate among themselves, but to excommunicate outsiders.[7] The book has made the word 'expert' into a target for suspicion ever since. Science has become a racket.[8]

The trend towards making science forceful and exclusive was perfectly exemplified by the progress of palæontology. Since dinosaur research had slipped into a quiet patch by the 1960s there was suddenly a conscious effort to reinvent the field. The new image was destined to be very different; dinosaurs were now to be promoted as warm-blooded, active and intelligent creatures. The new generation of reconstructions showed them bounding across the prairie or engaging in vigorous battles which usually showed one leaping into the air. They jousted and fought, roared and chased each other, slashed with their mighty tails and stamped with their huge feet ... these were monsters of machismo. Even though there was little scientific evidence to support all this, these new dinosaurs were far more exciting than the plodding reptiles of previous generations. Suddenly palæontology was the most exciting science in the world, and the funding started to flow in once more.

This new interpretation made little sense to me. Dinosaurs were far more massive than elephants, and they were the largest land animal we knew – even mammoths, which were bigger than elephants, rarely topped 10 tons. The *Amphicœlias fragillimus* that Cope had discovered in 1877 would have weighed 10 times as much as an elephant. To my way of thinking, as a simple-minded inquiring biologist, no such creature could have moved about comfortably on land. When I stood and studied the *Diplodocus* in London, the whole subject seemed to me confused, even more so after it had its tail raised and held aloft in 1993, rather than resting it comfortably on the floor. The skeletons of *T. rex* and *Iguanodon* were even more perplexing, because they had such diminished forelimbs. Any massive land animal would need four sturdy legs to support its bulk. There was no evolutionary pressure to cause the forelimbs to shrink away – yet the largest carnivorous dinosaurs of all were the tyrannosaurs, and their arms were reduced to mere stubby little claws. None of it made any sense.

One of the problems with the reconstructions of dinosaurs so popular in magazines and on television is that they so often show the dinosaurs in a present-day environment, with rippling streams and craggy mountains clothed with broadleaved forests towering over undulating verdant meadows. The world of the dinosaurs was dramatically different. When they existed, the continents as we know them had yet to form. Continental drift – the theory that became plate tectonics – had demonstrated how the continents have moved with time and had shown that, during the age of the dinosaurs, the continents of South America, Africa, India and Australia were joined with Antarctica as the supercontinent Gondwanaland. Eurasia, Greenland and North America were similarly once joined as the northern supercontinent Laurasia, whereas the land that became China and Japan lay near the equator. When we inspect the geology of a region, we can sometimes see the progressive layers of Arctic, and then tropical, vegetation as the entire area wandered across the globe from icy conditions to hot. Geological strata can often be read like a history book. If we wind the clock backwards, we can see how the continents drifted back in time, as we go towards the period when the Earth first formed. To early humans, the world had always been the same; Aristotle considered that it had existed unchanged throughout eternity. In Asia, both in India and China, there were beliefs that the world was created, destroyed, and then re-emerged in immensely long cycles lasting billions of years, though there was not the slightest evidence that this might be the case.

The first person on record who considered that the Earth originated at a specific date was the Venerable Bede (672–735). He was a monk and scholar in the north of England who lived at the monastery of St. Peter in the ancient kingdom of Northumbria of the Angles (today this is in Tyne and Wear). Bede wrote the first recorded history of England in 731, and has long been celebrated as the father of English history. His book was transcribed as a manuscript and was not published in book form until more than 740 years later.[9]

Bede's was a remarkable book, containing vital information on music of the early Middle Ages, between the sixth and eighth centuries. The Saint Petersburg Bede is an early illuminated manuscript of Bede's eighth-century history, the *Historia ecclesiastica gentis Anglorum* (Ecclesiastical History of the English People). It was moved to the Russian National Library during the French Revolution. In a paper from 725 entitled *De Temporibus*, Bede wrote about the Earth, discussed various calendars, reported on the changing Moon and the seasonal changes of a spherical world; and he worked through the Bible to calculate when the Earth began. According to Bede, this was 3952 BC; his record is the earliest recorded calculation of when our planet was formed. One of those ancient documents survives; though not contemporaneous with him, it was produced just 120 years after his death and is at the St. Gallen Stiftsbibliothek in Switzerland.[10]

The next calculation of the age of the Earth was undertaken by John Lightfoot, who was born in Stoke-on-Trent in 1602 and became Master of St Catherine's College, Cambridge, and Vice-Chancellor of the University. His method was methodically to accumulate a meticulous chronology of world history derived from the genealogies spelled out in the Old Testament, and he worked out that the world was created at 9:00 am on September 21, 3298 BC.[11]

Lightfoot was followed by the better-known James Ussher, born in 1581, who rose to become Archbishop of Armagh and Primate of All Ireland, and who painstakingly combed through all the Middle Eastern and Mediterranean histories he could find, as well as the Old Testament, before announcing his decision in 1658. There was no doubt about it, Ussher concluded; the world had been created on October 23, 4004 BC. It was a Sunday.[12]

So far, there were three separate ages for the Earth in the literature: 4,677 years in the time of Bede, 4,958 from Lightfoot, and 5,662 according to Ussher. All were based on ancient writings, which have a tremendous appeal as surviving texts, but had nothing solid and scientific on which to base a calculation. Some science was introduced for the first time by the English astronomer and mathematician Sir Edmond Halley, who became Astronomer Royal in 1720. He carefully considered geological and climatic considerations in calculating how the Earth had formed and set up a sensible line of reasoning. Halley said that, since salt accumulated in the oceans from the small amounts that had been washed down by rivers and streams, we could think of a way to reason the age of the Earth. Clearly, it could not be very young, because if it were, then the oceans would still be composed of predominantly fresh water. Conversely, it could not be incredibly old, or all the oceans would be saturated with salt, like the waters of the Dead Sea. The actual calculation he proposed to leave to others to complete.[13]

In France, Georges Louis Leclerc, Comte de Buffon, the distinguished naturalist, decided to tackle the task empirically. He carried out experiments that involved observing the rates of cooling of red-hot spheres of iron, and then scaled up the volume to give an idea of the time it would have taken the Earth similarly to cool down. This made little sense, even in his time. The radiation from a hot globe in space is unlike something in a study or laboratory here on Earth, and it takes no account of how hot the interior of the world might still be

– let alone considering heat generated from within the Earth (like the hot rocks of Cornwall, which are heated by atomic energy produced by radioactive granite). That aside, he did at least carry out experiments, and used objective criteria by which to assess the results; he may have been using ill-considered premises for his arguments, but he was at least attempting to be scientific. Buffon calculated the age of the Earth to be 75,000 years.[14]

Needless to say, the geologist Charles Lyell made his own attempt at the puzzle in 1828. This time he stated that fossilized shells were of 'vast importance' in working out the detailed history of the Earth, and he came up with a figure of 240 million years for the age of our planet.[15]

In 1862 Lord Kelvin – the British physicist William Thomson – returned to Buffon's pioneering, if muddle-headed, calculations. Although he was similarly handicapped in possessing a limited understanding of the physics of the early Earth, he produced a range of ages between 24 million and 400 million years. Once again, the figures were wildly inaccurate; but the logic behind his research was carefully considered at the time.[16]

Radioactivity was discovered by a French physicist, Antoine Henri Becquerel, in 1896, and he was awarded the Nobel Prize for this far-reaching revelation. One of the earliest applications of the principle was in the treatment of cancer. This application was introduced into medicine by a Dublin physician, John Joly, as long ago as 1914. Curiously, Joly had previously tried to estimate the age of the Earth by resorting to Halley's method. He used the known data for the salinity of the sea to propose that the Earth was 100 million years old. Although the figure was incorrect, it had some reasoning behind it.[17]

In 1905, two Cambridge physicists teamed up to work on this problem. They were Ernest Rutherford, a physicist from New Zealand, and Frederick Soddy, born in Eastbourne on the south coast of England. Working at the Cavendish Laboratory in the middle of Cambridge, they realized that radioactive decay in isotopes occurred at a constant rate, so they set out to apply this technique to the measurement of the age of rocks and various minerals. Rutherford, who had discovered that the decay of uranium releases helium, now felt

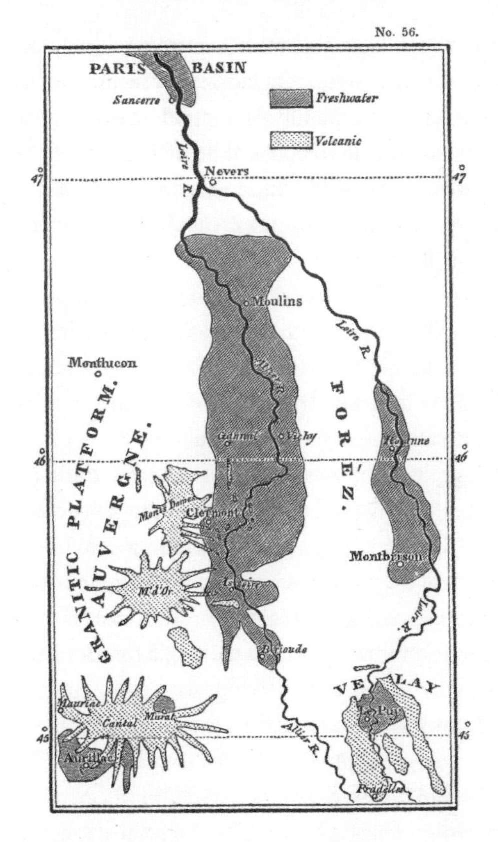

When Charles Lyell published his book *Principles of Geology* in 1830 he explained earthquakes and introduced the terms Palæozoic, Mesozoic and Cenozoic. He detailed the prehistoric volcanoes that had given rise to our present-day landscape.

that a valid age for the Earth could be estimated, and he worked out that the oldest rocks would be about 500 million years old. At the time, rates of geological sedimentation or the cooling of the Earth were the most popular models for calculating the planet's age, and most scientists still accepted Lord Kelvin's figures of 24 to 400 million years. On a visit to New England, Rutherford gave a lecture at Yale and his words caught the attention of a young chemistry graduate, Bertram B. Boltwood, who was transfixed and started to investigate on his own. Boltwood discovered that ratios of uranium to lead in rocks from the same strata were always the same, while in older samples of rock there was always proportionately less uranium and more lead. He concluded that lead was the ultimate product of uranium decay and realized that the age of the rock could be determined from the ratio of the lead (chemical symbol Pb) to uranium (U). He dated 26 mineral samples between 92 and 570 million years. It later transpired that ^{238}U naturally decays to ^{206}Pb with a half-life of 4.51 gigayears, while the other radioactive isotope ^{235}U decays to ^{207}Pb with a half-life of 710 million years. Boltwood looked again at the calculations and this time he came up with an age for the Earth of 1.64 billion years. This was an astonishing figure and put the birth of our world back to a much earlier date than anybody else had calculated.[18]

The other great proponent of radiometric dating was Arthur Holmes, who, as we have seen, was a passionate advocate of continental drift. By 1911, as a research geologist at Imperial College, London, he obtained a date for a rock sample from Ceylon (now Sri Lanka) at 1.6 billion years old, using lead isotopes. At the annual meeting of the British Association for the Advancement of Science in 1921, the audience heard that radiometric dating was emerging as a credible research tool, and they were told that the Earth was several billion years old. This was an astonishing revelation though it fitted well with the results that Boltwood had obtained.[19]

Although Holmes initially calculated that the most ancient rocks were 1.6 billion years old, by 1927 he had revised this figure to about 3 billion years, and in 1945 he increased it to 4.5 billion years. Many of his calculations compare well with our modern figures. For instance, Holmes calculated that rocky samples from the Devonian period were 370 million years old (we now accept that this geological era occurred between 419.2 and 358.9 million years). His results put the rocks from a Silurian period specimen at 430 million years (our latest calculations would say 443.8 to 419.2 million years), and his estimates for the Carboniferous period were around 340 million years old (now we would say 358.9 to 298.9 million years). These were extraordinary results. For his later figures, Holmes relied on techniques developed in America by Alfred Otto Carl Nier, a gifted nuclear physicist. Nier had designed a revolutionary mass spectro-

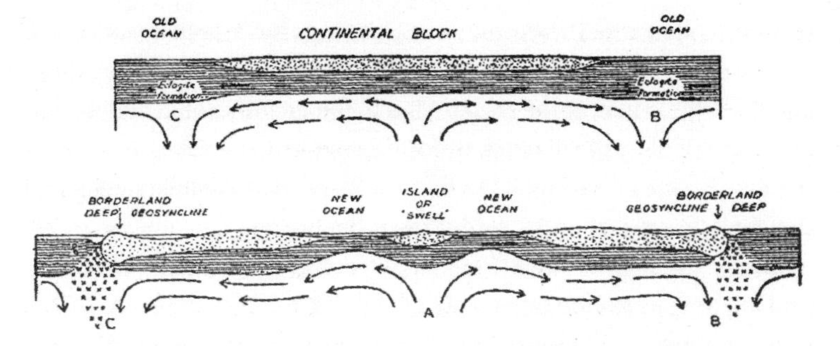

Arthur Holmes, son of a Newcastle cabinet-maker, revolutionized geology with his proposals for geochronology and continental drift. In his 1913 book *Age of the Earth* he advanced the concept of convection causing the continents to move.

graph that allowed him to prepare a pure sample of ^{235}uranium for Enrico Fermi, pioneer of the atomic bomb. Investigations at Columbia University revealed that this rare isotope could also undergo nuclear fission, like ^{238}uranium. Between 1943 and 1945 Nier worked with the Kellex Corporation in New York, designing and constructing mass spectrographs for the Manhattan Project, which developed the atomic bombs that the United States dropped on Japan near the end of World War II. Separating and identifying these isotopes was crucial for creating the atomic bomb, and it was also the vital step that gave Holmes the best way to estimate the age of the Earth. War often provides unexpected insights into the world of science, for the urgency of wartime is a far greater stimulus to science and technology than times of peace.[20]

We now accept that the age of the Earth is four and a half billion years – 4.54 ± 0.05 billion years (4.54×10^9 years ± 1%), a result based on radiometric dating. Holmes had it right by the end of World War II, an impressive achievement. The kind of life we see around us today is a very recent arrival on the scene.

We can see this vividly if we condense the whole of the Earth's existence into a single year. Scaling everything down to one year gives us the best way to appreciate the timescales of our planet. I have often resorted to this analogy, though my own calculations were called into question when I visited climatologist Bill Hay in Colorado. He recommended modifications, so my timings have been revised. For the first quarter of the year, the world is just a hot volcanic planet devoid of life. Then, at the end of March, the first microbes appear. For the first half of its life, Earth has nothing but tiny microbes alive on its surface. By the end of July, the first multi-cellular creatures evolve in the shallow seas, then in mid-November we see the emergence of trilobites. No sooner have they appeared, but the earliest vertebrates evolve – these are elongated, small fish, and it is November 20 by the time they appear. The earliest land plants date from November 22, and by the end of that month there are insects on dry land. The world begins to look vaguely recognizable to us only in the final month of the year, when the earliest amphibians evolve on December 1. By December 4, the single great continent of Pangea forms and dinosaurs first emerge on December 11. Then the early landmasses began to drift apart, and on December 23 the newly separated continents start to collide, as they are wafted along on the convection currents from deep inside the Earth and massive crumple-zones form mountain ranges from the Alps to the Appalachians. The dinosaurs, having ruled for over a fortnight, die out by December 26 as the Rocky Mountains rise up when two vast tectonic plates strike each other at speed and today's continents start to emerge. On December 29 apes evolve, but Neanderthals do not appear until it is half an hour before midnight. Not until 1 minute before midnight do the earliest human civilizations emerge. The chimes for midnight were already striking when Jesus Christ lived at 14 seconds to midnight. Then matters race ahead at increasing speed; the Age of Reason emerges 3 seconds before midnight, and the Industrial Revolution begins in Britain within the final 2 seconds of the year. So the whole of human existence fits into a single minute, whereas the era of the dinosaurs lasted more than 2 whole weeks. Thoughts like this give us a more realistic perspective. The dinosaurs endured for a long time in the spotlight, while humanity remains a flash in the pan. If we carry on as we are, foolishly wondering whether global warming is just a myth, we will be gone by the time the midnight clock has rung its second 'dong'.

Little progress was made over refining our calculations of the age of our planet in the 1950s and 1960s, and the earth sciences were equally unproductive at that time. This period has since been described by palæontologists as the 'dinosaur doldrums'. As we have seen, research became sparse, and fresh ideas on dinosaurs stagnated. Palæontology looked back nostalgically to the recognition of the existence of dinosaurs in England, those great dinosaur skeleton discoveries in Belgium, the acrimony of the 'bone wars' in nineteenth-century America, and the competitive 'bone rush' that had followed in the earlier decades of the twentieth century. Dinosaurs had been found in a variety of forms, but since then they had begun to slip from view. There were reasons for this: for over a century they had been regarded as a crucial category of magnificent monsters that had descended along lines that were independent from other reptiles. Since 1888 – the year that the Geological Society of America had been founded – they had been classified in two main groups: the Saurischians and the Ornithischians, distinguished through differences in their pelvic bones and joints, which became popularly known as the lizard-hipped and bird-hipped dinosaurs.[21]

And so the traditional idea that dinosaurs were a single large group was beginning to vanish, and it now appeared that there might have been two kinds of those gigantic prehistoric reptiles, rather than one. The proposal was investigated in the 1930s by Alfred Romer when he was a professor at Chicago and later at Harvard. He carried out extensive work on vertebrate evolution, drawing together data from embryology, comparative anatomy and palæontology to determine how aquatic organisms had slowly evolved to conquer land and had then specialized into the groups we currently understood.[22] Romer taught that the environment would influence form and function, and this had generated the evolutionary tree. He classified the dinosaurs in two disparate groups as though they were merely extinct descendants of ancient reptiles from which crocodiles and alligators had also evolved. Suddenly, dinosaurs were no longer as special as they had once appeared; the scientific classification of Owen's *Dinosauria* had disappeared.[23]

Once Charles Darwin's *Origin of Species* had appeared in 1859, and that celebrated fossil of *Archæopteryx* had been discovered, the distinguished British biologist Thomas Henry Huxley was led formally to conclude that birds had descended directly from dinosaurs. He used anatomical comparisons of skeletons to prove his point. Between 1860 and the 1920s this remained the scientific consensus, but it disappeared with a paper published in 1926. This was the work of a Danish amateur ornithologist, Gerhard Heilmann, who had studied medicine but became sidetracked and turned to work as an artist for a while. As Huxley had done, Heilmann started with *Archæopteryx* and compared it both with present-day birds and with fossil reptiles. He made a simple point: birds have a wishbone, and dinosaurs do not. The wishbone, formally known as the furcula, is indeed a feature of birds. It is modified from the clavicles. You will also find clavicles in primitive reptiles, creatures that had existed prior to the appearance of dinosaurs, but it had never been found in the dinosaurs themselves. Heilman's research was very detailed and the reasoning was simple: if lizards had lost their clavicles in the process of evolving into dinosaurs, then those bones could not have subsequently re-emerged. His conclusion was that birds could never have descended from the dinosaurs.[24]

Although Heilman didn't know it, a dinosaur clavicle had been discovered three years before his book was published. In 1923, Roy Chapman Andrews had recorded a well-preserved oviraptor fossil in the Gobi Desert, in which that elusive furcula was clearly visible. Clavicles are not easy to find in dinosaurs to this day – they are delicate bones and not given to preservation – but they do exist and have been found in a range of different types. The essential criterion upon which Heilmann based his argument was, simply, wrong. Even so, once he had dismissed the link between dinosaurs and birds, that view persisted through the 1960s. Here too, another important evolutionary role for the dinosaurs had been eclipsed.[25]

It seemed to me that we could learn far more about living organisms if we looked at them, not as collections of limbs and organs, but as colonies of separate living cells. I had thought little about applying this principle to dinosaurs – not my field of interest – but I did create a different approach to the way we looked at health and nutrition, so in my twenties I wrote a textbook to spread the word. It was based on looking at these disparate disciplines from the viewpoint of a single living cell, whether that cell was producing food (like yeast), damaging food (like *Salmonella*), being food (like the muscle cells in steak), or consuming food (like the cells of our body). Little did I realize how useful this would later be when I returned to look at three emerging realities of the dinosaur world.[26]

Change was soon afoot in the study of dinosaurs. The shadows were thinning, and the dinosaurs were about to emerge once more into the spotlight. The impetus for this revolutionary new approach stemmed from the enterprising research by John Harold Ostrom, an American palæontologist, born in New York in 1928, who set up a major excavation at the Big Horn Basin in Wyoming in the 1960s. Ostrom was a cheerful, thoughtful man with a broad smile and knowing eyes. He studied at Union College with a view to becoming a medical doctor like his father; but he was sidetracked by the evolutionary story and enrolled at Columbia University. After a year teaching, he transferred to Beloit College and then went on to the place upon which he had set his sights – Yale University. He was appointed Curator Emeritus of Vertebrate Palæontology at the Peabody Museum of Natural History, where all the collections from Othniel Charles Marsh were still held.

It was in 1964 that he made his great discovery. It would lead to one of the most important game-changing periods in twentieth-century palæontology. Ostrom established a full-time excavation at the Big Horn Basin in Wyoming, digging both there and at nearby at Rocky Hill. Later that year he was digging in sandstone rocks known as the Cloverly Formation when he came across his life-changing discovery. He excavated a few dinosaur fossils dating from some 110 million years ago that were beautifully preserved. In all, they retrieved more than 1,000 bone specimens representing three of four separate dinosaurs. As they were assembled, this dinosaur struck him as being different in key respects from the others that had been found, so he set about reconstructing their lives. It took him five years. These turned out to be smaller dinosaurs, measuring no more than 10 feet (3.5 metres) long and 4 feet (about 1 metre) tall. They would have weighed no more than a person, some 165 pound or 12 stone (about 75 kg). Nothing like this had ever been reported before. An extended line of inquiry subsequently showed that this dinosaur had been excavated previously – there were some bones from the same creature that had been retrieved by Barnum Brown in 1932, though he had not investigated them further and no report had ever been published. Ostrom decided to name this new genus *Deinonychus*, from the Greek δεινός (*deinos*, fearsome) and ὄνυχος (*onychos*, claw). He soon concluded that *Deinonychus* was a hunter that could attack by leaping and slashing with a vicious claw. Ostrom claimed to find evidence of elongated bands of muscle that ran the length of the tail, which he said would have made it suitable as a counterweight for running and leaping. His paper is wonderfully detailed. He describes the features of the skeleton, with clear line and stipple drawings of each component, showing how the claws could have flexed back and forth. He includes microscopic studies of cross sections of the bones, photographs of some of the details of the tail, and tables of measurements that compare his new dinosaur with others already known. Ostrom's published report was some 170 pages long, and it is one of the most thorough papers in its field.[27]

Ostrom's conclusion was that, if this dinosaur was so active, then it must have been a warm-blooded creature much like a mammal. It must be said that he was not the first to think of an active dinosaur that could jump high in attacking its prey. Charles Knight had launched precisely the same view with *Leaping Lealaps*, which he painted in 1897 and which predated Ostrom's conclusions by more than 70 years. There was just as much energy in Knight's later studies published in 1942, which showed a pair of *T. rex* 'locked in mortal combat' and sparring like boxers in the ring, with another towering threateningly over a stegosaur and standing almost on tiptoes. Those images had seemed fanciful when they were painted, but now they fitted the new view of dinosaurs that was emerging.[28]

What Ostrom sought to promulgate in the 1970s was the idea that dinosaurs were different to what everybody thought. They should no longer be considered as idle, sluggish, ponderous beasts, Ostrom insisted, but as active, fearsome and dynamic. This was the clue that was to trigger a revolution in our attitudes to the world of the dinosaur. Later he studied the trackways left by *Hadrosaurus*, and

One of Charles R. Knight's most memorable paintings is this remarkable study, entitled *Leaping Laelaps* and dating from 1896. Even at this early time, he showed these dinosaurs (since renamed *Dryptosaurus*) as energetic and dynamic.

concluded that they proved that these creatures must have been social animals. The duck-billed dinosaurs were not simply grazing creatures that wandered aimlessly across a primeval prairie, Ostrom insisted; they were associating in herds.[29]

For the first time in decades, dinosaurs were headline news once again. There were other changes afoot. In the late 1950s, Bill Hay, at that time a graduate student at Stanford University, had been collecting samples and studying the fossilized plankton from the sediments in the Gulf coastal plain of eastern Mexico. He thought he detected a sudden change in their populations between 60 and 70 million years ago and, looking further into the matter, he found that the same sudden transition had been reported from Italy and the Caucasus. Hay concluded that something dramatic must have happened to the plankton of the world's oceans at around that time, and he concluded that only a major, catastrophic event could have produced such a sudden and dramatic shift in the evolutionary trends of those tiny organisms. His doctoral dissertation committee was not convinced;

most of them were strict followers of Charles Lyell's idea of gradual processes of change in the geologic past. To those people, catastrophic events were unthinkable. Nevertheless, they let him squeak through, and Hay's new idea was narrowly deemed to be acceptable. Later that day the same committee heard the defence of a thesis by Roger Anderson, who was studying botany. He described a similar sudden change in palæoflora at the same period of time, ranging from trees to ferns, in the Raton Basin of New Mexico. Hay later presented his ideas at the International Geological Congress in Copenhagen in 1960, and at last they began to attract wider interest. The idea that a major cataclysm once rocked the whole Earth was supported in a paper published in 1970 by an Irish-born Canadian geoscientist, Digby Johns McLaren.[30] He was a kind and smiling man with a mop of dramatically wavy hair. He also had presence; a regal air of authority and a theatrical and charismatic way of addressing a gathering. In his spare time, McLaren was a connoisseur of fine wines, a practical joker and a keen grower of exotic orchids, while his professional life was concerned with campaigning for a better future. Although he was instrumental in making some of the major finds of petroleum in Canada, he later became a campaigner against the burning of fossil fuels and began to press for the need to limit the increase in the global population. McLaren devoted much of his life to a study of the first great extinction event – not the demise of the dinosaurs, but a more distant time some 375 million years ago between the Frasnian and Famennian stages of the Upper Devonian, when whole families of organisms had dramatically vanished. Nobody could work out why. McLaren, who was president of the Paleontological Society, devoted his inaugural address to the problem, and announced his startling conclusion: the normal functioning of the biosphere must have been disrupted by a massive meteorite impact. Some celestial body had surely slammed into the Earth, clouding the skies, polluting the seas, darkening the surface of the planet and causing the death of the Earth's communities of monsters. The idea was sound enough in theory, though he had no proof of how big this object might have been, nor of where it might have landed, nor of any geological evidence to substantiate his point. It was a fine speech, but it convinced very few people at the time. There is currently only a small Wikipedia entry on McLaren, and he is missing altogether from their entry on the Cretaceous/Tertiary extinction event, but his insight and his conclusions were prescient.

Within a decade, the meteorite idea resurfaced, this time in Mexico and with hard evidence to support it. Two geophysicists, Glen Penfield and Antonio Camargo, were surveyors for the Petróleos Mexicanos Company in 1978 when they noticed some remarkable geophysical data from Yucatán. The results were unmistakable: there was a vast circular shadow on the screen from something deep below the surface of the Earth. It seemed to be an arc measuring 40 miles (about 70 km) in diameter. They investigated further and turned up a gravity map made years earlier that showed another portion of the same arc. Together, they comprised a huge circle about 110 miles (180 km) wide, with the village of Chicxulub at its heart. Penfield and Camargo attended a meeting of the Society of Exploration Geophysicists in 1981, where they formally presented their findings. At the time, nobody knew what to make of them. Impact craters were known on Earth, indeed even in the 1970s more than 50 had been identified, though at the time the announcement seemed strange.[31]

While this was going on, a father-and-son team of physicists, Luis and Walter Alvarez, were speculating on meteoric impacts on the Earth. Luis had won a Nobel Prize for physics in 1968. He was a superb scientist and inventor, and was once described by the *American Journal of Physics* as one of the most brilliant and productive experimental physicists of the twentieth century. The pair now had data to back them. There was widespread evidence of a layer of iridium in the Earth's surface, dating back about 66 million years. Iridium is rare on Earth but is much more abundant in meteorites. Its presence was taken to indicate that an astronomical cataclysm had occurred, and the most popular suggestion was that the iridium could have originated in a nearby supernova. The two physicists became convinced that it had been an asteroid or a meteorite that was the cause, and they set out to disprove the supernova theory. They published their results in 1979.[32]

Theirs was a remarkable finding, because it seemed to coincide with the timing of the extinction of many types of life on Earth that occurred about 66 million years ago, at the boundary that marked the end of the Cretaceous and the dawn of the Tertiary period. This became known as the K-T extinction event.*

This was exciting news. Hay had speculated on a cataclysmal change, McLaren had thought that an asteroid could have caused such an event, scores of meteorite craters were being discovered, Penfield and Camargo were reporting what seemed to be a massive crater deep under Mexico – and suddenly here was a theory that tied everything together.[33]

Subsequent investigations at Chicxulub provided confirmation that a huge meteorite had indeed been the cause, and the conclusion was that an interplanetary body travelling at about 70,000 mph (30 km/sec) and estimated to be 6 miles (10 km) across had impacted to release energy amounting to 420 zettajoules of power (a billion times more than the atomic bombings of Hiroshima and Nagasaki combined). That is hundreds of times more energy than is released by the most violent volcanic eruption. The calculations suggested that it would have created a tsunami over 330 feet (100 metres) tall and it would have devastated the climate for a decade. This is what is now claimed to have wiped out the dinosaurs.[34]

All this added fuel to the fire, and discussion of dinosaurs was once again on the ascendant. The new generation of palæontologists was reinventing dinosaurs as dynamic and hot-blooded, with the predatory dinosaurs presented as fearsome warriors; and we now had a dramatic death designed for these denizens – killed off by a massive rock from space that decimated our entire planet. As we have seen, the previous phase of dwindling interest had been dismissed as the 'dinosaur doldrums', but now a new era was opening and new research

* Sometimes people ask why we call this the K-T extinction event. Since it occurs between the Cretaceous and Tertiary, should it not be the C-T extinction event? The term Cretaceous was coined by Belgian and English scientists, who called it respectively *crétacé* and *cretaceous*, and the Latin for chalk is *creta*; however, the German word is *Kreide*, and it is this German term that gives us the 'K'. In any event, this has changed, for 'K-T event' is now considered an obsolete term. Geologists currently prefer to use the term 'K-Pg boundary' (the Pg stands for Palæogene, a term most people know nothing about), so you should stick to this new version if you wish to be considered up to date with the jargon.

Coloured-coded strata in the geological map published by William Smith, the self-taught surveyor, in 1815 revealed a glimpse of what lies beneath our feet. His huge map was 8 feet (2.6 metres) long and is in breathtaking detail.

Mary Anning's discoveries were interpreted in this 1830 study by Henry de la Beche. The painting, entitled *Duria Antiquior* ('ancient Dorset') is the first picture to show a prehistoric landscape. Prints were produced by Georg Scharf for sale.

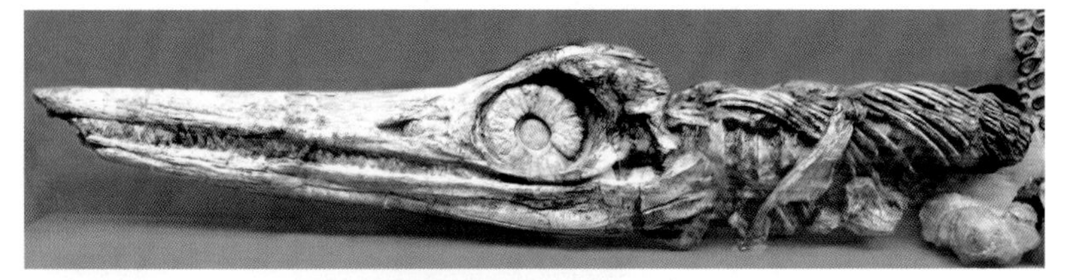

A fossilized ichthyosaur *Temnodontosaurus platyodon* from 190 million years ago was discovered by Mary Anning's brother Joseph in 1810. Mary later excavated the neck. The specimen is in the Natural History Museum.

After the Great Exhibition in 1851, the Crystal Palace was moved to Sydenham in South London and Benjamin Waterhouse Hawkins was asked to create models of extinct creatures, including these curious-looking iguanodons.

William Dyce's painting 'Pegwell Bay, Kent, a Recollection of October 5th, 1858' captures a typical scene of women collecting specimens from a beach at sunset. Popular interest in science is evident from Donati's comet, just faintly visible in the middle of the sky.

Heinrich Harder was a German art professor who studied natural history. In 1915 he began painting dinosaurs, like this *Diplodocus,* to appear on cards entitled Tiere der Urwelt ('animals of the early world') for the Reichardt Cocoa Company.

The horn which Mantell mistakenly placed on the snout of *Iguanodon* appeared on many editions of Edgar Rice Burroughs' novel. This example dates from 1918, but it was still there in 1963.

Stravinsky's 'Rite of Spring' was the music chosen to accompany Walt Disney's view of dying dinosaurs in his 1940 movie 'Fantasia'. This was the first cartoon to depict their demise.

Brachiosaurs were painted by the Moravian artist Zdeněk Michael František Burian in 1941. This concept could not work; submerged dinosaurs could not inhale, and would float to the surface.

There are few women dinosaur artists, and the Canadian painter Eleanor Kish shows extraordinary skill as she portrays *T. rex* attacking an edmontosaur. However, no 10-ton beast could ever move in this way. Tyrannosaurs were as heavy as elephants.

Although current accounts claim *T. rex* was covered with feathers, studies of its skin reported in 2017 by Australian investigator Phil Bell show that was untrue. This magnificent monster painted by Roger Harris now needs to move into the water.

Gigantic brachiosaurs were the first dinosaurs that visitors to 'Jurassic Park' encountered. This still from the 1993 movie, based on a novel by Michael Crichton and produced by Kathleen Kennedy and Gerald R. Molen, exemplified the unfeasibility of such terrestrial giants.

Zdeněk Burian's painting of *Brontosaurus* (actually an apatosaur) appeared in *Life before Man* by Zdeněk Špinar and featured in the Monty Python sketch with John Cleese playing the part of Anne Elk. In 1994 it reappeared on a Czech postage stamp.

Next to the Loch Ness exhibition centre, near Inverness in Scotland, is a parking lot surrounded by a wooden fence. Visitors who peep over the top will find this concrete effigy of Nessie.

Skipping like a gazelle near limpid pools, this is the most frequently reproduced image of *Edmontosaurus*. Fossils of this duck-billed dinosaur, which weighed as much as an elephant, are always found in strata from former swamps and shallow lakes. BBC documentaries show it grazing near pools.

The Russian edition of Wikipedia published this impression of *Apatosaurus* at the water's edge. The physical constraints of evolution have placed an upper limit of about 10 tons for terrestrial creatures. There is no scientific basis for interpretations like this.

In November 2009, this spinosaur skeleton some 26 feet (8 metres) long was unveiled at the Drouot Montaigne auction house in Paris. Because it is not possessed by a recognised museum or university, it has never been formally described.

The newly-discovered *Argentinosaurus* is the most dramatic display in the Fernbank Museum near Atlanta. From nose to tail it measures 125 feet (38 metres) – yet the only bones discovered are shown with natural colour superimposed. The rest is imaginary and made of plastic.

The 30-inch (80 cm) tuatara *Sphenodon punctatus* of New Zealand is a survivor from the age of the dinosaurs, and dates back 200 million years. Like dinosaurs, its skeleton features a gastralium, the bony rib-like structure that protects the abdomen.

Scale models were used by the author for volumetric analysis in this picture, which was featured by *Laboratory News* when the aquatic dinosaur theory was first announced. Critics ridiculed the idea, though it is the accepted method used by palæontologists everywhere.

A huge carnivore, *Torvosaurus tanneri,* was identified in Colorado in 1979. In 2014 its European cousin, *T. gurneyi* (above) was unearthed in Portugal. Measuring 33 feet (10 metres) long and weighing 5-10 tons, it is clearly incongruous on dry land.

Pulau Komodo, a tiny island east of Java, is home to the Komodo dragon *Varanus komodoensis* which, at 10 feet long (3 metres) is the largest lizard in the world. Studying it in the wild reveals the problems faced by large reptiles.

José Ignacio Canudo, a Spanish palæontologist, is snapped lying down by a dinosaur femur in Trelew, Patagonia, in January 2014. Several other bones were excavated and the dinosaur, *Argentinosaurus*, has since had the rest of its skeleton imaginatively re-invented by scholars.

Gigantic dinosaur footprints were reported by Steve Salisbury, of the University of Queensland, in 2017. An aboriginal elder, Richard Hunter, lay down to demonstrate their size – 5 ft 6 in (1.75 metres) long. The dinosaur would have weighed over 100 tons.

Researching in Colorado, the author encountered Bob Bakker, resplendent in beard and a jacket with the stars and stripes. Introductions were by first names only. Bakker popularized the dynamic, athletic dinosaur in a lucid and entertaining (though misconceived) book.

Amateur enthusiast Thomas Boylan used 5,796 dinosaur bones weighing some 50 tons to produce this unique cabin which opened as a museum in 1932. Situated beside Route 30/287, five miles east of Medicine Bow, Wyoming, it is long since deserted.

Skulls of sauropods are rare; the heads of decaying carcases were bitten off cleanly by therapods. This skeleton of *Dicræosaurus hansemanni* at the Museum für Naturkunde, Berlin – like most on display – has an invented plastic skull substituted for the real thing.

The fish-eating dinosaur *Spinosaurus*, with its dorsal fin and clawed extremities, was recognized by the author as aquatic in 2012. Although relatively few bones have been studied, we know it was 50 feet (15 metres) long – larger than *Tyrannosaurus rex*.

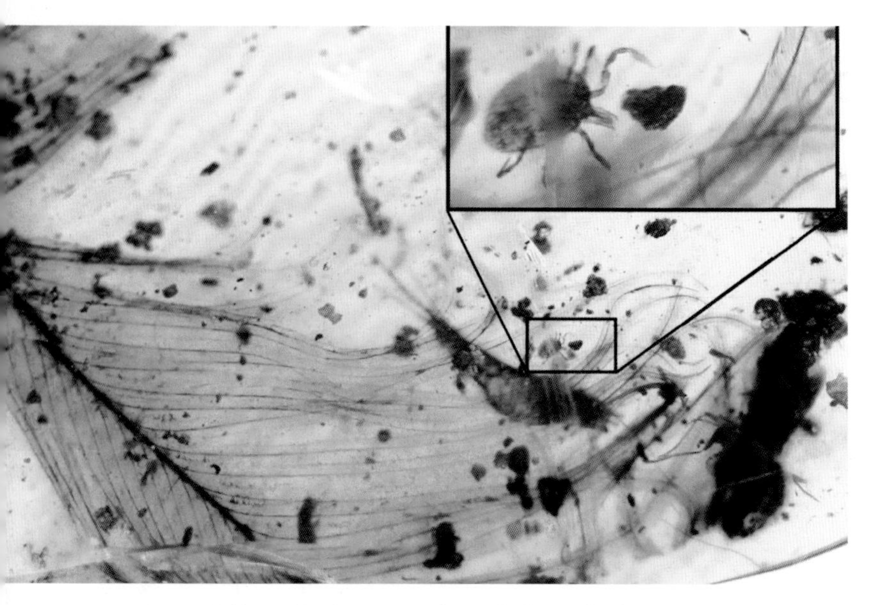

In 2017, a remarkable sample of Burmese amber studied in Madrid by Enrique Peñalver was found to contain a dinosaur feather and a tick engorged with blood from 99 million years ago. This has resonances in the story of 'Jurassic Park'.

Beneath our feet, Mesozoic rocks are bursting full of dinosaurs. This mass of bones, 500 feet (150 metres) long and 25 feet (8 metres) tall, was recently discovered in Yunyang, China. It is the largest 'fossil wall' yet discovered.

Kevin Ebi is an American wildlife photographer who captured this image of *Alligator mississippiensis* cruising in the Everglades. The plates on its tail are similar to those of a stegosaur, reminding us that the dinosaur legacy still survives.

Philip Lanoue, formerly a news and sports photographer, now specialises in photographing wildlife. He captured this menacing shot of an alligator in a pose that gives an appearance close to that of *T. rex* when seeking flesh to scavenge.

The first titanosaur from Africa was announced in 2018. *Mansourasaurus shahinæ* was excavated from the Egyptian desert and this vivid re-creation by Andrew McAfee of the Carnegie Museum shows it close to the water. It needs to be largely immersed.

The only attempt to mount dinosaur fossils copulating is at MUJA, the Jurassic Museum of Asturias, Spain. This pair of *T. rex* are on display – though the act would be more realistic if they were buoyant in water.

Copulation in dinosaurs was rarely discussed, though when Beverley Halstead, a British palæontologist, was interviewed for *Omni* in 1988, Ron Embleton was commissioned to illustrate the article. His vivid artwork conveyed acts that were entirely imaginary, not scientific.

funds started flowing in. With the momentum created by the push from Ostrom's conclusions, dinosaurs were suddenly re-emerging as sharp-eyed and calculating, intelligent, thoughtful and crafty. We have seen that today's Wikipedia actually publishes a warning against anyone imagining that dinosaurs might have rested in limpid pools, and the new palæontologists took this to extremes. Henceforth, dinosaurs were to be construed exclusively as dynamic and bright; all earlier views were dismissed as heretical. Says one book on the so-called dinosaur heresies: 'Dinosaurs didn't produce any swimming predators at all …' and goes on to assure the reader that *Tyrannosaurus*, *Allosaurus* and their 'ecological colleagues' definitely lived on dry land. A new terrestrial tyranny took over in palæontology, and no alternative view would be entertained. These new theories had to uphold the view that dinosaurs, in order to be so active, must have been warm-blooded. For every change in body temperature of 10°C, the rate of metabolic activity doubles. A reptile with a body temperature close to ours, 34°C (93°F), will reduce its metabolic activity by one half if its body temperature is reduced to 26°C (79°F); if it sinks to 17°C (63°F), its metabolic rate is reduced to a quarter of its initial value. On a cold day when its blood is at only 8°C (46°F), it will be one-eighth as active as it was originally. For the astonishing levels of activity that the reformist writers were now postulating, a dinosaur would need a high blood temperature all the time – and so, they reasoned, they must have been endothermic (i.e. created their own metabolic heat), just like us. This confers on the dinosaurs an extraordinary level of activity. Indeed, illustrations produced by Robert (Bob) Bakker – who studied under John Ostrom at Yale – show them leaping about the landscape like gazelles. Huge sauropods were shown standing on two legs, balancing balletically and looking out for any trouble; the *Megalosaurus* that Mantell had described was shown rearing up like a wrestler, towering at least 30 feet (10 metres) above its writhing prey, while two 'bull *Brontosaurus*' are shown dodging each other like boxers in the ring. The larger they were, the more often dinosaurs were being conceived as using their tails to supplement their hindlimbs as a tripod. They could rear up into the air, as tall as a tree, with their weight resting firmly on three huge supports.

Illustrations show a *Diplodocus*, coyly posing with legs akimbo, like a horse performing dressage, and they even envisage a stegosaur, a triceratops and a tyrannosaur standing poised on a single hindleg. We see formidable dinosaurs like *Triceratops* and even *Tyrannosaurus rex* portrayed as galloping across the prairie like super-sized racehorses. Bakker's account baldly claimed: '*Allosaurus* and *Ceratosaurus* were long-legged and nimble-footed ...' This was a new view of the prehistoric denizens – not sluggish reptiles, but intrepid and exciting conquerors of the world. All of this is fanciful and flies in the face of science.[35]

'Nimble-footed' is not quite the right word to describe a dinosaur weighing 10 times as much as an elephant – and here the new dinosaur writers held a card up their sleeves. They explained that the reason that dinosaurs were so agile is because they really weighed far less than everyone had previously claimed. Estimates were simply slashed; the *Diplodocus* originally estimated to have weighed more than 100 tons was conveniently reduced to less than 20 tons, purely to fit the otherwise unanswerable objections to the new approach. *Stegosaurus*, once a 20-ton giant, was expediently shrunk to 5 tons, and is currently little more than 2 tons, so its pivoting like an athlete became progressively easier to comprehend. As Bakker wrote in 1986: '*Stegosaurus* must have been a grand performer under attack, a five-ton ballet dancer with an armor-plated tutu ...' for which there is not a scrap of scientific evidence, though there is an abundance of objection. One intriguing dinosaur was the *Spinosaurus*, which had been described by the German palæntologist Ernst Freiherr Stromer von Reichenbach but which had been destroyed by Allied bombing in World War II. Bakker describes it vividly: 'A strutting *Spinosaurus* must have been a singular sight – striding on its long hind legs, its head twenty feet above the ground, turning broadside to dare its rival to test its potency.'[36]

Bakker's fanciful theories went further, and he even published an account of dinosaurs that danced. The fact that his extraordinary drawings of these athletic leviathans had no scientific evidence to support them and that there is not a single example of footprints left by dinosaurs performing such grotesque antics, were no discourage-

Chief proponent of the animated dinosaur is Robert Bakker, whose line-and-stipple cartoons vividly convey his ideas on their dynamism. Here he shows *Triceratops* and *T. rex* pirouetting on one foot, though they each weigh 10 tons. Bakker's drawings are imaginative cartoons rather than serious diagrams.

ment: he ploughed on regardless, and the eager palæontologists meekly followed.[37]

After the draft of this book was written, I chanced upon Bakker in the museum at Morrison, Colorado. We were introduced by forenames only, and no mention was made of this forthcoming book. If anybody is an archetypal dinosaur hunter it must be Bob Bakker. He has long hair and was sporting a grey straggling beard, a bush hat, and a denim jacket emblazoned with the stars and stripes. Were you to go to a costume party like this, as a palæontologist, you'd be in with a fair chance of winning top prize. Bakker is quick-thinking and bright, with steely grey eyes that don't miss a trick. I found him likeable but given to using his personable presence to dominate his colleagues. 'The family name isn't Bakker, with an "a"', he said, but Bokker, more like an 'oh'. It's Dutch.' I tried it, but he shook his head disdainfully. It was safer to call him Bob.

It was easy to see how his unchaperoned concept of a dynamic dinosaur would lead him into fanciful rhetoric about monsters that were impossibly balletic and senselessly hyperactive. With nobody to calm him down and restore reality, he would simply set off on flights of fancy. We discussed the reception his book had received. 'Ah yes, it got me into hot water when it first came out,' he mused, 'and that bothered me at first.' Then he added, 'And now it is taught everywhere as gospel – and that bothers me even more.' I said to everyone, just before we left: 'Today's visit is going to have repercussions. Dr. Bakker doesn't know what repercussions, but he's going to find out.' His response was typical: 'He can't pronounce Bakker quite right,' he said to the group, 'but he's close ...' He stood on the porch of the museum as I drove away, waving his cowboy hat and grinning broadly like the star of an old Western movie. Bob Bakker is likeable, knowledgeable, intelligent and ruminative; but – once he had seized upon his idea of terrestrial tyrants loping across the landscape and fighting with fury – his mind ran amok. He abandoned any claim to basing his conclusions on scientific evidence. Where dinosaurs are concerned, Bakker is a benevolent fantasist.

We can compare this attitude with the dinosaur sketch in Episode 31 of *Monty Python's Flying Circus*, when Miss Anne Elk (played by John Cleese) announced her theory on brontosaurs. She wasn't so keen to reveal any details, but sought to ensure that she was known as a great dinosaur innovator, emphasizing: 'This is my theory, it is mine and belongs to me, and I own it and what it is, too.' When pressed, she finally revealed her discovery: 'All brontosauruses are thin at one end, much, *much* thicker in the middle, and then thin again at the far end,' she said. The audience laughed loud and long at the absurdity of it all – though at least Miss Elk had more scientific facts on her side than the proponents of the dancing dinosaurs. The sketch later appeared on the album *Monty Python's Previous Record*, released in 1972,[38] and the interview itself is sometimes quoted in media-training literature to illustrate the behaviour of such interviewees.

Although the most exciting dinosaurs like *Spinosaurus* ate flesh, the majority of the dinosaur fossils that we find are plant-eating sauropods. The browsing herbivores outnumber the carnivorous theropod dinosaurs like *T. rex* by at least thirty to one. Clearly, the theropods used to feed on the sauropod species. The ratio is the same if we compare fossils of carnivorous to herbivorous mammals, and from this it has been deduced that – if the mammals were endothermic (i.e. warm-blooded) – then, obviously, so were the dinosaurs.

The 'Monty Python' TV show featured Graham Chapman interviewing John Cleese as Anne Elk about her new *Brontosaurus* theory. Between them is a print of a painting by Czech artist Zdeněk Burian which, in 1994, was used on a postage stamp.

One obvious and immediate objection to the notion of dinosaurs being warm-blooded and successful is that no reptile has ever evolved a method of maintaining a warm inner-body temperature. Another objection is that dinosaurs became extinct. If they had evolved so successfully, why would they disappear? The argument advanced was crisp and precise: a group of organisms with a high metabolic rate becomes much more vulnerable to a catastrophe. Creatures with a sluggish metabolism might struggle to survive, but the elegant and warm-blooded dynamic dinosaurs were destined to die.

This terrestrial tyranny revolutionized palæontology, and the analysis of footprints left by dinosaurs was now being used to fuel this new approach. Firstly, the trackways were said to show a habit of herding, as in the case of the duck-billed *Hadrosaurus*, and they were also used to substantiate the view that dinosaurs were sprinting along – size for size, faster than a human athlete. The stride of *Acrocanthosaurus atokensis* fossilized in mud was measured at 11 feet 6 inches (360 cm), while the average walking stride was a mere 5 feet 11 inches (1.81 metres). Results like this have been used to calculate that dinosaurs could have been running at 25 miles per hour (40 km/h). There was also a problem posed by the lack of young dinosaur fossils, particularly of juveniles. Plenty of fossilized eggs had been found over the years, but the only records of very young skeletons were in the Gobi Desert, at Djadochta and Iren Dabasu, until more were discovered in America, in the Two Medicine Formation in Teton County, Montana. In 1978, American fossil dealer Marion Nehring Brandvold discovered 15 hadrosaur skeletons in sedimentary strata of the Two Medicine Formation. Each was about a yard (roughly a metre) long. They were formally described by the expedition leader, Jack Horner, as being associated with a 'nest-like structure', and from this it was concluded that these duck-billed dinosaurs must have been social animals that lived in organized communities. More important was the discovery that there were several young dinosaurs in the area, and they were too mature to be hatchlings. The conclusion was clear: not only were dinosaurs active, intelligent and sociable, but they even cared for their young.[39]

More discoveries were soon made by Brandvold, who ran the TRex Agate Shop on the site of a former Catholic church beside Highway 89 in Bynum, Montana. She kept finding dinosaur eggs and had a long life as a fossil hunter; she died at the age of 102 in 2014. The area in which she made her finds was nicknamed 'Egg Mountain'. Eventually, more than 200 specimens were excavated, covering a range of ages. Then, in 2015, the skull of a mature dinosaur was retrieved by Laurie Trexler and it was given to Horner and a colleague, Robert Makela, who collects fossils in the summer, to describe formally. They named the new dinosaur *Maiasaura*, from the Greek μαία (*maia*, good mother) and the feminine form of the Latin *saurus* (*saura*, lizard) – hence, good mother reptile. Jack Horner and James Gorman published a book that described the finds, and the idea of social dinosaurs and the tender, caring dinosaur mother now became firmly established.[40]

Meanwhile, a team of Canadian and Chinese palæontologists had been excavating in Mongolia when they uncovered another group of young dinosaurs. These were identified by Philip Currie of the Royal Tyrrell Museum of Paleontology in Alberta as ankylosaurs, armoured dinosaurs of the genus *Pinacosaurus*. Five of the fossils, each the size of a sheep, were found together in a group, so the palæontologists concluded that this showed that they must have been living together as a coordinated community and had doubtless been cared for by their parents. The genus *Pinacosaurus*, from the Greek πίνακας (*pínakas*, panel), were vegetarians and had barrel-shaped, squat bodies covered with thick, reinforced plates. The end of the tail was equipped with a heavy, club-like weapon that was well suited for defence. An adult was 18 feet (5.5 metres) long and weighed at least a ton. The strata nearby made it obvious that the area at that time was covered extensively with shallow lakes, though the team still concluded that – when the young dinosaurs died – the area was dry land covered by sparse vegetation. They concluded that they may have been overwhelmed by rapidly shifting sand. In any event, the closeness of these skeletons gave Currie and his team the evidence they needed to promote the idea of gregarious dinosaurs that lived together in social groups.[41]

Subsequently, from Japan came findings of trackways that had been left by a dense group of 33 different *Toyamasauripus masuiae* dinosaurs. They were named from the Toyama Prefecture, where the fossils were found, and the name of Masui Hamuro, the wife of Toshikazu Hamuro, the palæontologist running the dig. No skeletal material was found in the area, but the fossilized footprints gave a clear indication of the dinosaurs that had once lived there. These were the most crowded set of dinosaur footprints ever found in Japan, and the scientists were certain that this proved that the dinosaurs must have been social animals. The conclusions were, once again, that dinosaurs were gregarious creatures. The notion of communities of dinosaurs was now flourishing as never before.[42]

Gigantic sauropods soon entered the social spectrum. In the Jurassic bone beds of Argentina, groups of dinosaurs of differing ages have been found fossilized together, and some trackways have been thought to show segregation according to age, with the younger dinosaurs gathering together separately from the adults. Microscopic studies of tooth wear have also been thought to show segregation according to age, so it was thought that the younger dinosaurs were congregating in places away from the adults, and herding separately to allow them to forage in peace. At least one survey has suggested that the evidence shows there was little parental care of the young, though that had little impact on the widely agreed support for the concept of friendly dinosaurs living harmoniously in a community.[43]

In the Cretaceous strata exposed in a quarry at Agrio del Medio in Patagonia, one team came across the remains of three sauropods that had died about 100 million years ago. One was an adult; the other two (found in a different part of the quarry) were smaller, and presumably younger. The skeletons had been broken up and disarticulated, as is usually the case with herbivorous dinosaurs, though there were few signs of transportation of the bodies – i.e. the skeletons seem to have been fossilized where they fell. From this several conclusions were reached: first, the three dinosaurs were social animals; second, they had all lived together; third, their deaths had been 'almost simultaneous'; and finally, the researchers claimed, the death of the adult had 'triggered the death of the two juvenile individuals.' These findings all go to support the idea of dinosaurs as active, thoughtful and caring creatures that lived in coordinated communities. A sceptic would be quick to point out that the existence of piles of bones says nothing whatever about the sequence of events that put them there. There is no scientific evidence that they must have lived together, and the proposal that they died at the same time is pure guesswork. Concluding that the death of one has caused the demise of the others is wild conjecture at best.[44]

This new era of the dinosaurs was proving to be so very different from the old. They had begun as mysterious monsters, had gone on to become the largest and most fearsome beasts, massive and lumbering, crashing across the landscape and occasionally collapsing to rest in weedy swamps, and had subsequently sunk – if not into oblivion – at least into the realm of idle curiosity. Dinosaurs had then lost their hold on serious scientists. Although children might have found them fascinating, the general era of excitement had evaporated. Yet now dinosaurs were back, and with a vengeance. They had re-emerged as spritely and sociable, thoughtful, warm-blooded and gregarious. They were clearly terrestrial, obviously successful, and manifestly important to our understanding of the world. To be a palæontologist was suddenly the height of fashion, and books on dinosaurs addressed extraordinary topics. The 'How do' series for children explored topics ranging from *How do dinosaurs say good night?* and *How do dinosaurs stay friends?* to *How do dinosaurs say I love you?* while *How do dinosaurs say Merry Christmas?* was rapidly mirrored by *How do dinosaurs say Happy Hanukah?* Paul D. Brinkman, Head of the History of Science Research Laboratory at the North Carolina Museum of Natural Sciences, launched a course called 'Dinomania! A Cultural & Scientific History of Dinosaurs' and tells me he is teaching it again in Autumn 2018,[45] while for 89 cents you can stream music about dinosaurs, ranging from 'Going on a Dino Hunt' to 'Liopleurodon Lament', and there are books including *Stegosaurus, the friendliest Dinosaur* and *The Dinosaur that Pooped a Planet*. The Disney movie about a friendly dinosaur, one that smiled affectionately and sometimes glanced coquettishly from side to side, suddenly seems mild by comparison.

The news media now love to feature bigger dinosaurs, and there have been many popular reports of massive 'titanosaurs' that would have been 120 feet (35 metres) long and weighed more than 100 tons. These are always promoted as the latest development and the most amazing discovery, but remember, this isn't so new. When Edward Cope discovered his vast sauropod dinosaur *Amphicœlias fragillimus*, he calculated that it measured 190 feet long (58 metres) and weighed up to 120 tons. That was back in 1877.

Dinosaurs that are similarly colossal continue to appear. *Mamenchisaurus* was discovered in 1952 during excavations for the Yitang Highway in Sichuan, China, and was analyzed by a Chinese palæontologist named C.C. Young. The genus literally means 'Mamenchi lizard' from the Chinese Pinyin 马 (*mǎ*, horse) and 门 (*mén*, gate), while chi derives from 溪 (*xī*, stream). The type specimen was a partial skeleton that Young decided to name *Mamenchisaurus constructus*. Its neck alone was found to measure some 50 feet (15 metres) long. Twenty years later another example was unearthed; it was given a different species name, *M. hochuanensis*. Its neck comprised 19 vertebræ and measured 30 feet (9.5 metres) long. After a further 20 years, a third species, *M. sinocanadorum*, was unearthed, consisting of a damaged skull and a selection of vertebræ. Some claimed this as a world-record dinosaur, which could have been 115 feet (35 metres) in length, with its neck alone measuring 56 feet (17 metres). A better skeleton was found in 2001, with a fine skull, and well-preserved shoulders and forelimbs along with part of the tail. Even so, the rest of the body has to be inferred by specialists, so we cannot be sure of the real size. From South America came news of a similarly vast dinosaur discovered in 1987 by Guillermo Heredia, a rancher in Argentina. At first, he assumed that the fossilized femur was a petrified tree trunk. It was located in the Huincul Formation of Neuquén Province, and when the area was explored, a single vertebra was excavated. It was said to be the size of a person. This was accepted as a type species *Argentinosaurus huinculensis* and was academically published in 1993 by two Argentine palæontologists. This gigantic dinosaur would have flourished within the late Cenomanian stage of the Cretaceous, about 95 million years ago.[46]

As with many other gigantic dinosaurs, not very much specimen material has been found. In Victorian England, as we have seen, collectors sometimes made good missing portions of a skeleton with painted plaster, as Thomas Hawkins used to do. Then it was regarded as indefensible; not any more. It is now common practice for most of the skeletons to be supplemented by detailed portions made of plastic that give the estimated impression of an entire skeleton, when only a few scattered bones have been found. Remember, very little of *Argentinosaurus* has ever been recovered. The bones that were eventually excavated amounted to about 10 vertebræ, some ribs, most of a femur and part of the right rear fibula. You might conclude that, from this alone, no reasonable estimate of size could be calculated. However, the single bones are said to tell the tale. Each vertebra measures 5 feet 2½ inches (1.59 metres) tall, and the fibula measures 5 feet 1 inch (1.55 metres) long. The femur measures 3 feet 11 inches (1.18 metres) around – at the thinnest part. How would the entire dinosaur have appeared in life? A graphical reconstruction was attempted by an American palæontological artist, Gregory S. Paul. He has produced large numbers of vivid images and is typical of the present-day artists who believe that dinosaurs were lively, dynamic, active, and so he is not likely to exaggerate the mass of a dinosaur. Paul concludes that

Argentinosaurus would have been 98–115 feet (30–35 metres) and weighed about 90–110 tons. This agrees reasonably well with the restoration that is on display at the Museo Carmen Funes in Plaza Huincul, Neuquén Province, Argentina. They have made it 130 feet (40 metres) from nose to tail and 24 feet (7.3 metres) tall. The first reconstruction of this monstrous skeleton to be put on public display is claimed by the Fernbank Museum in Atlanta. Their replica measures just over 123 feet (37.5 m) long and it loops across the vast central atrium of the museum, its body towering over visitors like a commemorative arch. I visited it with Rich Brown, a leading microscope specialist and a good friend, and we mused on its improbable energetics. Online there is a digital reconstruction of *Argentinosaurus* hobbling along, its neck and tail both held horizontal, though this is an absurd interpretation of the facts. The entire axial skeleton, from the end of the nose to the tip of the tail, is represented as being rigid like a steel girder. There is none of the flexure and counterpoise of a living creature, where sinuous movements of the neck would absorb vertical thrust forces and the tail would gently change its curve to neutralize the movements of the body. A child could create this simplistic picture, and it is ridiculously wrong.[47]

South America has continued to provide numerous astonishing discoveries in the last 20 years. In 2000, an enormous fossil dinosaur that would have measured 105–112 feet (32–24 metres) was unearthed at Barreales, some 55 miles (90 km) north of Neuquén in Argentina. It was discovered by Jorge Calvo, a palæontologist from the Universidad Nacional del Comahue in Neuquén, and was named *Futalognkosaurus*, not this time from the Greek, but from the local Mapudungun language: *futa* meaning giant and *lognko* signifying chief. Three similar fossils were found, and between them they provided about 75 per cent of an entire skeleton. This gave the most complete picture at that time of any of the titanosaurs.[48]

Calvo set up a base that tourists could visit, and recorded 10,000 visitors per year. Many were local executives, who sought a break from city life and were happy to join with the teams as volunteers digging for more fossils, while others came as sightseers from thousands of miles away just to drink in the sense of discovery. Although the visitors liked it all, other academics did not. The chief palæontologist at the nearby Carmen Funes Museum, Rodolfo Coria, insisted that the skeletons should be studied in a laboratory and disagreed strongly with the idea of making the excavations into a tourist attraction. Rubén D. Carolini, the director of the dinosaur museum in El Chocón, near Neuquén, was once said to have chained himself to the fossilized remains of a gigantosaur, of which he had discovered the first femur in 1993, campaigning for the return of South American dinosaur fossils to their place of discovery. Carolini was a character – before becoming a museum director he was originally a car mechanic and amateur fossil hunter who drove around the deserts in a buggy, sporting a bush hat like Indiana Jones. He knew that giants like *Argentinosaurus* and *Futalognkosaurus* had been making headlines worldwide as the biggest herbivores ever known, but he had already discovered the biggest of all the meat-eaters – *Gigantosaurus carolinii*. This monstrous carnivore measured 43 feet (13 metres) long and weighed around 15 tons, making it far larger than *T. rex*. Carolini had discovered this vast monster in 1993 and his find was of a skeleton that was more than 60 per cent complete. It was a truly impressive dinosaur, with a skull some 6½ feet (2 metres) long.

The discovery of the Argentine dinosaurs had begun back in 1882, and the early finds all gravitated towards the great museums in Buenos Aires and La Plata. It was Carolini, however, who brought the search into the modern era with his stupendous discoveries of the biggest dinosaurs, though he has been largely overlooked. There is no page for him in the current Wikipedia, and a web search currently reveals very little on the man.[49]

Another colossal dinosaur was found in 2013 by a farm worker named Aurelio Hernández on the Oscar Mayo Ranch in the desert near La Flecha, about 160 miles (250 km) west of Trelew in Patagonia. One of the ranchers subsequently called in to the Museo Paleontológico Egidio Feruglio in Trelew to ask whether this might be an important find. It took a team of technicians and volunteers 18 months to recover 225 bones representing the bodies of six separate titanosaurs. The most spectacular single find was a femur measuring 8 feet (2.4 metres) in length, and one of the expedition leaders, Diego Pol, lay down alongside to be photographed. This single bone dwarfed him. The museum authorities had moulds made, so that perfect casts of the bones could be manufactured, and the first complete replica was erected in the American Museum of Natural History in January 2016. It stands 122 feet (37.2 metres) long and cannot be accommodated even in the museum's largest rooms. Instead, its head and neck protrude through the entrance towards the lift shafts, welcoming visitors as they exit the elevators. So new was the discovery, and so hasty its reconstruction, that it had no name at the time it was unveiled in New York. Not until August 2017 was it formally identified – it was announced as *Patagotitan mayorum*, the name commemorating the place where it was found and the family who owned the ranch. The final appearance was deduced from the various skeletons, all of about the same size, and all apparently from the same species, though none was complete. As is usual, not one of them showed a complete skull, so the final version was created from the few traces that remained plus some creative imagination.[50] As we shall see, when the discovery was published, the facts were partly obscured by hyperbole.

Although most of the gigantic dinosaur skeletons that have been found are fragmented, the bones being widely scattered, a few that have recently been discovered are surprisingly intact. In 2005, palæontologist Ken Lacovara was excavating in Chubut Province, Patagonia, when he came across the remains of a sauropod. It was *Dreadnoughtus schrani*, and would have been at least 85 feet (26 metres) in length, weighing some 80 tons. The head, as is usual in these finds, was missing; but in this instance more than 70 per cent of the rest of the skeleton was present. This is highly unusual. The tallest complete sauropod skeleton in the world is the specimen of *Giraffatitan brancai* in the central hall of the Museum für Naturkunde in Berlin. This is all bone, without any plastic, and with nothing created out of imagination. But discoveries of entire skeletons like this remain rare. The great majority of sauropod fossils are known only from small portions of the skeleton, typically less than 10 per cent, and the rest is invented; so estimations of size and mass are problematic. Palæontologists usually have to be content with a few limbs, some ribs, a selection of vertebræ; they just invent the rest and construct it from plastic. How curious that, when Charles König went probing the skeletons assembled by Hawkins and showed some small parts had been re-created out of plaster, it created a scandal. Now it is considered perfectly proper to take a few bony fragments of a fossilized dinosaur and guess how to make all the rest from resin. How things change.

Perhaps the biggest of all dinosaurs have yet to be discovered. Gigantic footprints have been documented in Western Australia, for which no actual dinosaur has ever been found. Each of these footprints measures 5 feet 9 inches (1.75 metres) long. The site, dating from between 127 and 140 million years ago, has been described as 'Australia's own Jurassic Park, in a spectacular wilderness setting' by the team leader Steve Salisbury. A joint team of palæontologists from the School of Biological Sciences at the University of Queensland and the School of Earth and Environmental Sciences at James Cook University had worked along the coast of the Dampier Peninsula from 2011 to 2016, after the footprints were threatened by the $40 billion project to build a liquid gas processing plant. When this industrial scheme was announced in 2008, the indigenous people living in the area, the Goolarabooloo, turned to the palæontologists for advice on how their sacred sites could best be protected. Their chief, Phillip Roe, said simply: 'We needed the world to see what was at stake.' A variety of footprints was discovered during the five-year project, representing about 20 different dinosaur species including vast long-necked sauropods, bipedal ornithopods, and even some armoured dinosaurs. But the biggest footprints were made by a mystery monster that nobody has yet discovered. When one of the Goolarabooloo elders, Richard Hunter, lay down alongside one of the newly recognized footprints, it was almost as long as he was. The Australian government was impressed, both by the finds and by the solicitous attitude of the local communities. In 2013 the natural gas project was abandoned, and the area was granted National Heritage status. Salisbury, ever one to catch the attention of the headlines, supplemented his earlier comments by describing the site as the 'Cretaceous equivalent of the Serengeti National Park.' He could be right, if only we had the dinosaurs to go with the footprints.

Those endless revelations of massive dinosaurs that cared for each other, travelled together in social groups, laying eggs and tending their nests like brooding swans: it could all paint an idyllic image of reptilian cooperation. My own view was very different. The palæontologists, it appeared to me, were delusional and were simply seduced by the massive machismo of these lively leviathans. They now revelled in their public personae as extrovert characters, like Roy Chapman Andrews and Rubén D. Carolini (and our familiar image of Indiana Jones). Dinosaur science had become a cutting-edge area of research. No longer was palæontology a dry-as-dust fringe subject, for it had been reborn as a headline-grabbing topic. These were the scientists who dazzled everyone with the majesty of their research, and they alone could reveal the detailed stories of life a million years ago from a few fragments of rock. This was a new era, not just for dinosaurs, but for their mighty followers. In this new age (to coin a phrase) palæontology rocks.

7

HOW MICROBES MADE THE WORLD

When our Earth was young it had no atmospheric oxygen. Carbon dioxide abounded in the world's early history, when life first emerged 3 billion years ago. At that time, newly evolved primitive bacteria began to metabolize inorganic components in their habitat as an energy source. Cyanobacteria followed, tiny cells that lacked a nucleus, yet had developed the ability to capture solar energy – they were the first cells to photosynthesize. The process of photosynthesis is the green plant's way of capturing sunlight. Chemically, the solar energy is used to harvest carbon dioxide and water and combine them to produce carbohydrates. We can simply represent the synthesis of glucose from atmospheric CO_2 and H_2O:

$$6CO_2 + 6H_2O + E \rightarrow C_6H_{12}O_6 + 6O_2$$

The crucial component of this equation is *E*, which represents the input of solar energy. Science has still failed successfully to reproduce this ostensibly simple task, which is carried out every day by each small weed in your garden. Although science cannot mimic photosynthesis, this is the essential process that gave us life. We pride ourselves on the achievement of creating solar panels, though microbes have been relying on their own diminutive solar systems with greater efficiency than ours for billions of years.[1]

The cyanobacteria are the organisms that began to secrete oxygen into the air and laid down dead remains that (unlike the trees of today) did not decompose. Instead, they grew in large layered communities 2.5 billion years ago that have come down to us as fossils. These are stromatolites, which came to their greatest prominence more than a billion years ago. By the time larger grazing organisms had evolved, and during the Cambrian era, the stromatolites progressively declined. We see resonances of this around us. In today's world, we can observe layers of mucilage produced by microbial growths, which then accumulate small particles of debris and eventually form masses of limestone that contain high levels of stored carbon. Chemically, limestone is composed of a mixture of calcium carbonate $CaCO_3$ and calcium magnesium carbonate $CaMg(CO_3)_2$ and it is the vast amount of carbon trapped within these molecules originating in carbon dioxide that released the atoms of oxygen upon which we rely for our survival. Today we can observe growths of cyanobacteria forming thrombolites – these are today's equivalent of the stromatolites we know as fossils. It is these massive accumulations, added to the precipitation of carbonates in the shells of marine molluscs, that have stored away the carbon and released oxygen to the atmosphere. There are at least 50 million Gt (50 petatons) of carbon stored in limestone plus a further 15 million Gt in organic materials that we are now harnessing as shale gas. It is because of limestone locking away carbon that we have oxygen in the air.*

Although carbon dioxide is much in the news, it remains a rare gas in the air. If you ask people casually to name the third commonest atmospheric gas, hardly anybody has the correct answer. Most people say 'carbon dioxide' – but this is wrong. Nitrogen makes up 78 per cent of the air and 21 per cent is oxygen, leaving 1 per cent unaccounted for ... so what is it? In fact, the third commonest atmospheric gas is argon, of which most people know little. It was once used to fill lamp bulbs, and is now used to make lasers. Carbon dioxide is a much rarer gas and comprises only 0.04 per cent of the air – just 4 parts in 1,000. There are about 800 gigatons (Gt) of carbon in atmospheric CO_2 with about 600 (Gt) held in green plants, and an additional 1,500 Gt is stored in soil. There is far more CO_2 in the oceans: almost 40,000

* A gigaton (Gt) is a billion metric tons or 2200 billion pounds.

Gt, yet this is rarely considered. The oceans play the greatest role in moderating our climate and did so in the time of the dinosaurs.

For billions of years, it was microscopic organisms that laid down carbon in the form of carbonates in their shells, and this is what gave us those colossal beds of sedimentary rocks such as limestone. The chemistry is simple. Dissolved carbon dioxide in the oceans can be converted into carbonate (CO_3^{2-}) or bicarbonate (HCO_3^-), and CO_2 reacts with water to produce carbonic acid:

$$CO_2 + H_2O \leftrightarrow H_2CO_3$$

The reaction is reversible and in the oceans we find a chemical equilibrium between CO_2 and H_2CO_3. Ions of hydrogen and bicarbonate can be released:

$$HCO_3^- \leftrightarrow H^+ + HCO_3^{2-}$$

It is this reaction that serves to buffer seawater against random fluctuations in acidity. Many microbes and other minute marine organisms can fix bicarbonate biologically with calcium (Ca^{+2}) to create calcium carbonate, $CaCO_3$, which is produced in several different forms, including aragonite and calcite, and this process gives rise to their exoskeletons and microscopic shells that fall to the seabed, accumulate in strata, and eventually become sedimentary rocks. This form of limestone comprises 10 per cent of all sedimentary rocks, and our civilization is founded upon it. Almost everywhere you look there is limestone.

You may think that fossils are a specialized subject, but our society is founded on fossils. We burn fossil fuels (once it was only coal, now it is oil and gas) and have built our cities, hospitals, airports, schools … of fossils. Microscopic fossils gave us the limestone of which our buildings are made – our lofty cathedrals and churches are carved from limestone, and held together with lime mortar, itself made from limestone; it is limestone that is furnace-burned to produce cement and concrete; it underpins everything in our lives. Aggregate, used to create roads and massive embankments, is made from limestone;

acidic farmland is neutralized and made productive by the application of billions of tons of pulverized limestone, and the fumes from power stations are treated with limestone dust to make it safe. You find powdered limestone in paper and paint, plastic and toothpaste; it is added to flour and medicines, cereals and cosmetics, it boosts animal feed and is a crucial component in the manufacture of iron and steel. Chances are the room in which you may be reading these words is made with concrete (or stands on a bed of it), and the source of all these present-day products was all laid down, atom by atom, by microscopic creatures nobody ever sees. The microbes that made limestone created our environment as well as giving us the air to breathe.

It was microbial metabolic processes that locked away atmospheric CO_2 in a form we are now exploiting. About 90 per cent of the CO_2 in the oceans is in the form of bicarbonate HCO_3^- and the other roughly 10 per cent is as the carbonate ion CO_3^-. Only about 1 per cent is dissolved CO_2 gas. When the temperature of the sea increases, it encourages the replacement of the carbonate by the bicarbonate ion, and this makes the calcium carbonate dissolve more readily. Marine organisms find it harder to secrete their shells, and existing carbonate deposits become increasingly soluble. Since the solubility of gases is inversely proportional to temperature, less CO_2 is absorbed by the oceans as they warm, which further increases the problem. Today's world is getting warmer, and that includes the sea. The oceans are also becoming more acidic, and it is microbes that notice it first. If the oceans had been acidic in the time of the dinosaurs, there would be few fossils to find; dinosaur bones would more likely have dissolved away instead of being petrified. In today's world, limestone is dissolving. This is where sinkholes originate, and why the carvings on cathedrals are dissolving away. Rain has always been acid; but not *that* acid.

Oxygen remained at low levels in the air as geological ages unfurled; indeed, around 2 billion years ago those low levels may have fallen close to zero. That all changed when more complex plants emerged at the dawn of the Cambrian period, 540 million years ago. Photosynthesizing land plants poured out excess oxygen that accumulated in the atmosphere and soon rose to comprise 15 per cent of the air. Volcanoes were active throughout this time, spewing out huge amounts of carbon dioxide into the air. Thus was born the Carboniferous period, which as we have seen was named by William Conybeare and William Phillips in 1822. Suddenly there were growths of plants everywhere in the moist and soggy environment, taking advantage of the atmospheric carbon dioxide that they needed for photosynthesis to proceed at high rates, thriving in the water-rich low-lying landscape and basking in the optimized temperatures caused by the CO_2. For the first time in Earth's history, plants in the Carboniferous period could grow to gigantic proportions. None of them were flowering plants, which had yet to evolve, and many were of types that are long extinct, including *Cordaites*, a primitive conifer, and the gigantic tree fern *Psaronius*. The moist warmth and the high levels of carbon dioxide meant that plants then were larger and more luxuriant than any of those we see around us today. *Calamites* was a huge spreading horsetail (their descendants today are *Equisetum* plants that rarely grow more than 1 foot (30 cm) tall, while *Lepidophloios* towered high above the swamps reaching 130 feet (40 metres), whereas their present-day descendants, the club mosses such as *Lycopodium*, grow to a height of only 6 inches (15 cm). There were many species of scaly, cone-producing trees like *Lepidophloios* and *Paralycopodites*, and there are similar survivors today. The monkey puzzle tree *Araucaria araucana* is one example, and the Norfolk Island pine *A. heterophylla* is another. This tree survived only on Norfolk Island, a tiny speck of land between Australia and New Zealand, though it is now widely planted in parks and gardens around the world. Several other examples of the trees from the Carboniferous period have survived in tiny corners of the world; one is *Ginkgo biloba*, of which we have 270-million-year-old fossils and which survived only in a tiny region of Central China where it was grown ceremoniously in temples. The most extraordinary survivor from the Carboniferous must surely be *Wollemia nobilis*, the Wollemi pine. This plant was well known from fossils dating back 200 million years and had always been believed to be extinct, yet suddenly it was discovered in 1994, alive and well, luxuriating in a sandstone valley only some 90 miles (about 150 km) from Sydney, Australia. We will see how these are plants that could be grown today to produce a forest very like that in which the dinosaurs lived.

This unprecedented explosion of plant life poured out oxygen into the air. So rapid was the growth of plants everywhere that there was no time for them to rot away. They accumulated in deep layers, slowly degenerating under the action of fungi and bacteria, but never decomposing completely. Their carbon content remained behind, and slowly the accumulations built up to form the coal measures that we find today. This is what gave us the oxygen – the carbohydrates in the dead trees would have broken down to form carbon dioxide and water. Let us exemplify this simply by modelling the process with a simple carbohydrate, rather than the immensely complex molecules found in plants. Carbohydrates are composed of molecules of $C_6H_{10}O_5$ that are joined together in a polymer chain that contains thousands of the separate molecules. But, if we take just one molecule, we can instantly see what happens when it degrades in an atmosphere where there is too little oxygen. The monomer of carbohydrate will oxidize to produce plenty of water, a small amount of carbon dioxide, and a residue to elemental carbon. That is where your coal comes from. The great majority of the oxygen remains in the atmosphere, while the carbon remains locked up inside the coal seams.

$$C_6H_{10}O_5 + O_2 \rightarrow CO_2 + 5H_2O + 5C$$

The early Earth was dramatically changing; the air was becoming enriched with vital oxygen, and the land was being buried under massive growths of plants. This was the perfect environment for the evolution of animal life. By the end of the Carboniferous, some 300 million years ago, the levels of oxygen were at an all-time high, reaching some 35 per cent of the atmosphere. This would kill us. We could not breathe so rich an atmosphere without suffering damage to the lungs and eyes. In 1976, I concluded that this explains why there were vast insects, including a dragonfly *Meganeura* with a wingspan of more than 2 feet (65 cm), and giant amphibians like *Sclerocephalus* which was almost 6 feet (175 cm) long. Insects obtain oxygen through fine tubes that carry air through spiracles, tiny rounded perforations in the exoskeleton; amphibians take in oxygen from the water in which they live through feathery gills or their wet skin. Neither is efficient, and the additional oxygen would have encouraged the evolution of the mightiest of their race.[2]

Everywhere we travel there are the remains of the great eras of the past. Coal measures, the remains of those vast forests, are scattered around the globe. Where microscopic foraminifera once proliferated in incalculable numbers we see their remains in gigantic chalk cliffs. Similarly, when diatoms flourished and constructed astonishing glass shells for their cells, they deposited their remains in the towering diatomite deposits in California. The Blue Lias beds, which gave rise to so many of the early dinosaur discoveries, are exposed at the surface for miles along the coast of Wales, from the south coast of Sussex to the north coast of Yorkshire, and right across to York – then, as we have seen, they occur again over much of Europe. The deep bone beds of the United States, from which the most remarkable discoveries emerged, lie in the Morrison Formation, the bed of a vast fossilized sea extending from Alberta and Saskatchewan in Canada right down to Arizona and New Mexico in the USA. The sandstone strata at the Dampier Peninsula, Australia, are just one part of that primeval kingdom. Extensive shallow seas covered the whole of Central Australia in the time of the dinosaurs. Japan is now giving up more fossils, while more than 400 different genera of dinosaurs have already been recorded across Asia. And remember, that range of different dinosaurs has been excavated even down to Antarctica.

Because palæontologists deal with the fossilized remains of long-dead creatures, they have to re-create reality from their own interpretations. How much flesh goes on the body, where the appendages are deployed, whether the feet were webbed, how fast they might have moved; all these are conjecture based on ancient remains. As a result, this branch of science has uniquely tended to bend the facts to suit the desired result. When the gigantic sauropods were first being described, there was a wish to exaggerate their size for the sake of dramatic effect. The artists' impressions showed them to be enormously rotund, hugely muscled, towering high and threatening. Their weight was repeatedly overstated. Remember that when Cope announced his discovery of *Amphicœlias* in 1877, it was claimed to have been almost 200 feet (60 metres) long and to have weighed 120 tons. Few of the largest dinosaurs discovered in recent decades would be so prodigious. Cope's aim was to present the dinosaurs in the most amazing fashion he could – so increasing their vital statistics would only help his campaign, even though he had very little upon which to base any calculations. When he published his description in 1878, he was relying on the find of a single broken vertebra: just one partial bone. Cope stated that it measured 5 feet tall (1.5 metres), and no vertebra of that size has ever been seen, before or since. Even the vertebræ of the colossal *Argentinosaurus* are not so large. The bone that Cope described has never been seen again, and 40 years after his formal description, a detailed survey of all Cope's sauropod fossils at the American Museum of Natural History could not find any trace of it. All we have is Cope's measured drawing, and – even though his diagrams were known to be meticulously accurate – it may be that the measurements were overstated. Woodruff and Foster, two American investigators, have recently written a report. They say: 'By deciphering the ontogenetic change of Diplodocoidea vertebræ, the science of gigantism, and Cope's own mannerisms, we conclude that the reported size of *A. fragillimus* is most likely an extreme over-estimation.'[3]

In Cope's time, everybody wanted to make dinosaurs as massive as they could. When the new era of territorial tyranny came to rule the world of palæontology, that could no longer apply. Then the need was to remind everyone that dinosaurs lived on land, and thus they cannot have been too heavy. And so palæontologists started systematically to scale down their estimates of weight.[4]

If we go back to the 1995 edition of *Encyclopædia Britannica*, brachiosaurs were stated to weigh at least 80 tons; by 2009 this had been slashed to a mere 25 tons. Similarly, the weight of a *Diplodocus* was variously quoted, sometimes around 100 tons, but even if we go back no further than that 1995 encyclopædia it is given as up to 80 tons. The latest estimates (to make it look more acceptable as a terrestrial creature) state that the weight was closer to 10 tons. These leviathans have had to slim down to match the fashions of the age. Scientific evidence plays no part in these imaginings.

The more palæontologists insisted on dinosaurs as terrestrial creatures, the lighter their dinosaurs have become. Although I believe that this was done to reinforce the macho notion of a dynamic dinosaur pounding across the pampas, palæontologists would not agree. They find other justifications, the best of which still makes me smile. Jan Peczkis, who works at the Department of Earth Science at Northeastern Illinois University, studied the slow shrinking in estimated weight and eventually concluded that it was due to current collectors preferring the smaller specimens. 'This suggests that early researchers tended to collect giant dinosaurs,' he concludes. Priceless.[5]

That would only work if we imagine surveyors busily exploring bone beds and discarding anything that was considered too large. In fact, it has been the resurgence in funding and public interest that has caused the current craze for dinosaur studies. Since the mid-1990s, after Peczkis had published his paper, the greatest dinosaur skeletons ever discovered keep coming to light. We now have far more gigantic specimens than ever before. Even if those lighter weights were true, persistent problems still remain: the meat-eating theropods were bipedal. What would be the evolutionary imperative that would impel them to lose their massive forelimbs and move only on their hindlegs? The red kangaroo *Macropus rufus* has done that, it is true; but they did so to increase their running speed. In Australia I have watched a kangaroo running at full pelt, when it can leap 30 feet (9 metres) in a single bound while travelling at 45 mph (70 km/h) – an astonishing speed for an animal weighing 200 pounds (90 kg). Remember that, in 1925, the Pathé Review film *Monsters of the Past* (p. 205) showed a bipedal tyrannosaur leaping about on the hindlegs just like a kangaroo. But dinosaurs could not possibly have emulated that. Even with the lower weights recently claimed, that dinosaur is still 100 times heavier than the mightiest of all kangaroos. When we turn to the sauropod dinosaurs, the evolutionary pressure towards a massive bulk becomes insurmountably problematic for those who espouse them as terrestrial creatures. They became progressively larger over a period extending from 230 million years ago until they had become the dominant vertebrates 200 million years ago. The gigantic *Argentinosaurus* emerged around 95 million years ago, and all the dinosaurs were finally extinct 66 million years ago. These largest species represented the extreme of dinosaur development and must have weighed at least 100 tons. Palæontologists who are still determined to claim that the likely weight of lesser dinosaurs was not so great (as they did for *Diplodocus*, which shrank from 100 tons to 10 tons) are still trying to apply the same principle to the unmistakable giants, the titanosaurs, which must have weighed at least 100 tons and perhaps more. The most recent claim for the 'biggest animal ever to walk on Earth' was announced to an astonished world's press in August 2017. It was published in a paper for the Royal Society by José L. Carballido and his team from the Egidio Feruglio Paleontology Museum in Argentina and was reported in news outlets around the world.[6]

'Meet What May Have Been the Largest Land Animal Ever, "Patagotitan mayorum" dino may have been as long as 7 elephants,' exclaimed the Newser website.[7] Echoed *National Geographic*: 'New Dinosaur Species Was Largest Animal Ever to Walk the Earth.'[8] For the physics.org website, Seth Borenstein wrote it up with this eye-catching headline: 'Patagotitan mayorum: New study describes the biggest dinosaur ever,'[9] while the *USA Online Journal* proclaimed: 'Meet Patagotitan, the Biggest Dinosaur Ever Found.'[10]

A report for Independent Television News (ITN) in London began with wording that was typical of the other reports: 'Scientists have found the fossilized bones of what they believe is the biggest ever dinosaur. The creature could have been as long as 35 metres from head to tail – more than the length of three buses. Named Patagotitan mayorum, it is thought to have weighed the same as a Challenger 2 tank – around 62 tonnes.'[11]

Within one week there were 274,000 sites online that mentioned the 'new' dinosaur. People were talking about this exciting new discovery in the pub. What they weren't aware of was the full story. This wasn't a new discovery after all. This was the skeleton originally discovered on La Flecha Farm in the Chubut Province out in the Argentinian desert by an old shepherd named Aureliano Hernandez,

One of the largest dinosaurs, *Titanosaurus*, featured on a BBC programme with David Attenborough, and the *Radio Times* compiled this artwork to illustrate its size. An elephant is roughly as large as a land animal can sensibly be.

who had found the vast femur in the rocky ground back in 2013. He mentioned it to the Mayo family, the farm owners, and then Hernandez died of old age before he knew how important his find would prove to be. More than 90 per cent of the news stories omit to mention his name, but he is the person who made the discovery and is the key figure in it all. When excavations began, it took just two weeks for the team to unearth the rest of the *Patagotitan* skeleton, and it proved to be remarkably complete. In spite of those lurid headlines, it was not a new discovery; it already dated back 4 years. Even more surprising, it had already been put on public display. The original skeleton was erected in Argentina, and this was the monster that my distinguished friend Sir David Attenborough had discussed in detail in a BBC television programme made in 2016.[12]

The dinosaur had already been seen by many millions of people, because it had been on public exhibition in the American Museum of Natural History for 20 months before the news announcement. Theirs is the 123-foot (37.5 metre) long plastic cast (the original must remain in Argentina). Once again, palæontology had conspired to create an image which the facts did not substantiate. The newly published report claimed that the dinosaur was slightly shorter, at 115 feet (35 metres) long. They claim that it could have weighed 76 tons, though this may be an attempt to make it look more feasible as a terrestrial creature. Like the other titanosaurs, it would probably have weighed more than 100 tons. An artist's impression was created in Argentina, which shows the dinosaur holding its neck aloft, towering above the spectators. This is also wrong. The caudal vertebrae are not constructed in a way that allows dinosaurs to reach up in this manner, and even if that were not the case, then (as Roger Seymour of the Department of Ecology and Evolutionary Biology at the University of Adelaide has shown) a giant dinosaur would consume half its metabolic energy simply holding up its head.[13]

This dinosaur was announced by the Royal Society as being 'the largest species described so far'. One would assume that the referees would ensure that statements are properly justified, but the facts suggest something very different to me. What is the scientific evidence for it being the biggest? The femur may be huge, but some other long-bone dimensions are smaller than they are in other giant genera. We need scientific evidence, so let us look at published details for the length of the humerus (the upper forelimb bone) of other sauropod dinosaurs. The longest of all is *Giraffatitan* at 84 inches (213 cm), followed by *Brachiosaurus* at 80 inches (203 cm) and *Turiasaurus* measuring 70½ inches (179 cm) long. The humerus of *Patagotitan* measures a mere 66 inches (167.5 cm), which is 1 foot 2 inches (35.5 cm) shorter than the same bone of *Giraffatitan*. The widely reported *Dreadnoughtus* had a humerus measuring just 63 inches (160 cm) long. On that basis, *Patagotitan* is well down the list.

Now, the humerus is just one of the long-bones, and it could be that a dinosaur had conspicuously long forelimbs, so the length of the humerus is not sufficient to draw a final conclusion. A better comparison might be the circumference of the slimmest part of the femur; this must take the weight of the dinosaur – so, the heavier the creature, the more substantial the bone. José Luis Carbadillo and his team at the Egidio Feruglio Museum report that in *Patagotitan*, the circumference of the femur is 40 inches (101 cm).[14] The femur of an incomplete *Argentinosaurus* was measured at 46½ inches (118 cm), which suggests that this dinosaur could have been considerably more massive than *Patagotitan*.[15] Proponents of *Patagotitan* will point out that this particular femur was not complete, and so a diameter cited by Gerardo Mazzetta and his team at London's Natural History Museum might only be an approximation.[16] However, there is a smaller – but complete – *Argentinosaurus* femur on record which has a confirmed circumference of 44 inches (111.4 cm), so even this lesser specimen is clearly from a heavier dinosaur than *Patagotitan*.

Wait – there is another indicator: the diameter of the core of each vertebra. Since it is the backbone that distributes the mass of these creatures, we can assume that the dimensions of each vertebra will correspond to the size of the creature from which it came. The published papers give a diameter of 23½ inches (60 cm) for gigantic sauropods like *Puertasaurus* and *Argentinosaurus*, whereas *Patagotitan* vertebral cores measure a maximum of 23 inches (59 cm). So the load-bearing bones of *Patagotitan* are slightly smaller than these of some other dinosaurs. We always need to base our conclusions on scientific evidence. In this case, both the vertebrae and the femora measure less than its rivals, and so does the humerus. Yes, it was a massive beast; but no, it was not the largest dinosaur ever discovered. The evidence for that exaggerated claim does not exist.

So, what was actually being announced? Read the title of their article again, and you will see that the Argentine team were really writing about ways of calculating the weight of dinosaurs. No new dinosaur had been discovered, in spite of what the headlines proclaimed. Their summary emphasized the one aspect that they could not answer but which this book has set out to do: in their words, 'there are still many unknown aspects about their evolution, especially for the most gigantic forms.' This major source of perplexity has dogged the heels of palæontologists for centuries, and the single item of news in their paper was the choice of name: they had decided to call it *Patagotitan*. The team explained that this was derived from Patagonia (the region of Argentina where it was found) and also from Titan, which they said was the Greek for large. That's incorrect; 'titan' does not mean large, and never did. The mythological titans (Τιτάν) were the gods of Ancient Greece. The word is often used to connote something of god-like magnificence, and 'titanosaur' is one of the names used in dinosaur taxonomy, so that 'Patagotitan' works well enough – but, if you are going to describe the etymology of a term, it is better to have your facts right. So all the claims for newness and novelty, for unprecedented size and all the other details that featured in the news, even the way they chose their name, were either exaggerated or erroneous. Once again, the lure of exaggerated palæontology had seduced editors from the Royal Society in England right across to *National Geographic* in America.

What can we conclude? The largest dinosaurs all attained a similar size and evolved to be as large as a creature can and still remain intact. They must have reached about 120 feet (37 metres) in length, weighing at least 100 tons. Some were perhaps bulkier, others may have been slightly more slender, but longer; and these are clearly approaching the largest dimensions that an animal could attain.

Edward Cope's underwater dinosaurs, breathing through a snorkel, made an appearance in Zdeněk Burian's painting of submerged brachiosaurs. Kenneth Kermack was the person who spelled out that water pressure would make it impossible to breathe. A point he missed is that they would float to the surface.

Fifty years ago, in 1968, I was walking through the dinosaur displays in the National Museum of Wales in Cardiff with the children, when suddenly we came across a glass-fronted display case. It was over a yard (about a metre) wide and contained a vivid diorama display of diminutive model dinosaurs. They were featured in a desert plain which faded into a blue-grey haze to give the impression of distance and with lilac-tinted mountains painted as the background. I showed the family. We crouched in closer to take in the view at eye level. Then I could see so clearly how the bodies of the dinosaurs were too bulky to be supported on their stubby legs – and, even if they could be supported, there was no notion of their being dynamic and lively. The painted blue haze seemed to supply a clue, and in a dramatic revelation I could suddenly see the answer to all these perplexing problems. If the dinosaurs had evolved in water, every paradox could be resolved. It took my breath away. I stood up and thought – if they had always lived in an aquatic habitat, then so many of the outstanding problems would surely be solved. I said to the children: 'See how the lower part of this scene is painted blue? Now, imagine that as water. If the bodies of the dinosaurs were partly submerged in the water, it would be easy to see how they could have moved about.' Even they understood. It made perfect sense. I left the museum that day with my mind buzzing with ideas.

I mused on this further from time to time, and whenever I took visitors around the museum we'd pass through those galleries, and once in a while I would point out how that scene made far more sense if we imagined the landscape as a lake, rather than a desert. Occasionally someone would say: 'That's really interesting; you should do some work on your ideas and publish it as a new theory,' but I always dismissed the idea. This wasn't my field. Somebody more central to the science was bound to reach the same conclusion. My research work was with living cells, not gigantic dinosaurs, and the last thing I wanted was to be identified with a new dinosaur theory. Could you imagine, turning up at a conference to give a speech about microscopic cells, only to be greeted by people saying: 'Here he is! The man with the dinosaur theory!' Dinosaurs make far bigger headlines than cells – it was the last thing you'd want if you were an outsider to the field. However, I did resolve to keep an eye on the published scientific papers on dinosaurs and I began to watch the research that was being published. When some palæontologist eventually decided to spell out that dinosaurs must have evolved in water it would surely make a good read. I wanted to see the approach of that day.

We have seen that some of the nineteenth-century palæontologists proposed that the biggest dinosaurs might have retreated to swamps to rest their limbs, but that's not the interpretation I have in mind. Dinosaurs did not merely resort to pools of water but owed their existence to their habitat. They were not languishing in a bog but must surely have evolved only because they lived in shallow seas and lakes. The water was not simply there to take their weight – it alone provided the habitat that was responsible for the development of massive monsters out of smaller, terrestrial reptiles. Water is 800 times as dense as air, and would provide the buoyancy that heavy animals need, and it is 60 times as viscous as air, providing the thrust from the tail and permitting slow and stately progression.[17]

One of the greatest dilemmas in palæontology is the cause of gigantism; the motive that impelled dinosaurs to keep evolving larger and larger. None of the explanations was satisfactory. I soon concluded that this aquatic habitat provided the imperative driving dinosaurs towards gigantism and this environment had always defined how they lived their lives. It must also have been because of their evolution in shallow water that the originally four-legged carnivorous tyrannosaurs lost the use of their forelimbs. Their buoyant bodies were propelled along by their muscular hindlegs reaching the bottom, for which hindlegs alone sufficed. Evolutionary pressure always leads to the abandonment of a superfluous feature, as we see in the loss of eyes in the blind salamanders found in deep, dark caves, and the dwarfing of the pelvic girdle in cetaceans. The environment in which the salamanders evolved meant that sight was no longer important, and so they lost their eyes. In precisely the same way, the tyrannosaurs evolving in water meant that four limbs were no longer necessary to carry their massive bulk, and so their arms dwindled away to vestigial proportions. Their fossilized forearms are as tiny as twigs. There is no other explanation, anywhere in the world, which fits this scientific evidence. It must similarly have been their aquatic habitat that led to the development of all the features that we see in those gigantic dinosaurs and – above all – to their assuming enormous dimensions, as big as a whale. In the quest to find an evolutionary reason for their prodigious size, theories for the gigantism of dinosaurs have been a popular topic for debate among palæontologists in recent decades. They show that larger terrestrial animals tend to develop proportionally in larger areas of land and remind us of 'Cope's Rule', which states that any group of organisms tends to evolve to a larger body size with the passage of time. This dubious concept is named after Edward Drinker Cope, though Cope never stated it himself in any of his writings. David W.E. Hone of the Bayerische Staatssammlung für Paläontologie und Geologie in Munich and Michael J. Benton of the Department of Earth Sciences at the University of Bristol put together the mathematical relationships used to predict animal size and concluded that the dimensions of the gigantic dinosaurs are far greater than they should be, according to the standard mathematical models. Hone and Benton listed the perceived advantages in becoming larger:

- Guarding against predation
- Increase in predation success
- Greater range of acceptable foods
- Increased success in mating
- Greater success in intraspecific competition
- Increased success in interspecific competition
- Extended longevity
- Increased intelligence (with greater brain size)
- At very large size, the potential for thermal inertia
- Survival through lean times and resistance to climatic variation and extremes

This is an impressive list of advantages, though 'guarding against predation' as an imperative driving the evolution of gigantic herbivores would be counterbalanced by 'increase in predation success' on the part of the therapods that devoured them.

- Increased vulnerability to predation
- Increased development time (both pre- and postnatal)
- Increased demand for resources
- Increased extinction risk because of:
    - Longer generation time gives a slower rate of evolution, reducing the ability to adapt
    - Lower abundance (i.e. small genetic pool, also reduces ability to adapt)
    - Lower fecundity through reduced number of offspring[18]

Some of these are questionable. The benefits of increased intelligence might not apply to sauropods, if they had extremely small skulls and very tiny brains, and, in any event, brain size does not equate with intelligence. If it did, then an elephant would be vastly more intelligent than a mouse. The idea of increased success in mating seems absurd – indeed, as we shall see, it is the physical inability to copulate that is one of the greatest problems faced by a science that insists on dinosaurs being terrestrial. You could also doubt whether 'increased vulnerability to predation' was a real disadvantage. An elephant, rhinoceros or hippopotamus – like a whale – will be infrequently killed by natural predators, solely because of their massive size. Perversely, the authors claim that 'the hindlimbs were considerably longer in sauropods than the forelimbs,' whereas the converse is often the case. In the prosauropods, which came first, the hindlimbs were longer – but the gigantic sauropods which came to rule featured forelimbs that were longer than their hind legs. It is the curiously longer front legs that gave sauropods their characteristic profile, generally like a giraffe, but bigger, much bigger (a gigantic dinosaur weighed more than 50 giraffes). One gigantism project kept 15 top German palæontologists busy for years, analyzing the mechanisms and reporting on the various theories. They admit at the outset that there is an 'innate human interest' in finding the largest of any type of specimen (try telling that to Peczkis, who imagined that only the pioneers troubled themselves to find the largest specimens). Those palæontologists account for the phenomenon of gigantism by concluding that there were several 'evolutionary innovations' that triggered a 'remarkable evolutionary cascade'. The most important of these 'innovations' was the very long neck, their most conspicuous feature. This, they say, allowed the sauropods to reach vegetation from all around them, and gave them a great advantage over other herbivores. The long neck reaching upwards worked only because there was a tiny head, which weighed little, and because the bones of the neck were filled with air spaces, like the bones of a bird. This is unlikely to be sensible, because the neck bones of dinosaurs are constructed in a way that precludes their reaching upwards. The German team also makes the obvious claim (though with no scientific evidence in support) that sauropods produced numerous small offspring. And that's it.[19]

There are few real reasons in any of these studies to account for the gigantism of dinosaurs. These explanations, with the factual mistakes and internal contradictions, don't amount to a viable theory. If those considerations applied, then (even though the dinosaurs became extinct) other massive, long-necked animals would since have evolved. What palæontology needs is a reasonable theory that explains why, in the era of the dinosaurs, they had reptiles on land that were as big as today's whales are in the sea. There is only one possible explanation – dinosaurs evolved in an aquatic habitat that conferred buoyancy on their bodies. They must have evolved in a watery world.

So, let us now think back to other known giants of the animal world and see how their figures compare. Remember, the heaviest creature is the blue whale, *Balænoptera musculus*, which may weigh as much as 150 tons and measures up to 90 feet (about 30 metres) in length. This is an extreme, of course; it is the greatest animal that we have known to exist. There are reasons for suggesting that this has evolved to the limits that a physical body can sustain and it has been able to do so purely because it is always weightless in its marine habitat. Dinosaurs, even though they evolved in an aquatic habitat, must have been able to drag themselves onto land to lay eggs. It cannot have been easy, but this exacting process is equally difficult for other creatures, like turtles, that haul themselves from their accustomed habitat in order to procreate.

Other whales are less massive, like the bowhead whale, *Balæna mysticetus*, weighing 130 tons, and the fin whale, *Balænoptera physalus*, that weighs about 80 tons. On land, animals must be lighter to walk. The largest land animal is the African elephant *Loxodonta africana*, 13 feet (4 metres) tall at the shoulder and weighing about 7 tons. The biggest elephant ever to have existed, was *Palæoloxodon namadicus* from 24,000 years ago which was 14 feet (4.4 metres) tall and weighed 10 tons. Bones excavated in Kent date back 400,000 years and were found close to stone tools made by *Homo heidelbergensis*, so it seems that early humans used them for food. The largest land mammal ever was probably *Paraceratherium* from 20 million years ago, once thought to weigh 30 tons but known through only a few bones and now believed to have an average weight nearer 11 tons. So the most likely weight of the largest successful terrestrial mammal is close to 10 tons, whereas the greatest weight of a successful marine creature is ten times larger, at about 120 tons. Sauropod dinosaurs weighed as much as a whale.

All this evidence drew me towards my own, very different, view. The giant dinosaurs had always seemed to me ill-equipped for a life on land. Their colossal size, their surprisingly shallow footprints, the occurrence of their skeletons only in the sedimentary rocks that formed at the bottom of shallow water … this is what had convinced me that they could only have evolved in an environment of shallow water. Since they had first emerged on the scientific scene, dinosaurs had been portrayed as terrestrial animals. Their vast bulk had made a few earlier investigators speculate that they must, surely, have sometimes rested in swamps and, as we have seen, there are images that show them in this position. The most extreme example was that idea that the giant sauropods actually walked about on the bottom of a lake, using their long necks like a snorkel. But we are agreed that this was a non-starter – the pressure of water on the thorax would have prevented an animal in this position from inhaling. In the drawing by Charles Knight, and in the painting by Zdeněk Burian, the dinosaurs' lungs are shown to be at least 20 feet (about 7 metres) below the surface, which would have imposed an additional pressure of 20 pounds per square inch (1,000 Torr). Over the entire thorax, this would amount to several tons pressing the chest wall inwards. In spite of this very obvious objection, at least eight leading palæontologists continued to believe this interpretation, and it remained current for decades. Dinosaurs would bob to the surface and, even if they could stay down, Kermack had shown they could not inhale. This is basic science that a child could grasp. It seems remarkable to me that such an untenable idea had flourished for so long.[20]

In present-day science, once a theory becomes fashionable and attracts funding, scientists search around for any snippet of information that might further substantiate it. They do not obtain evidence and, from that, deduce the truth; instead, they select only the evidence that supports the current pet hypothesis. Having resolved to prove that dinosaurs were terrestrial, scientists have sought out every trifling item that might fit their preconceived view, while ignoring any inconvenient truths and inventing the rest to fit. This is not how science is meant to work. What we need to do is look at every aspect of dinosaurs – their skeletons, anatomy, ecology, size, footprints … every aspect we can analyze. If they all fit a specific theory, then that theory is viable. You cannot simply select the few scant items that work for your own fashionable hypothesis. Objectivity should be the key – and a wide-ranging and self-critical approach is the only way ahead. If dinosaurs were aquatic, then there will be evidence of modifications in dinosaurs that are similar to those observed in other water-dwelling creatures, and this is indeed the case. Many of the most widely discovered dinosaurs are hadrosaurs – duck-billed dinosaurs. This group includes dinosaurs like *Edmontosaurus* and *Parasaurolophus*, which were widespread throughout the late Cretaceous period all around the world. They had mouths like ducks, which we now know was equipped with a shovel-like beak, a shape of mouth evolved specifically for eating wet water-weed. The clue is in the name – duck-billed dinosaurs lived, unquestionably, in water like ducks and shovelled water plants for food. Yet there is no image anywhere that suggests this is the case. Whenever these hadrosaurs feature in movies or television documentaries, they are portrayed wandering along through meadows or trudging through forests. Even if there is a stream shown in the background, the dinosaurs themselves are invariably portrayed as terrestrial creatures walking by the water. Because dinosaur fossils are found in rocks that were laid down in water they are often portrayed with puddles of water nearby. And (because they must always be seen as terrestrial) they are shown sipping water at the edge of a pond. No matter how obvious the aquatic habitat might be, the terrestrial imperative rules. Curiously, an interview about dinosaurs on BBC radio is burbling away in the background, just as I am writing these words. It is between broadcaster John Humphrys and Nick Longrich of the Department of Palæontology at Bath University about a dinosaur that has just been discovered in a Moroccan mine.

> John Humphrys: 'A lot of Africa was under water at that time, of course.'
>
> Nick Longrich: 'Yeah, it was pretty warm back there and the sea levels were pretty high, and so huge areas of Africa were flooded by shallow seas, so we don't get a lot of dinosaurs because they were terrestrial, but through an extraordinary coincidence we have a dinosaur that got washed out to sea and was found among marine deposits.'
>
> Humphrys: 'That is unusual, isn't it? To find dinosaur fossils at sea?'
>
> Longrich: 'Yeah, it's kinda like going looking for whale fossils and finding a sabre-toothed cat or a lion. It's *Chenanisaurus* … a fairly large one, about 8 metres long [that's 26 feet].'[21]

You see? Dinosaurs, says Longrich, were definitely terrestrial. He admitted that the fossil had been discovered in strata that had formed under water, but (instead of conceding that this connotes an aquatic species) he dismisses that salient fact as 'an extraordinary coincidence'. There have been a great many such 'coincidences'.

Now that they are so very fashionable, the rate at which dinosaurs are being discovered is astonishing. On average, a new species of dinosaur is currently being named every 10 days. How many more are there? So ideal were the conditions in those shallow lakes and seas of the late Jurassic and Cretaceous periods that there could easily have been 1,000 different kinds of dinosaur alive at any one time. But dinosaurs were rapidly evolving during their time of dominance; old species died out as new species emerged. If we accept that dinosaurs ruled the Earth for about 160 million years, and propose (for the sake of argument) that a single species lasted for a million years, then you could claim that there were 160,000 different types of dinosaur in the prehistoric world. If the rate of extinction and emergence was faster, which you could easily imagine in that fertile and fecund, verdant paradise, then the total number of different dinosaurs might have been higher – possibly 500,000 different species. Compare that with humanity, with a dozen or so species spread over a mere 2 million years, and the dinosaurs were a fabulous success story.

This leads us to a persistent paradox. If the dinosaurs were so successful, why did they die out? The meteorite theory did much to stimulate a renewed interest in dinosaurs, and it certainly triggered an enduring fascination with the way they became extinct; yet it cannot have been the answer. Dinosaurs were on the decrease prior to the arrival of that asteroid, and the variety of species was already dwindling. As we shall see, the date of extinction is not coincident with the date of the impact. Not only that, but many other reptiles that were contemporaneous with dinosaurs did not suffer the same fate. Crocodiles, alligators, tortoises, turtles, snakes and lizards … they were subject to precisely the same constraints as the dinosaurs yet they did not become extinct. If the asteroid brought about the end of the era of reptiles, they would all have disappeared at the same time. Although the Chicxulub event was held as the likely cause, there were many other impacts about which we do not hear so often. The Earth was once scarred by countless craters caused by meteorites, but they have mostly long since weathered away. Craters survive on heavenly bodies like Mercury and the Moon only because there is no weather to cause them to wear away. The total of known impacts on our planet within the last 70 million years has risen to more than 60 (10 more than were known about in 1970) and there are doubtless many more waiting to be discovered. And they still occur. In 1908 a meteorite slammed into Tunguska, Siberia, and devastated thousands of hectares of forest. In 2013, a meteor measuring about 40 feet (13 metres) wide burned up over Chelyabinsk in Russia, sending shock waves that caused damage to buildings and is said to have injured 1,000 people. Three years later, on September 7, 2016, an asteroid 1 mile (1.6 km) wide named 2004 BO41 made a near Earth pass, missing us by 7.3 million miles. In December 2017 an asteroid measuring 3 miles across (about 5 km) past near Earth, missing us by 6.4 million miles. It was named asteroid 3200 *Phaethon*, after the Greek god who was killed after he almost destroyed the world. This was the closest pass since 1974, reports NASA. Every year there is a sizeable body that misses Earth by a narrow margin. One day in the future, one won't.

My theory of the aquatic dinosaur also solves another enduring paradox – were dinosaurs warm-blooded or cold-blooded? Warm-blooded creatures are endothermic: their warmth comes from within. We also call them homeothermic. Humans are like this; we burn energy (largely in the liver) and this warms the blood. The rate at which energy is consumed is carefully adjusted to keep us at a constant temperature, and we are also cooled by the latent heat of evaporation through perspiration from the skin. In this way our bodies normally remain at a constant temperature, day in, day out, of about 98.6°F (37°C). This is the temperature at which our metabolism functions best, and allows us to keep our senses and metabolic mechanisms – especially our enzymes, which work best at these temperatures – functioning at optimum capacity. The body temperature varies a little from morning till night, and sometimes from day to day, but remains remarkably constant so long as we are well. Cold-blooded, ectothermic creatures (like lizards and crocodiles) are very different. They depend on their surroundings for their body temperatures. These we also call poikilothermic, from the Greek ποικίλος (*poikilos*, varied), because the temperature of the blood goes up in hot sunlight and down on chilly nights. This means that reptiles, insects and amphibians have to find some way to warm up every morning before their bodies can function properly. Analyzing the bones of dinosaurs under the microscope showed that they must have grown at the rapid rates only found in endothermic animals, and this – coupled with the new image of active, sociable dinosaurs – gave impetus to the idea of a warm-blooded dinosaur. In biology, we have the Q_{10} rule, which states that, within an animal's preferred temperature range, growth rates double with every increase in temperature of 10°C (18°F). The rates double, in line with the rate of metabolism. This is important. Temperature matters.[22]

No reptile has ever evolved endothermy, so to many biologists the idea of a warm-blooded dinosaur seems counterintuitive. Some palæontologists put forward a kind of halfway house and decided that dinosaurs must have been 'mesothermic' from the Greek μέσος (*mesos*, intermediate). This immediately attracted further controversy. Scientists retorted that the estimated growth rates in dinosaurs that were being quoted fell wide of the mark. None of the explanations was satisfactory. There was no way to answer this essential paradox: how could dinosaurs show a relatively consistent internal body temperature, if they lacked a metabolic mechanism to keep it constant?[23]

The more I reviewed all this, the more it seemed that my aquatic theory would solve the problem. A large ectothermic animal that lived in water would inevitably equilibrate at the same body temperature as the water, since that provides a thermal buffer and does not itself become significantly warmer or cooler as night changes into day. The aquatic dinosaur would thus have a relatively constant body temperature – without needing an internal mechanism to control it. Climate simulations reveal that the temperature of the water was close to normal human body temperature, so it would have been absolutely ideal for a dinosaur. One paper says: 'We find little variability in the sea-surface temperature records' even during a period when the temperature was formerly believed to have changed.[24]

You might be wondering whether the temperature near the poles would have thrown doubt upon this theory, but the sea at that time – even in the Arctic – was much warmer than we think. One report even concludes: 'polar waters were generally warmer than 20°C during the middle Cretaceous (~90 million years ago)', which is when the giant dinosaurs were at their peak. There was no polar ice on Earth throughout this period, so even at the North and South Poles a temperature of 70°F (20°C) might have been feasible. There was a problem in that the continents at that time are seen by climatologists as being cold and dry – though I postulated that the world of the Cretaceous was actually different from what the climatologists had claimed. I began to speculate that our models for the climate of the Cretaceous period were certainly due for radical revision. Our present ideas needed replacing.[25]

There is another curious problem that is peculiar to the carnivorous theropod dinosaurs; the question of yaw. You will have experienced this when you try to change the direction of a trolley in a supermarket. At the end of an aisle, you may wish to change direction, but your trolley doesn't. It has momentum and its tendency is to continue to travel in a straight line. Even if you turn it sideways, its momentum still wants to propel it straight ahead. Actually turning it around to travel on a different heading takes a surprising amount of force. Or imagine you are carrying a ladder on your shoulder and you suddenly need to turn a right-angled corner. It is virtually impossible because Newton's first law of motion comes into play. This sideways turning is known as yaw, and it is a perpetual problem for palæontologists to try and explain. Every time you see a filmed reconstruction of a bipedal dinosaur (the giant carnivorous dinosaurs, like *T. rex*, were always supported on two hind legs) it looks ungainly when it tries to turn to one side. In most situations, yaw would cause it simply to fall over. As a terrestrial animal, a bipedal dinosaur is fundamentally unstable – whereas in a watery environment it would, with a gently controlled movement of its tail, be able to turn from side to side

A new dinosaur, *Neovenator*, was discovered on the Isle of Wight in 1978, and dinosaur artist John Sibbick created this orthodox image of its stance. As a terrestrial creature, turning sideways would be near-impossible because of yaw.

with the minimum of effort. This is another example of a major, intractable problem that my aquatic theory solves in an instant.

As we have seen, most of the dinosaur skeletons that make palæontologists famous are discovered by members of the public; but there is a convention that a new dinosaur can be formally described only when it has been acquired by an academic institution – a museum or a university. For this reason, dinosaur skeletons usually end up in the hands of the professionals. Some of them pose peculiar problems, and it is the aquatic theory that offers the answer. On January 7, 1983, a part-time fossil hunter named William J. Walker was digging around the Smokejacks Pit, a clay quarry in Surrey, England, when he suddenly discovered a huge claw. It looked like a massive tooth. Nearby was a phalanx (a toe bone) and a piece of fossilized rib. When he cleaned up these finds at home, Walker realized that part of the claw had been broken off; so the next weekend he returned to the pit and found that too. He took all the fragments to the Natural History Museum in London and asked for an identification. The museum specialists to whom he presented his discovery were Angela Milner and Alan Charig who were astonished at what they saw. After a few discussions, it was agreed that this had the makings of an important discovery, so a small team was sent back to the pit to carry on digging. Within a month, they had found more bones, but dangerous conditions in the quarry halted the excavation for several months. Eventually, a team of 8 was back on site, and by June 1983 they had unearthed the remainder of the skeleton. Most of the bones were sheathed by nodules of siltstone and the mineral siderite ($FeCO_3$) in a matrix of fine silt and sand, while others were buried in clay. The bones were all confined to an area measuring only 17 x 7 feet (5 x 2 metres) and, although some had been disturbed by a bulldozer, many of the bones were roughly in their original, articulated positions. Altogether the team listed a damaged skull with teeth, numerous vertebræ extending from the neck to the tail, the ribs and sternum, pelvis, limb bones and some claws. This was a highly important find. Work proceeded for three years, and in 1986 the dinosaur was officially announced: it would be named *Baryonyx walkeri* from ancient Greek βαρύς (*barys*, strong) and ὄνυξ (*onyx*, claw), while the species name identified the discoverer. This large theropod dated from the early Cretaceous period – it was a world's first. Nothing like this had been previously discovered. Palæontologists were excited, and it was even described as the most important dinosaur discovery of the twentieth century. The world's media came to report the announcement, and there was a television documentary devoted to *Baryonyx*. This creature was a streamlined dinosaur measuring some 30 feet (10 metres) long and weighing several tons. Digital reconstructions showed it prowling through buildings and approaching a river bank – and there it halted, cautiously dipping its snout into the water. This seemed to me to be wrong; in my view, *Baryonyx* clearly had the anatomy of an aquatic creature. To begin with, its head was similar to that of a crocodile. The shape of the skull was long and tapering, and it had a bulbous snout like that of a gharial. This was a clear pointer to me that these dinosaurs were aquatic, and more was to follow.[26]

Further research showed that, hidden among the abdominal remains, were some small bones that had previously been overlooked. There were some fish scales, and a few small teeth. Under a low-power microscope it could be seen that these fossils had been damaged. They had been etched away by acid – stomach acid! These were the remains of the dinosaur's last meal. The fish turned out to have been *Scheenstia mantelli*. I was immensely excited by this revelation. Charig and Milner were concluding that *Baryonyx* must have fed on fish by dipping its powerful claws into a nearby stream. To me, the evidence revealed a completely different story – this was a dinosaur that evolved in an aquatic environment and always lived in water. This was why it fed exclusively on fish; everything was pointing in that direction.[27] Even so, all the palæontologists persisted in saying it must have been terrestrial.

Fossilized bones are not the only remains that dinosaurs left behind, of course. We have repeatedly seen how many times they left their footprints behind and how (in places like Australia and Japan) we can work out the dinosaur populations, even when no dinosaur fossils have been discovered. We know that the spacing of dinosaur footprints has been used to estimate the speed at which they walked, and – in places where they abound – palæontologists have concluded that they have evidence of a dinosaur stampede. In other cases, people have decided that dinosaurs must have walked side by side, interacted socially, or even strolled along in family groups. This is dangerous territory. Footprints in mud endure far longer than we imagine, particularly those in mud at the bottom of a lake. The fact that two sets of tracks lie side by side cannot allow us to infer that the dinosaurs were there at the same time; they might have been days apart, or many weeks. The existence of numerous footprints cannot imply that this was the sign of a stampede; it may have been very few dinosaurs walking on the same mud over a prolonged period of time. Finally, although the spacing of footprints has been taken to prove that an animal was hastening along at an unfeasible speed, if this was an animal propelling itself along by wading in water, then the spacing reveals a radically different story. The results then show that the rate of movement was leisurely and slow, far more in keeping with a dinosaur's size.[28]

A survey in 2017 showed that large dinosaurs cannot run fast, and *T. rex* would have been physically unable to chase its prey as the standard accounts suggest.[29] These findings echo conclusions that were reached many years ago at the Royal Veterinary College of the University of London. Tyrannosaurs were slow-moving. They could never have run.[30]

However, I could see that they were perfectly evolved for a life in shallow water. So far, we have considered the spacing of the footprints, but what of their depth? Dinosaur tracks are always shallow, and this is where my aquatic dinosaur theory offers the only explanation. If the conventional views were correct, then dinosaur footprints would be endlessly varied, and their depth would be a function of the consistency of the substrate and the mass of the dinosaur. On firm and consolidated ground, they would leave an impression that was slight. In wet and more soggy conditions, their footprints would be more deeply impressed; while in deep and soft mud, a large terrestrial dinosaur would have sunk up to its armpits. Their trackways would vary constantly, initially dependent upon the changing consistency of the substrate on which they walked, and the depth of each set of footprints would be a function of the size of the dinosaur: the bigger the animal, the deeper the track. When we look at the innumerable dinosaur trackways that have been discovered, this is not what we find. All of them are roughly the same depth. From the biggest dinosaur tracks (made by creatures which would have weighed, on dry land, 100 tons) down to the smallest (left by dinosaurs the size of a child), they are mostly of the same proportionate depth.

The trackways left by stegosaurs in what is now Colorado prove my point. These footprints are rare, but revealing, for they substantiate my view that they existed in an aquatic habitat, their weight being taken by water, rather than by their limbs. The rocky strata containing the footprints consisted of fluvial sandstone that forms part of the Morrison Formation of the upper Jurassic period, about 150 million years ago, and were excavated from Dinosaur Ridge, just 14 miles (22 km) west of Denver, Colorado. This was a site that had been intensively excavated for Othniel Charles Marsh by Arthur Lakes in the 1880s. When their work was finished, Marsh ordered Lakes to dynamite the entire site, purely to prevent Cope from exploring there subsequently – but Lake did not do it. It seems that he could not bring himself to destroy a site that might still contain potentially important fossils, so he had some random rocks scattered across the diggings instead. The strata he left intact. Nobody carried out further work until 2002, when palæontologists from the nearby Morrison Natural History Museum began to carry out investigations of their own. That is when the footprints emerged into the daylight.

On my visit to the Morrison Museum, their resident natural history artist, Chenoa Ellinghaus, was keen to show me those intriguing stegosaur tracks. One slab of rock had a remarkable array of footprints made by infant stegosaurs. They were running about on wet mud and had sunk in deeply; on another rocky piece from nearby are footprints from the adults. These are much shallower, even though the mature dinosaurs were so much heavier. Most intriguing of all, even though there are clear prints both large and small, none of the footprints was made by forefeet. 'We have not found a single front foot!' she told me. The Morrison team had found only one possible answer to the mystery. The sacral vertebræ of sauropods at the base of the spine tend to fuse in later life. This tends to happen to all of us, though in sauropod dinosaurs the effect is to produce a solid mass of bone supporting the pelvis. The standard explanation has always been that this supports a female dinosaur when a massive male is mating, though to me that makes little sense. The sacral vertebræ fuse in later life, whereas breeding is always most active in young adults. And, whether they are fused or not, the female dinosaur would have to bear the male's weight anyway. Ellinghaus explained their thoughts to me:

> It does answer a weird postulation that goes way, way back when looking at sauropod dinosaurs. As the animal ages its sacral vertebrae fuse, and the suggestion has always been that, if you stick a giant dinosaur on top of another giant dinosaur you break its back. So maybe this is a mating adaptation. But what if the fusion actually has more to do with allowing counterbalance? Makes sense, if you're putting 80–90% of your weight on your hind feet. It opens a range of possibility.[31]

This is bending the facts to fit a preconceived hypothesis: it cannot work. If 80 per cent of the weight of a stegosaur is being borne on its hindlimbs, then the front feet would leave impressions only 20 per cent as deep – but in fact they never touched the ground. Only the hind feet left any impression at all. Only one answer fits: that the dinosaurs lived in shallow water. In that case, the hindlimbs propel the adult animal along and the forefeet never touch base. Crucially, the most well-defined – and proportionately deepest – footprints are the smallest. The newly hatched stegosaurs ran about on their hindlegs on wet mud near where they had hatched, but above the waterline. Because of the lack of buoyancy, these footprints are proportionately deeper. Everything fits. The larger dinosaurs were buoyant in water. It was their shallow water habitat that bore their weight, and their hindlimbs served only to propel them along.

Walking is a controlled fall. You might imagine that the way you start to walk is to place one foot in front of another and then lift up the other foot; but it isn't so. Try it. If that is all you do, you will fall over backwards. No – the first thing you do when you start to walk is to lean forward. Your foot is then put out to prevent your falling. As we walk along, we are perpetually falling forward, and placing down our feet to prevent us dropping headlong to the ground. A bipedal creature must have moved like this on land, flopping down a foot in front of itself as its body fell forward. Once we envisage it in water, of course, then the situation is immediately obvious: the dinosaur is buoyed up by its surroundings, and its feet on the bottom of the lake serve only to move it along.

That may be the case for the two-legged carnivorous dinosaurs, but what about the massive herbivores? Those huge sauropod dinosaurs present the strongest evidence for an aquatic lifestyle, for they could not easily have moved – and could never have evolved to such an improbable size – without the buoyancy of their watery environment. Their shallow footprints (a couple of inches, or a few centimetres, in depth) are roughly as deep as those of their lighter carnivorous competitors, precisely as we would expect if they existed, and had evolved, in an aquatic habitat. We often find that the footprints of these four-footed leviathans represent only two of their four feet. The tracks may be left only by the front feet, or alternatively by the hind-limbs. When only the hindlegs have left footprints, it would be possible to argue that they were walking on their hindlegs, raising the forelimbs as they moved like a trained dog begging for biscuits. This is possible, perhaps, but not feasible; to have a dinosaur weighing 50 tons or more supporting its bulk on two legs is hardly realistic. The problems are compounded by the fact that other trackways show that only the front feet were in contact with the mud. Various explanations continued to be proposed to account for this, such as the notion that, because the weight of a dinosaur was mostly resting on the forelimbs, the hind legs would not have been exerting enough pressure to make any impression. This is a ridiculous fantasy and is a classic case of terrestrial tyranny replacing robust reality.[32]

My aquatic theory provides the obvious answer to these peculiar predicaments. The longest limbs reached the bed of the lake, and propelled the creature along, while the shorter legs did not reach down quite far enough, and thus left no prints at all. The body of the dinosaur would be supported by the water in which it lived, and it did not always need its four feet touching the bottom. Searching through the literature for papers on this subject, I came across a 2002 description of 105 perfectly preserved footprints that had been unearthed in South Korea. Only the front feet had left prints. These authors did not prevaricate but concluded: 'The tracks … were made by sauropod dinosaurs while swimming.' They could not have been 'swimming' or there would be no footprints at all. They were wading, precisely as my theory proposes.[33]

Trackways studied by footprint expert Michael Hawthorne in Texas around the same time were similarly found in a region which was 'a broad area of shallow-water' He said: 'Deposition occurred primarily in shallow water' in an area that, during the Cretaceous when the trackways were deposited, 'included streams, marshes, tidal flats, lagoons and shallow subtidal marine environments.' This is precisely in line with my predictions.[34]

Wherever we look, there are accounts that skate round the obvious interpretation. Here is another account from the discovery of gigantic dinosaur footprints in France some years ago:

'The dinosaur footprints in Plagne are circular depressions surrounded by a fold of limestone sediment. These depressions are very large, up to 1.50 m in total diameter, suggesting that the animals were larger than 40 tonnes and 25 meters in length. The limestone dates to the Tithonian stage [upper Jurassic, 150 million years ago], a period during which the area was covered by a warm and shallow sea.'

Indeed so – enormous dinosaurs were producing mere 'depressions' in the bed of a 'warm and shallow sea'. Precisely my point.[35]

There is also one curious observation about sauropod footprints that provides an undeniable piece of evidence. The study of footprints is a complex business but can reveal much about dinosaurs that otherwise we might never have discovered. Palæontologists prefer to call them trackways, rather than footprints, and the term they use for such traces is *ichnofossil* from the Greek ιχνος (*ikhnos*, track). From these we can deduce something of the way that dinosaurs behaved.[36]

Care is needed: calculations carried out by R. McNeil Alexander came up with the following expression of the rate of movement of dinosaurs:

$$v = 0.25g^{-0.5}\,\delta^{1.67}h^{-1.17}$$

where v = velocity, g = the acceleration of free fall, δ = stride length and h = hip height. Other investigators eagerly followed, and it was soon agreed that smaller bipedal dinosaurs would have been able to run at 25 mph (40 km/h), with the fastest travelling up to 50 mph (80 km/h). Run-of-the-mill theropods would travel at 12 mph (20 km/h), while sauropods would make about the same speed, or a little faster. Stegosaurs would move no faster than 4 mph (6 km/h), a brisk walking speed for an adult human. Calculations like these spread around the world of palæontology and, with variations, became generally accepted, until an anatomist, Peter Dobson, ran the figures again and showed that, if you applied the same formula to a person, then humans would normally be moving around at about 15 mph (23 km/h). Even when running for a train, most of us wouldn't match that.[37] The truth is that the spacing of the footprints should be interpreted with the animals largely immersed in water. They moved far slower than anyone realizes.

Throughout the writings of palæontologists one idea kept recurring: they insisted that large dinosaurs could not survive in water. They would be unstable and must inevitably tip over. Who first said so? Where did this idea originate? What was the scientific evidence? This conclusion was formally published in 2004 by Donald Henderson of the Department of Biological Sciences at the University of Calgary, Alberta, Canada. His paper used what he called a 'mathematical/computational model' to show that sauropod dinosaurs would tip over if ever they tried to go into the water. It is a poor paper and assumes that living creatures operate like an inflated beach toy or a rigid plastic mannequin. His models have the heads fixed upright on rising necks and show fixed points (like centres of buoyancy) which, in real life, continually change as an animal raises or lowers parts of itself. His calculations show how far divorced a theoretician can be from the living, breathing realities of an evolving organism. The paper deals in detail with the stability of dinosaurs created as computer models, and Henderson then calculates precisely how deep into the water a dinosaur could wade before it was doomed to tip over. His diagrams show that *Apatosaurus*, standing with its stiff neck upright, would measure 40 feet (12 metres) tall, and could have waded into water no more than 12 feet (3.7 metres) deep before it tipped over. Henderson claimed to show that a 60-foot (18-metre) tall

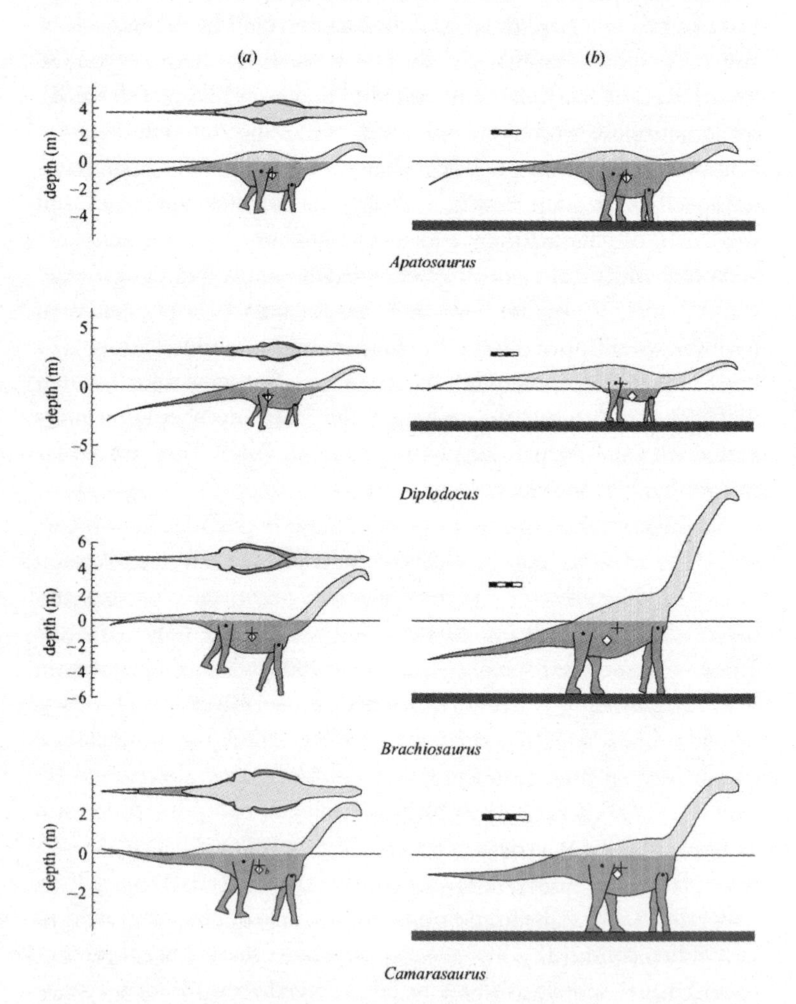

Dinosaurs wading in water were analysed in 2004 by Donald Henderson. His computer modelling utilized solid outlines, not responsive creatures; he claimed that they would inevitably tip over if they tried. A change in water level, or a lowering of the neck, could confer stability.

Brachiosaurus could have ventured into a depth of just 14 feet (4.3 metres), while *Camarasaurus*, reaching up some 50 feet (15 metres), could have managed up to 10 feet (3.2 metres), and *Diplodocus*, at 40 feet (12 metres) tall, would have to confine itself to water less than 8 feet (2.4 metres) deep. The computer modelling is meticulous, and the conclusions are unambiguous – it is just the models that are bizarre. This paper has been used throughout palæontology to reinforce the idea that dinosaurs could never have ventured far into water, yet it is fundamentally wrong. The premises on which his conclusions are founded are made-up models and the conclusions are based more on subjective speculation than objective realities.[38]

When you build algorithms or use computer models, you must ensure that the basis for your argument is sound, and Henderson's isn't. His diagrams of dinosaurs are crude and mathematical, lacking any of the flexible vitality of a living organism. His paper states that he estimated the basic tissue density for the axial body to be the same as water, 1 kilogram per litre. It was thus of neutral buoyancy. Because of the higher proportion of bone that we observe in the limbs, he set the density of the legs at 1050 grams per litre.[39]

This is just his guess. It is erroneous, because the density of solid bone is actually almost twice as much as water, and if a dinosaur's limb is half bone and half muscle, then its density is going to be 1,500 grams per litre. Henderson increases the density by a mere 5 per cent, but he is wrong; it is closer to 50 per cent greater. It is even worse when Henderson looks at the thorax; he gives this part of the dinosaur a density of only 850 grams per litre. That's as light as a piece of apple wood. On what does he base this figure? He uses the calculations from a book by Noble S. Proctor and Patrick J. Lynch, who obtained that result not from reptiles, but from birds.[40]

Proctor and Lynch found that, on average, the thoracic and abdominal air sacs of birds account for approximately 15 per cent of the volume of the trunk, so Henderson adopted this figure for his dinosaur calculations without evidence to support it. We know that the relationship of birds as descendants of dinosaurs was postulated as long ago as the 1880s, but to use data from present-day birds and transplant those into gigantic dinosaurs is an arbitrary flight of fancy.

This is Henderson's personal assumption and is not the kind of figure on which you can base an extensive computer model. He had created a simplistic computer re-creation of a dinosaur, provided it with absurdly lightweight legs, added the body of a bird, and (as if this weren't enough) stuck on top an upright neck and head that further unbalanced his rigid model. His simulations, according to the computer algorithms, would fall over. Sure. And bumblebees cannot fly. Because Henderson had created mythical monsters with impracticably low-density legs, impossibly flimsy bodies and absurdly unsupportable necks, his models would definitely keep away from the water. Real dinosaurs, my dinosaurs, would not.

Not only are his assumptions for the body of the dinosaur wildly inaccurate, but Henderson was wrong in giving his models heads held high. As in the case of their massive tails, this would have been an anatomical impossibility as well as requiring a conspicuous consumption of metabolic energy. The standard picture of dinosaurs is that they strode along with heads aloft on necks held erect. A paper published in 2009 by Michael Taylor and his team at Portsmouth University even used a photograph of the bones from a chicken neck to illustrate this erect posture.[41] This seemed logical. Robert Bakker once asked: 'Why evolve a 30-foot neck if all feeding is to be done on the ground?'

Were the dinosaurs all terrestrial creatures browsing on tall trees, that would be a valid point to make, but they weren't. Picture instead that they evolved in an aquatic habitat, and evolved with the neck resting at the water surface (like the tail), then the elongated neck would have been used to reach into the lush, deep growths of leafy plants, not high into the trees. Meanwhile, palæontologist Tracy Ford from San Diego, argued that dinosaurs must have held their necks horizontally. He had even co-authored a paper which concluded that at least one dinosaur might possibly have been aquatic.[42]

At around the same time, Roger Seymour at the University of Adelaide calculated that the blood pressure of a dinosaur holding its head erect would place unmanageable demands upon the dinosaur's heart. His argument put Taylor's into the shade. Seymour pointed out that the long necks of gigantic sauropod dinosaurs had long been

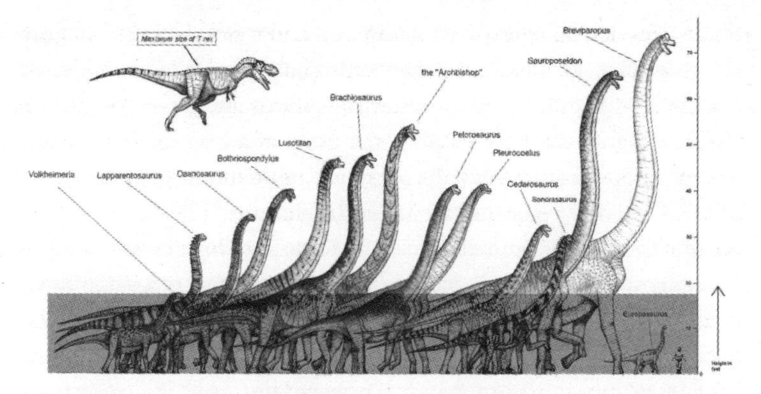

A range of dinosaurs was brought together on the Paleoking blog in a diagram created by Nima Sassani. To the original, a shaded layer representing the depth of a shallow lake has been added. Even the most massive are buoyed up by water.

assumed to be used for high browsing to obtain enough plant material for food, so he decided to look at the question from the point of view of the energetics involved. He argued that the energy cost of circulating the blood could be estimated accurately from two physiological findings that would relate metabolic rate and blood flow rate to arterial blood pressure. His reasoning was that the metabolic rate of an animal is proportional to the rate of blood flow and that the workload imposed on the heart is proportional to the product of blood flow rate and blood pressure. He calculated that – because a neck held high would have required a greatly elevated arterial blood pressure – the animal would expend approximately half of its energy intake just to circulate the blood. He concluded that it would therefore be far more realistic to use a horizontal neck to enable wide browsing because this would keep the blood pressure within manageable limits.[43]

Seymour's model was of dinosaurs grazing, like cows, and keeping their heads low. My refinement would provide the answer: because they were aquatic creatures their necks would be buoyant in the water. Rather than stretching up to reach vegetation, they were primarily reaching into vast and verdant thickets with their necks floating at the water surface, like a colossal snake. My calculations had suggested that a dinosaur would use about 15 cubic feet (400 litres) of oxygen every minute to hold its tail above the ground; Seymour had now calculated that a dinosaur would waste half its energy by holding up its head. Since both extremities are of comparable size, it's reasonable to assume that the head and the tail impose similar metabolic burdens. Thus, if the head and the tail are each consuming half the animal's energy, that leaves nothing whatever to power the dinosaur's body! Not only does my aquatic model provide the perfect answer, but the calculations show that the conventional conception cannot conceivably be correct. Even so, in an era of terrestrial tyranny, Henderson's bizarre proposal continues to be taken seriously. Palæontology decreed that dinosaurs *had* to be terrestrial at all costs; Henderson's computer calculations conveniently confirm that they could never have ventured far from dry land, even if they wanted to. Everyone has agreed with Henderson's findings: if dinosaurs dared to step into deep water they would topple over and drown.[44] Dinosaurs could not swim.

Tyrannosaurus rex poses a particular paradox with its absurdly tiny forelimbs. This dinosaur balanced on two legs, and could easily have fallen over. That may be true if it is terrestrial – but none of this applies when the normal habitat of *T. rex* is shallow water. The aquatic dinosaur theory alone demonstrates why a tyrannosaur had evolved such diminutive forelimbs. Evolving in water, it didn't need them, and so they became grotesquely reduced; its hindlimbs propelled it along, cruising with its jaws ready to consume its food. What other purpose could those tiny arms have performed? In 1970, British palæontologist Barney Newman proposed that the diminutive forelimbs were used to help stabilize the dinosaur after it fell, providing support as it tried to get up from the ground. To Newman, this evolutionary adaptation explained why tyrannosaurs needed those little limbs – they were simply struts, to prevent it sliding forward as it tried to stand.[45]

An engineer in California, Rob Reyes of Los Angeles, wrote to me with an intriguing idea. He thinks that one crucial use of those fragile forelimbs was to break the vacuum under the body when a dinosaur was resting in mud. A heavy body lying on a soft and sticky surface would find it hard to rise, simply because it was sucked into the surface and air could not easily penetrate beneath its heavy body when it chose to rise to its feet. Moving those tiny forelimbs could break the vacuum and let in the air, Reyes suggested, so that the dinosaur could extricate itself from the clinging mud.

Those small forearms of *T. rex* are absurd in any other scenario. Even if the dinosaur's hands held food prior to eating, they could not have conveyed food to the mouth because they were too short reach its jaws. The wrist joints were weak, and the elbow was limited to little more than 90 degrees of movement – it could never be straightened – and it had very reduced movement, both side-to-side and up-and-down. Fossils have sometimes been found with healed bones, showing that the forearms were broken in life, so these dinosaurs could manage without using their forearms for periods of at least one month (the time taken for a broken bone to heal). Only the aquatic theory provides the answer we seek.

What about the dinosaur's huge tail? The most telling evidence for my aquatic model comes from the tracks left by dinosaurs – not from the marks that the dinosaurs left behind, but those that they didn't leave. Among the tens of thousands of examples of fossilized dinosaur trackways that have been found, very few have revealed any sign of the tail touching the ground. They should be abundant. There are fossilized examples of crocodile tails leaving a dragging mark across the sand, and in the tropical jungles in Australia, Africa and Central America I have watched them do this in life, but marks left by dinosaurs are vanishingly rare. This poses an immediate problem: the tail of a terrestrial dinosaur was heavy (for some, at least 10 tons) and that would consume prodigious amounts of energy to hold up high – half the dinosaur's daily intake of food, if Seymour is to be believed.[46]

To make sense, the tail must have rested on the ground for most of the time because, as we have already seen, no animal could evolve to waste prodigious amounts of energy in holding its tail aloft. This had always seemed to me a powerful argument against terrestrial dinosaurs. The two-legged types, like *Iguanodon* or *Tyrannosaurus*, would have needed their tails as the only means available to support them erect while resting, or feeding if they were terrestrial creatures reaching upwards for leaves – but there were no marks to suggest they ever did so. Four-legged herbivores, like *Brachiosaurus* and *Diplodocus*, would have dragged their tails along precisely as Gertie the Dinosaur had done in her cartoon manifestation of 1912, and so there should be as many tail marks from dinosaurs as footprints. But there are hardly any anywhere in the world. This seemed to me overwhelming evidence in support of the aquatic theory, and it flew in the face of all those mounted dinosaur skeletons in the museums. In the 15-page chapter on dinosaurs in the 1995 edition of *Encyclopædia Britannica*, for example, there were five photographs of dinosaur skeletons, all of them with the tail resting on the ground. From a *Brachiosaurus* in Berlin to the *Triceratops* from New York, not one had its tail held aloft. The fossil tracks showed that this did not fit the facts.[47]

The final print edition of that great book (volume 10, dated 2010) added some new colour plates which showed dinosaurs obeying the terrestrial convention, and had them holding their tails up and parallel to the ground, though the older monochrome photographs in the pages (with the tail dragging on the ground) still remained. Yet the fossil evidence shows that this could not have been right. The lack of tail-dragging marks proved the point, and eventually the palæontologists at the Natural History Museum decided they had to change their skeleton and raise the tail high into the air. That was why they hoisted its tail up with steel bars holding its weight in 1993 and it stayed like that ever since. The metal supports were shipped with the skeleton when it was dismantled in 2017 and taken on a tour of all the major British museums that were willing to hire it. The strength of the museum's steel bars holding up the tail showed how heavy were the casts of the bones in porous plaster, let alone the full-bodied fleshy tail of the living dinosaur. My aquatic dinosaur theory solves that riddle. If they spent their days in an aquatic habitat, the dinosaur's tail would be floating at the water surface and they would have used it to help propel them along, much as we see in a present-day crocodile.

The evidence supporting my view was beginning to stack up. There were times when it seemed unfortunate that I was working with the majesty of microscopic living cells, when there was so much waiting to be said about the dinosaur world. Yet I still felt no urge to publish

The Independent Television documentary 'Dinosaur Britain' in August 2015 featured a re-creation of *Baryonyx*, a cousin of *Spinosaurus*. With partly digested fish scales in its abdomen it was clearly aquatic but was seen merely paddling in the Thames.

anything. This wasn't my field of research and I felt confident that the palæontologists would work out the answer. The last thing I wanted was a label of 'dinosaur enthusiast' hanging on my reputation. I had more than enough on my plate.

And yet, the strange saga of *Baryonyx* remained in my mind. To me, this was obviously an aquatic dinosaur, though it was not the only one of its type. The drawings and photographs of *Spinosaurus*, which Ernst Freiherr Stromer von Reichenbach had described in 1915, had shown a dinosaur that was very similar – though it was bigger than *Baryonyx*.[48]

When Stromer's fossils had been excavated, they were taken back to Munich to be carefully conserved. Eventually they were mounted on the walls of the museum, and Stromer begged the authorities to move them into safe storage as World War II approached. Nobody took any notice and, on the night of April 24, 1944, an Allied bombing raid on Munich destroyed the Bavarian State Collection of Palæontology, and all Stromer's original specimen collections were blown to smithereens. Some of the detailed drawings he made at the time were later recovered, and a few photographs were discovered in the archives. It would be decades before another *Spinosaurus* was found.[49]

The desert Arabs have known about fossils for thousands of years. Most of the extraordinary trilobite species have been found by Arabs, who clean them carefully and sell them to collectors. They know that a worthwhile discovery could be worth a lot of money from European or American palæontologists. In 2003, the remains of a spinosaur were found in the Kem Kem Beds in Morocco. It was little more than a skull, and it had been broken into several pieces. The remains were purchased for the Museo Civico di Storia Naturale di Milano (the Natural History Museum in Milan) and were examined by Cristiano Dal Sasso and his team. They found that the nasal bones alone measured a remarkable 39 inches (1 metre) long. Reassembling the skull was a difficult and demanding job, but when it was done they announced the result. From the size of the head, they estimated that the entire dinosaur would have been up to 60 feet (18 metres) long and must have weighed at least 10 tons. The restored skull was extraordinary – it was bigger than a person, measuring 6 feet (1.75 metres) long. This made *Spinosaurus* far bigger than the largest *T. rex*, previously the most fearsome theropod ever discovered. To me, the most interesting point about the skull was that (like the skull of *Baryonyx*) it was shaped like that of a crocodile. The nostrils were set high on the head, which is just what my notion of an aquatic dinosaur would require. Not only that, but the front of the snout was exactly like that of a crocodile or alligator: it was pitted with indentations, showing where sense organs had been situated. Organs like these are used by the crocodile family to detect the movement of fish in the water. The evidence substantiated my view that *Baryonyx* and *Spinosaurus* were taxonomic cousins, and there was no way in the world that they could have been terrestrial dinosaurs.[50]

While this was being announced in Italy, a remarkable auction was being held in France. This time, a privately owned *Spinosaurus* skull, which had apparently been smuggled out of Morocco in the middle of the twentieth century, was offered for public sale at the Drouot Montaigne auction house in Paris. It was listed as a *Spinosaurus* skull dating from the upper Cretaceous period (some 100 million years ago) and it was over a yard (105 cm) long. It had been in a private French collection whose owner had commissioned Flavio Bacchia, a restorer at the Italian company Stoneage, to prepare it for show. The result was pleasing to the eye, but was not scientifically rigorous, and experts believed it to be a chimera – it was composed of fragments from more than one skull. The Muséum National d'Histoire Naturelle entered a pre-emptive bid of €81,000 (£71,000 or $95,000), which would have been a record for French palæontology. Eventually they backed down: 'The piece was very beautiful on the catalog, but we had no guarantee that all its components came from the same individual,' said their director, Christian de Muizon. The British Museum in London had also been offered the skull, but also declined to buy it. This is curious; because the ownership of this skeleton is private, it is considered lost to scholarship. Collectors often have deeper pockets than museums, and in February 2019 a campaign began to restrict private ownership of dinosaur skeletons. Palæontologists want priority; they want their names on the find, not a wealthy collector.

Little has been reported of an extraordinary entire *Spinosaurus* skeleton measuring 26 feet (8 metres) long, which was unveiled at the Drouot Montaigne auction house in November 2009. The sale took place on December 1, 2009 and I have been told by Sylvie Tissot of Drouot Montaigne that the auctioneers who organized the event were the prestigious Cornette de Saint Cyr company. Like so many fossils, this one is believed to be comprised of bones from more than one skeleton, and a range of specialists was brought in to adjudicate, including palæontologist Andréa Valli, and Gilles Pacaud, the honorary chief of conservation at the Musées d'Histoire Naturelle in Paris. They agreed that about half this skeleton came from the same spinosaur, the rest being from nearby skeletons, though since they were all of similar size, the result was reasonably accurate. Many entire dinosaur skeletons have been imaginatively reconstructed from just a few bones, and this one was far better than most. There has been little written by anybody about this sale in Paris, and the skeleton is absent from the innumerable publications concerning *Spinosaurus*, and so, with eager interest, I asked them what had happened. The company owner, Bertrand Cornette, wrote back to me to say: 'Malheureusement ce spécimen n'avait pas trouvé preneur' – the lot had not found a buyer. I asked him what had happened to the skeleton since, and all he would say was that he was not concerned with what had happened to it after the auction.[51] I have recently been told that it is being reclaimed by the Moroccan authorities, since it was illegally taken to France.

Because palæontologists conventionally decline to publish descriptions of dinosaurs in private hands – a curious convention – this one remains lost to science. Only if it ever comes into the hands of an academic institution (a museum or university, for instance) will a formal description become possible. Until then, this important fossil skeleton has disappeared.

Academic interest was triggered again when Nizar Ibrahim, a postgraduate student from the University of Chicago, purchased some more spinosaur specimens. Ibrahim came from a Moroccan family and he visited the country on a quest for more fossils in April 2008. He was based at Erfoud, an oasis town deep inside the Moroccan desert and famous as a film location. (The 2010 movie *Prince of Persia: The Sands of Time* was largely filmed there, and so were the desert scenes in the James Bond film *Spectre*, released in 2015.) Ibrahim announced his intention to purchase some fossils, and a nomadic fossil hunter approached him a few days later with some. There were not many – a few vertebræ, part of the hip and leg, and some fragmented bones from the front of the skull – but enough to give a hint at the rest. Ibrahim bought the bones, and he took them to be examined by Samir Zouhri, a palæontologist at the Université Hassan II in Casablanca. Once his fossils became news, Ibrahim was painted in the Western media as a brave desert explorer, scouring the arid desert and digging in caves for long-lost leviathans; that is certainly more exciting than the reality of buying a bag of bones in a box from a Bedouin.

Hearing of the research already under way at Milan, Ibrahim went to Italy for his specimens to be examined by Dal Sasso. Not only were the dimensions similar, but the hue seemed identical and it was concluded that – by an extraordinary coincidence – these separate bones had come from the same specimen. Confirmation would have come from XRF analysis. Dal Sasso and his colleagues decided to submit their specimen to a CT scan, and this revealed a crucial component of the unravelling story. We have already seen that the snout of the *Spinosaurus* skull was pitted with indentations, and the scans now confirmed that these were probably where the dinosaur had sense organs like those of present-day crocodiles. In other words, it fed on fish like its distant cousin *Baryonyx*, in which partly digested fish scales had already been found. This is why *Baryonyx* was always portrayed as marching up and down the river bank, dipping its snout into the water to snatch any passing fish that it could seize.[52]

Further results added even more evidence for me to ponder. I knew that Romain Amiot at the Université de Lyon, France, had set up a team of some twenty scientists to analyze spinosaur bones, and they set out to measure the ratios of oxygen isotopes in the teeth of the dinosaur. There are two isotopes of interest: ^{16}oxygen and ^{18}oxygen. Of the two, ^{16}oxygen is lighter and is more readily released from the body. In aquatic creatures, the continual immersion in water means that the levels of ^{18}oxygen are lower than the readings obtained from terrestrial animals. Amiot and his team took readings from more than 130 Cretaceous fossils and they found the ratio measured in *Spinosaurus* was a perfect match for aquatic creatures (such as crocodiles and turtles). They concluded that, even if these dinosaurs were terrestrial creatures that simply pulled fish from the water, then they would still have had the signature typical of a terrestrial animal.[53]

When I read these results, the conclusion was unambiguous – spinosaurs must have lived in water. This provided the final link in the chain of the argument for dinosaurs being aquatic. Now it was obvious, and the implication was clear: *Spinosaurus* lived and fed in water.[54]

I was astonished that all the palæontologists still adhered firmly to their terrestrial model. 'I don't doubt the isotope data,' reported Paul Barrett, a palæontologist at the Natural History Museum in London, 'but if they were living in water, I am perplexed by why spinosaurs do not have limbs modified for aquatic propulsion or the flexible, propelling tails typically seen in aquatic animals.' This was illogical. Plenty of animals living in water did not have webbed feet, or other adaptations; crocodiles and alligators were prime examples. Many have clawed hands and feet, not webbed. And who said the spinosaurs did not have flexible tails, suitable for swimming? Their tails seemed fine to me. The protestations of the palæontologists seemed overstated. Even Hamlet's mother knew that.

Although a few early workers had speculated whether dinosaurs could swim, the idea was never much liked by palæontologists. It is hard to see why: most animals can swim. Horses and cattle, snakes and lizards, even raccoons and elephants, are able to swim. It is often said that the camel is the only mammal unable to swim, but they certainly can. Some armadillos find it difficult – but, apart from such very rare examples, vertebrates are capable swimmers.

Evidence finally proving that large dinosaurs could never have swum was provided by computer-modelling their profiles (pp. 363, 380). This research concluded that, if a dinosaur waded into water, it would tip over and drown. As we shall see, the reasoning is faulty, yet – so keen were dinosaur specialists to promote their terrestrial ideas – these findings were being accepted without demur. It was official: dinosaurs did not swim.

Even if they were not able to swim, they could wade. An environment of warm shallow seas, lakes and lagoons would support their weight and provide the evolutionary imperative needed for dinosaurs to develop into gigantic monsters. Here was an exciting possibility that solved the outstanding questions surrounding dinosaur development. At the time it was just an idea. All it needed was scientific evidence to prove the point.

8

WADING WITH DINOSAURS

After decades of contemplating that dinosaurs lived in shallow water, I was now celebrating the emergence of large amounts of scientific evidence that substantiated my new view. Everything pointed in the same direction. By now you will be wondering, as I was, two things. First, why were the vertebrate palæontologists obdurately refusing to announce that dinosaurs might have evolved in water? Secondly, why was I not publishing my own theory? Every now and again, perhaps at a university dinner or a museum visit, an academic colleague would wonder why I wasn't publishing any of my conclusions on dinosaur evolution. At Cardiff University, my professor of zoology James Brough had encouraged me to investigate further. Dilwyn Owen, Director of the National Museum of Wales, had offered access to their collections of dinosaur specimens. More recently in America, Curt Bonk, noted authority on distance learning, urged me to think about writing a book. Sir Sam Edwards, with whom I had energetic discussions when dining at high table at Gonville & Caius College, Cambridge University, loved the idea of challenging the establishment, and he was intrigued by these newly published findings. It was certainly true that the scientific research seemed to substantiate what I believed. When I was lecturing at the Cheltenham Science Festival I explained the gist of theory to my young intern, Katie King; she too said the theory was so clear that I should certainly have it in print. Everyone who knew about it said the same; but, to me, it still didn't seem appropriate. Palæontologists were now faced with all the evidence needed for an overwhelming case that dinosaurs must have been aquatic animals, and I felt that it was inevitable someone would soon publish this as the official view. Even so, it didn't happen.

Then, one day in 2011, I was sipping wine at Cambridge University when someone said casually: 'I see that *Spinosaurus* is on BBC television next week. I expect you will be watching?' This was important – I would make sure that I saw this documentary. I checked it out and, just as I had been told, the programme promised to set out the very latest research into *Spinosaurus*. At last, I was certain that the latest findings would finally prove the point, and we would surely see the emergence of the aquatic dinosaur. It was going to offer me immense inner satisfaction. That night my wife Jan had the television on in good time and a bottle of wine perfectly conditioned. There were delectable nibbles on a plate and the scene was set. The documentary was airing and there on the screen we saw *Spinosaurus*, beautifully re-created. With its crocodilian snout and pointed teeth, it was almost as if they had filmed a living creature. The script reflected recent research: '*Spinosaurus* is icthyophagous,' said the commentary: 'a fish-eater.' And then we saw its prey, giant sawfish named *Onchopristis*, measuring some 27 feet (8 metres) long, digitally cruising through the computer-generated water. The fish moved like dark shadows through a swiftly flowing river, darting from side to side and swimming with constrained power. From time to time we caught glimpses of the spinosaur, prowling along the dry riverbank with trees in the background (they looked like eucalypts, and had nothing in common with trees that would have grown during the dinosaur era, but dinosaur documentaries never have the vegetation right). The dinosaur was shown looking towards the water, as though seeking food. The rivers, we were told, were thronged with sawfish which, the commentary emphasized, provided the 'perfect opportunity for *Spinosaurus*'. The dinosaur was seen strutting along by the river, then standing in the shallows, wetting its ankles. Then, with a sudden darting movement, it dipped its jaws into the water and snatched a passing fish from the shallows. It threw the sawfish onto the dry earth and stamped on its head. Said the commentary, as kettle-drums throbbed in the background to increase the tension: '*Spinosaurus*'s conical teeth evolved to grip prey rather than tear flesh.' The sawfish thrashed about on the ground, as the spinosaur took a swipe with its forefeet. 'For that, it needs powerful arms and claws.' We saw the dinosaur peering down at its victim, like a vulture with a rat. Soon the *Spinosaurus* ripped off small portions of the fishy flesh and tossed them into the air, throwing its head back as it tried to catch them, like a puffin. This did not seem to be a very efficient method of eating. Said the commentary: 'With prey plentiful, *Spinosaurus* can afford to be wasteful ...' Oh, that's all right then.

On droned the commentary, which next reported the CT scans of the skull made in 2008 which had shown those curious pits around the snout. 'They looked just like those of crocodiles,' said the commentary. 'It's thought these contained pressure sensors; sensors that, like those of a crocodile, can detect prey, making it perfectly adapted to hunting in water. This gives us the best evidence of exactly how it hunted. Able to hold its snout in the water, because of its high nostrils, it can strike without even seeing its prey.'

And indeed, this is what we saw: a spinosaur standing with its feet at the edge of a river, dipping its snout just a couple of inches (a few centimetres) beneath the surface, as if testing the temperature. Then it snatched at another unseen sawfish, tossing it into the air and catching it again like a heron with a dab. I watched this, stupefied. For decades, we had endured the barrage of misinformation about the terrestrial dinosaur, and now we had a major television documentary determined to promulgate this pernicious propaganda. It had been proved that these dinosaurs had the morphology of crocodiles, the sense glands on the snout, the fish scales in the gut from their food, the oxygen isotope ratio and even the nostrils of aquatic creatures; yet these palæontologists had it portrayed as a land animal that simply dipped its snout in the water to catch fish. The recently published evidence was distorted to support the absurd terrestrial theories, even when this left us with the vision of a huge predator balancing on the edge of a river as if performing a stunt like a sea lion on a stool. For decades, I had assumed that palæontologists must have been working towards an aquatic model for the evolution of dinosaurs, and at last we had one example in which every aspect we could experimentally test proved the point beyond doubt. And what did the expert team of dinosaur experts present to the BBC? Why – the spinosaur grabbed fish while it stayed resolutely rooted to the riverbank as though frightened to fall.[1]

This had gone beyond a joke. I had been convinced that a rational assessment of the recent research would have given the programme's advisers the evidence they needed to present a realistic view of dinosaurs, and yet the viewer was being fed more of the absurd fiction that had maintained palæontology in its macho image of dominance since the 1970s. Next day I reviewed it all and realized that there was only one course of action: I would have to bite the bullet and publish my own theories. The topic was topical, there was no time to waste, and I needed a scientific magazine that could guarantee to print it without months of delays. Conventional peer-reviewed publications were out of the question – it would take many months to work through the system and the paper would certainly be rejected. I had first published in *Laboratory News* back in 1991, and this was a journal that allowed for speedy publication. A copy of the magazine was sent automatically to every laboratory in Britain, which was the ideal way to spread the

The BBC released its major series 'Planet Dinosaur' with a programme entitled 'Lost World' on September 14, 2011. Although it included the latest research on *Spinosaurus*, the producers still showed it as terrestrial, fishing from near the river bank.

word. Previously I had published articles in that magazine on a range of my academic interests – microscopes, plagiarism, science policy, mad cow disease, the origins of human society and genetically modified crops – surely they might allow me space for a rebuttal of these pervasive myths? In September 2011, immediately after the BBC had launched their lamentable show worldwide, I outlined the theory to the young editor at *Laboratory News*, Philip Prime, and he agreed that starting a debate would be a timely idea. 'I'd love a feature on your theory of dinosaur life!' he promptly replied. He promised that they could have it reviewed and printed with the minimum of delay. He just wanted to have the copy and proposed an October deadline.[2]

This immediately focused the mind – although I had maintained my interest as the dinosaur research had continued to unfold, my position remained that of an outside observer. Publishing a theory in print meant that I would be saddled with the consequences. I would also have to undertake some serious research but I was already overstretched with research projects and lecture tours. It was weeks before I could do anything, so I wrote to Phil Prime to say I would take longer to submit than I had intended. I also brought him up to date: 'We have been making scale model measurements and working out limb loadings,' I said. For a while the theory rested; there were other experiments fighting for space in the diary and I had recently had a book published, so spare time for dinosaurs didn't exist. There was a major charitable building project under way too, for which I was responsible, and I was due to appear as guest speaker on a voyage to Russia. Privately, I was still certain that some eminent palæontologist would launch the aquatic theory first and I hoped that would be the case. So I did not rush to print, and left it till the New Year. Eventually, I did not submit my paper to *Laboratory News* until the following spring and they soon approved the contents, promising three generous pages for the paper. It would appear in their edition for April 2012. The theory would explain that the gigantism of dinosaurs would make their evolution impracticable as land animals, and it would use the recent data from *Spinosaurus* as an example of a dinosaur that had been exhaustively investigated. To illustrate my theory, I put together a somewhat inexpert image of a spinosaur with an indication of how it might appear as an aquatic creature, but I didn't understand how to add reflections to a digital image, so I asked my friend Richard Hubbert, a concert performer of classical music and website designer who has far greater artistic skills, if he could prepare a picture showing a huge sauropod supporting its weight in a shallow lake. I sent him a few standard images of dinosaurs and some sample backgrounds. The pictures I provided were not inspiring, but Hubbert made far better use of them than I could have done and his renditions were fine. The finished article ran to 3,000 words, with 10 colour illustrations. The proofs were marvellously designed and extended across three large-format pages. The editor decided to promote it with a full-page front cover illustration. 'Are we wrong about dinosaurs?' said their banner headline; 'New theory suggets [*sic*] an aquatic habitat for these prehistoric beasts.' Yes, it might have been improved had it said 'suggests', but the message was powerful – and inside was my expansive article. The length of the text was far greater than many science news journals could countenance, and the graphic design was vivid. *Laboratory News* wanted to promote it with an online video, and we were all confident that a wide-ranging discussion would be triggered by the appearance of the theory in print.[3]

My *Laboratory News* article began by pointing out the essential paradox that so many popular ideas about dinosaurs do not make sense. They posed a huge problem (I described it as the biggest problem that you could encounter as a land animal) because the immense size of the largest genera would place colossal loadings on their extremities. A single limb might have to support tens of tons, which would not be compatible with the agility associated with dinosaurs. We accept their footprints without objection, I wrote, although for such gigantic creatures the imprints that we observe in rocky strata make no sense. The prints are roughly as deep as ours might have been, although the enormous mass of an adult dinosaur would cause it to sink up to its thighs. The footprints seem to be those of an altogether lighter creature.[4]

Children, I wrote, were traditionally fascinated by dinosaurs, and in recent years, new dinosaur discoveries had refocused popular attention. I pointed out that, had you said the word 'raptors' to a teenager a generation ago it would have signified birds of prey. But try it today and you'll find it has become those clawed and vicious creations from movies like *Jurassic Park*. Once we celebrated *Tyrannosaurus rex*; now youngsters were as likely to know about *Spinosaurus* and *Velociraptor*. Books on dinosaurs had been flourishing like never before, from the great National Geographic *Dinosaurs* to Dorling Kindersley's *Dinosaurs Eye to Eye*. Television, always eager to capitalize on any commercial trend, had been quick to follow, with the BBC producing *Walking with Dinosaurs*, which had now been followed by *Planet Dinosaur*. There had been an upsurge of enduring interest with the discoveries in North America by John S. McIntosh of Wesleyan University and recently in China by Dong Zhiming of the Institute of Vertebrate Paleontology and Paleoanthropology in Beijing, so publishers and programme producers were sinking millions into the subject.

The text emphasized that animatronic dinosaurs were not merely manufactured, but now they went on tour like rock stars. I reminded readers that the first such exhibition had been of those concrete dinosaurs made by Benjamin Waterhouse in 1854 for the Crystal Palace exhibition, with Waterhouse being personally instructed by Richard Owen, still the greatest dinosaur expert of the Victorian era. And now there were computer games like Dino Hunt. There seemed to be no end to the current fascination for the great dinosaurs. Yet dinosaurs made no sense. The most obvious problem to me was the sheer mass of the dinosaur and the modest plantar area of the lower limbs. Compounding this was the bipedal stance that is typical of those colossal creatures. I reminded readers that today's heaviest land animals are the elephant, the hippopotamus and the rhinoceros; these animals sensibly spread the load across four limbs rather than two – an obvious evolutionary necessity. Dinosaurs compounded the paradox by possessing a tail that is proportionately massive. I admitted that it was not clear how one could accurately calculate the expenditure of metabolic energy required to hold such a huge member horizontal and clear of the ground, but the lack of tail drag marks associated with footprints in fossiliferous strata substantiated my view that little ground contact for the tail was normal for dinosaurs. In every reconstruction, they held their massive tails aloft. This would have consumed a high proportion of the energy input from their diet, and this ran counter to the evolutionary imperative. We thus had a set of factors that made the largest dinosaurs seem impractical as a product of evolution. They were massive, whereas their surviving spoors suggest that they cannot have been so heavy, or they would have sunk into the mud across which they walked. They also developed massive tails which, in conventional portrayals, are held aloft and do not drag on the ground. Finally, and crucial to my understanding of how dinosaurs are supposed to have functioned, they typically evolved to walk erect, which doubled the load upon the hindlimbs. The latest estimates of an adult *Tyrannosaurus rex* reckoned that it weighed 10 tons, which must have been supported by a single limb when the animal moved. That seemed impractical.

In the article, I argued that we could find substantiation of my views on the role of evolutionary pressures in the development of load-bearing anatomy by reflecting on today's largest land animals, because the elephant and, to a lesser extent, the hippopotamus conform exactly to the argument I was proposing. Their great mass is supported by four legs rather than two and they have tiny tails that impose no metabolic burden to hold erect. Crucially, the trunk of an elephant is an additional member whose weight has to be supported, and you could analogize the elephant's trunk to the tail of a dinosaur – yet the elephant's trunk normally hangs downward and does not consume metabolic energy for support. Indeed, the hippopotamus and (to a lesser extent) the elephant both have one practical answer to supporting their body mass, for they often resort to partial immersion in lakes or rivers where the displacement of water can help to reduce the load on their limbs. I pointed out that the hippo is categorized as a semi-aquatic mammal. Although both of these creatures have diminutive tails, there were other large animals that are equipped with tails of dinosaur-like proportions – present-day reptiles. Monitor lizards, culminating in the Komodo dragon *Varanus komodoensis*, have substantial tails, though even the largest is far smaller than a dinosaur's. Dinosaur footprints hardly ever show the tail touching the earth. I have observed wild monitor lizards in Borneo, Central

America and in zoos elsewhere and they do not always hold the tail free from the ground. The trails of the Komodo dragons I have observed in the wild in Indonesia are often characterized by a groove left by the tail dragging behind them. Crocodilians, notably the saltwater crocodile *Crocodylus porosus*, can weigh up to a ton and possess truly gigantic tail structures which can be up to one-third of the body mass. These tails also drag along the ground, for they are too massive to be held aloft. I had been observing crocodiles in Costa Rica; the spoors that the animals leave behind them feature the clear signs of this massive tail dragging across the mud. Crocodiles, I wrote, were the only present-day creatures that compare with dinosaurs, and for most of their lives, apart from times when they haul themselves out onto the bank, they inhabit an aquatic environment. It is the displaced water that bears the weight of the tail, for this is the organ that primarily helps the animal to swim. This, I postulated, provided the answer to the paradox of the ungainly gigantic dinosaurs – the only reasonable solution to the problems that they otherwise pose. And then I stated my basic premise: the giant dinosaurs were essentially aquatic creatures.

The physics stacked up in support of my idea. An African elephant weighs around 8 tons. It keeps three feet on the ground when walking; so, each limb would support some 2¾ tons. Of the species of rhinoceros, *Ceratotherium simum* is the largest and weighs up to 4½ tons. Each limb when standing thus has to support about 1½ tons. The

The lines of parallel plates of a stegosaur are not as unique as people say. This fine study of an American alligator *A. mississippiensis* was taken by Kevin Ebi of Living Wilderness Nature Photography. Its tail is remarkably like *Stegosaurus*.

adult hippopotamus *Hippopotamus amphibius* weighs some 7 tons and thus each limb comfortably supports a weight up to 2½ tons. Our experiments showed that the effects of immersion in water were dramatic. Animal tissues, being predominantly composed of water, may be regarded as of neutral density for this discussion, and a large mammal that is 90 per cent below the water level will exert proportionately reduced loading on the limbs – from about 8 tons down to 800 kg for the elephant, and from 7 tons to 700 kg for the hippo. Similarly, a large crocodile that is scarcely able to stand on its own feet on land becomes effectively weightless in water. In the real world, a crocodile is most typically observed with only the top of the head exposed. In terms of metabolic efficiency, walking on land for a crocodile would be a costly indulgence.

I then turned the clock back to the Cretaceous. Volumetric analysis of scale-model dinosaurs - provided a helpful indication of the effects of an aquatic habitat on the physics of a large extinct reptile like *T. rex*, where the mass of an adult was claimed to be no more than 7 tons. Using scale models, I found that the head and shoulders occupy roughly 15 per cent of the volume of the entire animal. Partial immersion in the aquatic surroundings that I was now postulating sets the figures into an interesting context; the loading per limb is reduced to 1050 kg in total, or 525 kg per limb when standing. The largest dinosaur yet discovered, *Bruhathkayosaurus*, was a quadruped weighing some 120 tons. Each limb must have supported 30 tons when standing still. Yet that creature, if immersed in water so that its head and neck are exposed, would have exerted less than 5 tons per limb when standing on all fours, rising to 6½ tons when wading. The anatomical adaptations we would see in these species were consistent with what I envisaged: they have a large and bulky body with a massive, muscular tail. The mass of the abdomen is immaterial when it is buoyant and submerged, whereas the nature of the tail would fit it for use as a powerful organ of propulsion and steering for a swimming dinosaur. The diminished forelimbs are equally well accounted for by this view. I pointed out that every large land animal, like the elephant, hippo and rhinoceros, has four load-bearing limbs, and evolutionary pressure has been against the reduction in forelimb size that we observe in flesh-eating dinosaurs like *Tyrannosaurus*. If the large dinosaurs are conceived as primarily aquatic, however, then the specialization of the forelimbs would be towards manipulative dexterity. The fact that the limbs became foreshortened is entirely reasonable. Animals like to inspect their food as they eat and holding it closer to the face is normal behaviour. Conventionally conceived, the reduced forelimbs of *T. rex* make no sense. However, if we envision the animal as an aquatic carnivorous species, this adaptation becomes entirely reasonable. Later I reviewed in detail the dimensions of the forelimbs of a tyrannosaur. They were so rudimentary they could not reach anywhere near the mouth.

Long-necked genera such as the brachiosaurs and the more recently investigated *Bruhathkayosaurus* were herbivorous, I explained, and their evolutionary constraints were different. Since plants do not move away when attempts are made to prey upon them, and because leaves would have been eaten *en masse* and not painstakingly pulled apart prior to consumption, there is less evolutionary impetus towards increased manipulative facility for the forelimbs. Our modelling suggests that a 40-ton brachiosaur could have weighed a mere 3–4 tons above the level of the shoulder. The body mass in water is of near-neutral buoyancy and can thus be provisionally discounted, so this reduced mass is all that has to be borne by the limbs. Had only two supporting limbs evolved, as in the case of the carnivorous genera we have already considered, this loading on the limbs would have been as little as 1 ton per limb – less than that of a present-day elephant. The mass of the elongated neck, necessarily evolved for an animal specialized for grazing on the foliage of tall trees, would have impelled these genera to evolve towards a four-footed stance. Only through these means could the load factor on the individual limbs remain within the constraints that can be calculated for the other heavy animals we have already considered.

I felt that this concept offered an interesting revision of the many artists' impressions of large dinosaurs with which we are familiar. A typical picture from the Madriz website shows brachiosaurs wandering in the middle of a vast and arid plain.[5] They are depicted standing on dried, cracked mud in a desert wilderness. Now apply my own view and envision the scene instead as a shallow lake in which the water supports the weight of the animals. The mud of the lakebed would eventually form the layers of mudstone or shale, in which state it is certainly solid; but at the time it was, in my view, mud at the bottom of those interminable shallow lakes. With water flooding the scene in your mind's eye the picture suddenly makes sense and the scene as originally depicted suddenly seems unrealistic and impracticable. We had been calculating the likely loading upon dinosaur limbs with and without water, and in each case the realities revealed by physics become far more plausible when the bulk of the animal is supported by the buoyancy of partial immersion. In my *Laboratory News* article I acknowledged that my colleague Richard Hubbert had used his skills as a graphic artist to provide 'before' and 'after' views that set my proposals in context. The scene was far less incongruous when the dinosaurs were half-submerged. Set in this new context, the scene suddenly made sense. As we have seen, this revised hypothesis also rationalized the paradoxically shallow nature of dinosaur footprints in soft mud. Investigations have modelled the consistency of mud in which such shallow footprints can have been impressed, and they force us to conclude that closely limited constraints must have applied: the mud needed to be of exactly the correct consistency for footprints successfully to form. It has been concluded that normal alluvial mud would have been so soft that large dinosaurs would sink in deeply and become trapped. In fact, there are widespread dinosaur footprints from large and small species and this variety of depth of impression is not seen.

These considerations perfectly fit the concept of a dinosaur that was supported by the buoyancy of water. I went on to explain that there had been controversy over whether dinosaurs were poikilothermic (their body temperature equilibrating with the environment) or homeothermic (controlling their temperature through their metabolism). Some research led to the conclusion that large dinosaurs had a constant body temperature – but this did not mean they were homeothermic, as current opinions would have it. Conceived as primarily aquatic creatures, I pointed out that they would have been buffered against rapid thermal change and their bodies would be close to the temperature of their watery environment. Thus, dinosaurs could have had a steady internal temperature without any need for its metabolic regulation. The mean water temperature during this era adds a final substantiation – it was almost 37°C (98.4°F), the present-day metabolic temperature of us homeothermic humans. The latest scientific research fitted my proposals perfectly, I wrote. *Spinosaurus* was a 15-ton dinosaur with a large sail-line fin running along its back. If it were aquatic, as I was proposing, then the fin would have acted as a thermoregulator if bodily heat needed to be shed. I reported the investigations of *Spinosaurus* in Milan in 2009 that had subjected the snout to X-ray computed tomography which revealed that the dinosaur seemed to have been equipped with pressure sensors just like those found in crocodiles. Curiously, the researchers had concluded that the dinosaur must have dipped its head into water, using these sensors to catch swimming prey. I mentioned the recent BBC television reconstruction that shows *Spinosaurus* wading along and dipping its snout into a stream to catch fish in exactly that way. Clearly, I argued, it made sense only if the dinosaurs truly were aquatic and scooped up fish as they swam. I wrote that it had been in 2010 when an international group based in China had analyzed the composition of isotopes of oxygen in the phosphatic remains of *Spinosaurus* and found the ratios to be close to those seen in present-day crocodiles and turtles, which had led me to the inevitable conclusion that they might not have been land-dwelling dinosaurs at all but should have been envisaged as semi-aquatic. I added that I took it further still: they, and all gigantic dinosaurs, had evolved to live their lives supported by the buoyancy of water.

In conclusion, I argued that dinosaurs were not lumbering monsters, teetering about on an arid landscape and burning huge amounts of metabolic energy to support both their bodies and their tails, but had evolved when the world was largely covered in vast shallow lakes, using the water to support their mass, buoy up their tails, regulate their temperature and provide a habitat for their food. Without a watery environment, dinosaurs did not make sense. And I explained that this might also have provided a reason for their extinction. The era following the age of the dinosaurs was the period when the continents drifted towards their current positions. Mountain building was active; the vast shallow lakes were at an end and the dinosaurs' aquatic environment disappeared. The Cenozoic era that followed was also a time of cooler climates, which weighed against large reptiles without an aquatic thermal buffer in which to survive. During the Cretaceous period when the largest dinosaurs flourished, the enormous plants of the time would have favoured the production of atmospheric oxygen, and this would have aided the metabolism of gigantic creatures. Although there were many controversies that remained, I was certain that most of the paradoxes surrounding the study of the dinosaurs were resolved by making this change in concept. Dinosaurs looked more convincing in water, and the physics stood up more soundly. And I ended with a clarion call for change: all the while we had been speculating in science on those remarkable creatures, this single, crucial factor eluded palæontologists – dinosaurs were aquatic.

The article summarized many of my views and when it appeared in print I was confident that a debate could now start. Of course, I could not hope to encompass all the arguments and recent findings. This book itself contains more than 50 times as many words as the *Laboratory News* article, so I ended by reminding readers that there were many controversies still outstanding, yet most of the paradoxes that surround the study of the dinosaurs would be resolved by making this change in concept. My crude mock-up of *Spinosaurus* in shallow water exemplified the new approach. On the first page of the feature we printed the standard view of this dinosaur as a land animal: 'This interpretation of *Spinosaurus ægyptiacus*, posted on the jurassicpark web forum, is a typical reconstruction of a large carnivorous dinosaur. Each limb would have had to support up to 15 tons. It seems incongruous on dry land.' Beneath this description appeared my retouched picture with the shallow water added. Ran the caption: 'In this revised image, I have shown it immersed in water to the shoulders. The tail becomes buoyant and the mass of the body is supported by the water. The loading per limb would be less than 1 ton – and it would be reduced to zero if the dinosaur floated in water like a present-day crocodile.' On the following page, I reproduced a screenshot taken from the BBC documentary, with the dinosaur poised at the water's edge and ready to dart its snout through the waves to seize a passing fish. It bore the crucial caption: 'Current interpretations of *Spinosaurus* confirm it lived near water, plunging in its head to find food. In my view it was an aquatic dinosaur and lived largely immersed in water where its fish diet was abundantly available.'[6]

This was revolutionary science. Not only was there plenty of evidence to show that dinosaurs must surely have evolved in an aquatic habitat, but – in the case of *Spinosaurus* – the interpretation was clear. I had unequivocally stated: 'in my view it was an aquatic dinosaur' and thus the die was cast. There could now be a sensible debate.

The public response was immediate and enthusiastic. The phone rang with messages of congratulation. British newspapers, from the *Daily Telegraph* and the *Metro* to the *Daily Mail*, covered the story at length and the global press were equally positive with reports ranging from the *International Business Times* right across to *Fox News*.[7]

Large numbers of reports began to appear in countless countries right round to Australia, and blogs were soon claiming that the new theory had made a global impact. One announced the theory as: 'Dinosaur News! Were Your Favorite Prehistoric Beasts Actually Aquatic?' and went on to say: 'Regardless of where the science lands on this one, Ford's article is a fascinating read and an interesting exercise in changing one's imagination viewpoint … it's worth watching Ford's adorable video introduction.' After that I was not sure whether to smile or squirm.[8]

The BBC's science correspondent Tom Feilden heard about my theory from John Humphrys, who has been a friend of mine since the 1960s. Feilden came to conduct a lengthy interview for the *Today* programme, to be broadcast on publication day. Feilden opened with a clip from one of the BBC dinosaur documentaries imbued with an aura of mystery and magic. 'Somewhere on the edge of a Jurassic forest, a mother dinosaur is laying her eggs,' it began, the portentous words wrapped in subtle sound effects of mysterious life forms uttering strange and distant sounds. Feilden interjected: 'The classic, modern interpretation of life in the Mesozoic, as depicted by the

BBC's award-winning series *Walking with Dinosaurs*.' There were more mysterious sound effects, as the documentary sound-track continued: 'She is a 25-ton *Diplodocus* ...' before fading away in the distance as Tom Feilden said: 'Relying heavily on the latest scientific evidence, as well as the most advanced computer animatronics,* the series was as much natural history as it was prime-time entertainment.'

At this point the programme turned to my demonstrating the incongruities: 'Let me just call up this clip, from one of the BBC dinosaur documentaries,' I was saying: 'Look at them – how they're walking!' Feilden interjects: 'But for all its diligent research, and authoritative voice-over, Professor Brian J. Ford believes something is fundamentally wrong with this picture.'

I was saying: 'Well, every time you see these images, they're always the same; you have these huge dinosaurs, crunching across the arid desert, they are holding up these huge tails – the tail of a dinosaur could weigh 10, 20 tons – which they are holding erect, and they're looking around for prey in this sort of sunny landscape. It makes no sense.'

Feilden explained: 'The problem, Professor Ford says, is that these larger dinosaurs are simply too big; too big to support their own weight, too big to consume enough food to maintain their vast bulk, too big to be able to move around with anything like the agility other aspects of their physiology imply – unless, that is, all that bulk is somehow supported.'

At this point I continued: 'Just imagine that the landscape was flooded with water. Now it suddenly makes sense. His huge tail is buoyant, it's floating in the water; it doesn't cause any energy drain. If he waves it then it's going to help him swim; it becomes a swimming aid and suddenly he's in an environment which is sympathetic to him. That "arid" mud wasn't arid at all – that mud was the bottom of a shallow lake. The water took their weight, not their limbs.'

* This is not quite right; 'animatronics' describes the construction of a functioning model, whereas what Feilden means here is computer-generated imagery or CGI – there were no animatronic dinosaurs in the series, only computer-generated digital re-creations.

The press reports were all encouraging, though not all were accurate. Some, ranging from the *Daily Telegraph* in England to the *Irish Independent*, reported that my theory suggested dinosaurs lived *under* water, rather than in shallow lakes. They were quickly notified, and their later reports were corrected. Most were accurate, and some were extensive. In Germany the science magazine *PM* soon published a feature by their editor, entitled 'Saurier waren Amphibien'.[9]

Now the theory had been launched worldwide and the ground was clear for a sensible debate. We could all air the various strands of scientific inquiry, and eventually there would surely be a new consensus on how dinosaurs evolved.

That is what I thought. But palæontologists would have none of it. We had a hint of the establishment's response when Tom Feilden followed his interview with a comment, by way of balance, from the Natural History Museum. Feilden had commented: 'Brian Ford is, he freely admits, *not* a palæontologist. He's a cell biologist, but with a polymath's delight for overturning apple-carts. So, what do the experts have to say? I brought his paper, which is published in the latest edition of *Laboratory News*, to the Natural History Museum [in London] – where else? – to find out.'

'This actually harks back to a lot of the views that people held about dinosaurs earlier in the twentieth century, and even in the nineteenth century,' said a voice identified by Tom Feilden as 'Dr. Paul Barrett, one of the world's leading experts on non-avian dinosaurs.' Barrett continued: 'So, in a number of respects, this is if you like a recapitulation of exactly those same kind of arguments our colleagues were having almost 100 years ago.'

Mused Feilden: 'Nothing, it seems, is ever truly original in science. And sadly, that's about as far as Dr. Barrett's willing to go in endorsing Brian Ford's new aquatic dinosaur theory. Extraordinary claims, as the saying goes, require extraordinary evidence, and the Natural History Museum is not about to flood its dinosaur display any time soon.'

'Things have moved on quite a lot,' said Barrett, 'and there's been a huge body of work, in particular since about the 1960s, when people started getting interested in dinosaurs in a big way again, that show really quite conclusively that most of these animals are actually perfectly well engineered for a life on dry land.'

Feilden added: 'So the idea that – because they are so large, so heavy, they must have had some assistance supporting them – you're not buying it.'

Barrett concurred: 'I don't think we'll be re-writing any of the textbooks just yet.'

Tom Feilden closed his report with these words: 'Somehow, I don't think that's going to be enough to persuade Professor Brian Ford. As another famous scientific dissenter, Galileo, was reported to have muttered under his breath, when forced to deny that the Earth revolved around the Sun: *eppur si muove* – "and yet, it moves".'[10*]

When I heard his reports being broadcast, the mention of Galileo made me wince. The comparison between my wish to change palæontology and Galileo's struggle against the Church was like comparing an ant to a lobster. When I visited Florence shortly afterwards I went to the Museo Galileo (formerly the Istituto e Museo di Storia della Scienza) in the Piazza dei Giudic to apologize personally in front of Galileo's marble bust.

Barrett's protests about my views were only the beginning. The world of palæontology was incensed; they even insisted that the BBC should never have broadcast that interview. At Bristol University, the Department of Palæontology issued a statement:

> As everyone now knows, last week the respected and trusted *Today* programme on BBC Radio 4 ran an absurd nonscience piece on Brian Ford's wild, ignorant, uninformed speculation that all dinosaurs lived in shallow lakes because that was the only way they could support their weight ... plenty of people have shown what utter, contemptible nonsense this is. During 24 hours, 20 palaontologists [*sic*] signed, and so this is what was submitted at 3 pm on Thursday 5th April:

* The BBC report on the *Today* programme can be accessed through their website. The aquatic dinosaur recording has recently been deleted, though an off-air recording is available: https://youtu.be/ZsX1ux3HSvU

> 'Dear Radio 4, The *Today* Programme for Tuesday 3rd April 2012 contained a science piece by Tom Feilden regarding Professor Brian J. Ford's "theory" that dinosaurs did not live on land but in shallow lakes which supported their weight.
>
> 'Professor Ford's theory was published in a magazine rather than a peer-reviewed journal, and is wholly unsupported by any evidence whatsoever. It contradicts all evidence from dinosaur anatomy, biomechanics, sedimentology and palæoenvironments, and does not even qualify as fringe science. It is unsupported and uninformed speculation which Ford could have disproved had he taken just ten minutes to look at the readily available literature representing a century of consensus. By giving air-time to this speculation, even comparing Ford with Galileo, Radio 4 has unfortunately lent it a credibility that it has not earned, introduced a time-wasting controversy where there is not a controversy, misled the public, and maybe most importantly compromised its own credibility as a trusted source of science reporting. No listener with any knowledge of palæontology will have been able to take this report seriously; will they believe the next science report you broadcast?
>
> 'To mitigate this damage, we recommend and request that you broadcast a formal retraction.'

Although this had been left open for signatories to support for only 24 hours, it attracted support from all over the world. Many distinguished palæontologists wanted to be signatories: Mike Taylor, from the Department of Earth Sciences at the University of Bristol, England; David Marjanović, Museum für Naturkunde, Berlin, Germany; Silvio C. Renesto, Associate Professor of Palæontology, Department of Theoretical and Applied Sciences, Università degli Studi dell'Insubria, Italy; Grant Hurlburt, Department of Natural History, Royal Ontario Museum, Canada; Michael Balsai, Department of Biology, Temple University, Philadelphia, USA; Bill Sanders, Museum of Paleontology, University of Michigan, USA; Stephen Poropat, Department of Earth Sciences, Uppsala University, Sweden; Oliver Wings, Curator of Vertebrate Palæontology, Museum für Naturkunde, Berlin, Germany;

Jon Tennant, Independent Researcher, UK; John R. Hutchinson, Department of Veterinary Basic Sciences, The Royal Veterinary College, UK; Lorin R. King, Department of Science, Math and Physical Education, Western Nebraska Community College; Scott Hartman, Paleontologist and Scientific Illustrator, SkeletalDrawing.com; Neil Kelley, Department of Geology, University of California at Davis, USA; Matteo Belvedere, Department of Geosciences, University of Padova, Italy; Andrew R.C. Milner, Paleontologist and Curator, St. George Dinosaur Discovery Site, Utah, USA; James I. Kirkland, State Paleontologist, Utah Geological Survey, USA; Jerry D. Harris, Director of Paleontology, Dixie State College, Utah, USA; Andrew A. Farke, Curator, Raymond M. Alf Museum of Paleontology, Claremont, California, USA; Daniel Marty, Palæontology Editor of the Swiss Journal of Geosciences; and Manabu Sakamoto, School of Earth Sciences, University of Bristol, UK. This is a catalogue of the most eminent palæontologists working today – and all of them insisted that the BBC must admit that it should not have broadcast their report. They wanted a retraction.[11]

The BBC were quick to respond. Within a few hours, Mark Roberts from their complaints department wrote back at length. Among his comments, he said:

> I understand that you were unhappy with the inclusion of a report by Tom Feilden on a theory proposed by Professor Brian Ford regarding how dinosaurs lived. I note you believe the report gave credibility to this theory and compared the professor with Galileo.
>
> Your concerns were forwarded to the programme who explained in response that the item in question was a light-hearted feature looking at an outlandish new idea about the dinosaurs and which was clearly signposted as such. They added that the item even included one of the world's leading experts on dinosaurs, Paul Barrett, exposing its flaws and ridiculing it and that it was very clear where Brian Ford's article was published since *Laboratory News* was clearly mentioned. They also added that the reference to Galileo was simply an aside about the importance of dissent in science [suggesting that] Brian Ford was unlikely to be put off by

> the condemnation of the established experts, and not, as you suggest, a comparison between Brian Ford and one of the greatest scientists of all time.

The petitioners were not remotely satisfied with this anodyne response from the BBC. They found it 'completely unsatisfactory' and said: 'Trying to pass the segment off as "a light-hearted feature looking at an outlandish new idea about the dinosaurs and which was clearly signposted as such" just won't fly: its page on the BBC site is entitled "Aquatic dinosaur theory debated" and there is nothing about it that signposts it as any less serious than, say, the piece they did … on Brontomerus, or on sauropod neck posture.' The author added: 'As it happens, my mum called me for a chat a couple of days ago, asking me whether I'd heard "the new theory" on the *Today* show. It was pretty painful having to let her down. She obviously didn't hear it as "a light-hearted feature". It's going to be harder now for her to accept other science reporting on *Today*.'

My plan for a reasoned debate was proving to be a forlorn hope. I had imagined that the 'aquaticists' would line up with the 'anti-aquatic movement' and the known facts would be aired, so that we could see which view was probably correct. This would be the only way to reach a consensus. But instead, I was immersed in a rising tide of opprobrium. The entire world of palæontology was up in arms, and they all dismissed the theory as rubbish. One headline shouted: 'Who the Hell is "Prof. Brian J. Ford"?' while another ridiculed the reports simply as 'Bad Science Journalism' and one writer asserted: 'No, dinosaurs were not aquatic.' Another headline was: 'Sinking the Silly Idea of Aquatic Dinosaurs,'[12] while 'Brian J. Ford's Aquatic Dinosaurs Claim Holds No Water,' retorted another.

On April 3, 2012, the Smithsonian blog published: 'Aquatic Dinosaurs? Not So Fast! A cell biologist says dinosaurs spent their days floating in lakes, but his idea doesn't hold water,' and they returned next day with: 'Paleontologists Sink Aquatic Dinosaur Nonsense; tales of aquatic dinosaurs have proliferated through the news, providing one more sad example of failed reporting and the parroting of fantastic claims.'[13]

There were also email messages. On June 23, from the office of the curator of Palæontology (Vertebrates) at the Sedgwick Museum of Cambridge University, came these words:

'Brian,' said the message from David Norman, 'I am sorry to say that I think that you are completely misguided in both your approach and understanding of these matters.' The advice he offered was clear – I should *not* continue publishing in this vein. 'You have generated some "publicity" that, in Lord Beaverbrook's opinion, is "a good thing" but don't get carried away with this …' the email continued. It concluded by pointing out that my ideas were being 'widely ridiculed' by the authorities on-line.

The attacks seemed endless. Soon I began to receive messages and calls from people who had been following streams of messages on Twitter, saying that they were full of invective. Those kind folk all sympathized, saying how much I must have been wounded by these ceaseless insults. I was quick to explain that I wasn't in the least offended by it all, just surprised. So often this has happened to others who tried to innovate in science. That was a subject on which I had often lectured and sometimes written about, and here I was in the middle of just such an episode. I've never known what it was like to be lambasted for a new scientific theory, though I was finding it fascinating and was astonished not only at the unanimity of the condemnation in the field of palæontology itself, but also by the warmth of the wider scientific community who could see that the aquatic dinosaur theory brought together answers to so many paradoxes, and offered a single, succinct solution. This was going to provide endless material for my lectures, and I found it all fascinating.

At *Laboratory News*, editor Philip Prime was being inundated by complaints from palæontologists. He sent me a message to say that, in the interests of balance, he proposed to include an article by Naish of comparable length in his next issue to present the opposition expressed by the world of palæontology. I agreed entirely; it seemed to me that this might offer the start of a debate. Most of the rebuttal he published simply restated the known facts about dinosaurs, referring to the gigantic size of the largest, 'over 30m long and around or over 100 tonnes,' reminding readers that dinosaurs 'were mostly neglected as objects of scientific research' for decades, and concurred that the research was reinvigorated when palæontologists accepted that 'dinosaurs were "warm-blooded", that they lived on in today's world as birds, and that their story was one of success and innovation.' The article reported that the responses to my article had been 'aggressive' and they had questioned the 'abilities of a non-specialist to declare expertise in an area where there is no evidence of prior experience.' The response reminded readers that palæontologists believed dinosaurs to be 'unstable and prone to tipping' (this was Henderson's amusing paper rearing its head) and misguidedly claimed that 'rivers, lakes and ponds were rare, small or even wholly absent' from places where dinosaur skeletons had been excavated. Adaptations to an aquatic habitat would include 'paddle-like hands and feet, tails specialized for sculling, dense bones that contribute to buoyancy control, and eyes and nostrils positioned high up on the head,' it stated. The article even claimed that 'literally millions' of fossilized dinosaur eggs had been discovered. Millions?

Some of the points were constructive – it had to concede that dinosaurs 'may have been good swimmers' – but most were muddle-headed, like the bald assertion that the spinosaurs had 'crocodile-like jaws' because they were unquestionably 'waterside predators that waded in the shallows, dipping their snout-tips into the water to grab large fish.' Everyone assumed that was how they behaved, but I had shown this was wrong.

The weight of dinosaurs was an intriguing issue to raise. As we have seen, ever since 1970, palæontologists had been scaling down their estimates of dinosaur mass, simply because it was impossible to show how such vast animals could ever have moved about easily on dry land. However, the article referred to dinosaurs like *Amphicœlias*, which could have weighed 120 tons. There is nothing new about this – remember that *Amphicœlias* had been named by Edward Drinker Cope way back in 1878. This dinosaur is likely the largest ever to be discovered, and had been named from the Greek αμφι (*amphi*, on both sides) and κοιλος (*koilos*, concave) in allusion to the shape of its vertebræ. The few bones known to science had been excavated by Oramel Lucas, one of Cope's fossil hunters, in 1877, and estimates suggest that this would have been a colossal dinosaur. With recent discoveries of titanosaurs being regularly reported, palæontologists have now had to concede that the largest creatures must have been more than 100 tons in weight, and the earlier estimates are now being reluctantly revised upwards again.

Paul Upchurch, Professor of Palæontology at the University of London, was the only palæontologist in the world to attempt to start a discussion, and he said: 'I enjoy turning an issue upside-down and thinking the unthinkable.'[14]

His comments did not, in the event, show any inclination to agree. Among his points were the familiar claims that the predatory (in reality, scavenging) dinosaurs lacked anatomical adaptations for hunting in water: 'T. rex and allosaurs do not have flippers …' he said, before pointing out that they had been shown to be unstable in water. Curiously, among his objections he did include one of my own conclusions: 'Sauropods would have to float with their necks along the surface,' he insisted. In my view, that's exactly correct. Upchurch listed this as an objection to my theory, but in fact this is precisely how I believe they must have lived.

Throughout this time the notion of dinosaurs as birds was regularly raised in the media – but remember that the idea that birds were closely related descendants of dinosaurs had been published in 1887 by Harold Seeley, so this idea is already more than 130 years old. Not only that, but asserting that 'birds are dinosaurs' is wrong. Birds and dinosaurs share common ancestors. Birds are descended from an evolutionary line that involved the dinosaurs; but to say that they 'are the same' is as ridiculous as claiming that humans are lemurs, or butterflies are beetles. It is groundless to claim that shallow water in ancient areas where dinosaurs are found is 'rare, small or absent' because every dinosaur skeleton ever found has been retrieved from mudstone, from siltstone, or from sandstone. There is an argument that some of them have been found in strata that represented wind-blown sands, but there are plenty of reasons for that. A lake bed can easily dry out and become desert-like in the fullness of time, just as drifts of desert sands can be blown into a water-course. In any event, these shallow water zones are neither 'rare' nor 'small'. They extended over vast areas of the early Earth, and it's in those areas that we find fossilized dinosaurs – nowhere else.

One morning, an extraordinary proposal arrived. Several media folk had thought that this whole project would make a great television documentary, though nobody had produced a solid suggestion. It was now suggested that the work on this new theory would be suitable for the big screen – the saga of the aquatic dinosaur should be made into a movie for the cinema. The film would be produced by Fernando de Jesus, a successful producer with Independent Television in London, and directed by his partner Daisy Lilley, who has produced a range of television programmes including the *I'm a Celebrity* series. Over the years I have made a great many television programmes and have hosted documentaries and, for two seasons, even hosted my own game show called *Computer Challenge* when I was young and irresponsible. The idea for a film was intriguing, and we agreed that a short promotional sampler should be filmed at my house. Neither of these people wanted to use any of the existing footage, including interviews and lectures, and so they arranged for a full cinema shoot in 4K digital video, with a team using a movie camera the size of a suitcase and lights everywhere like Harrods at Christmas. They filmed discussions and demonstrations and gathered facts and figures. At times we felt as if the crew from *Downton Abbey* had invaded our home. Over the following months, we received emails excitedly mentioning the various theatrical outlets that were showing interest – but, in the end, nothing came of it. There is some fascinating footage (on file, no doubt), but the idea of a movie has so far come to naught. Although it always seemed a far-fetched proposal, it was an intriguing experience.

Palæontologists, as a group, are not very good at zoology. Many of them argued against my theory because they said that, if dinosaurs were aquatic creatures, they would have evolved limbs that were paddle-shaped, possessed webbed feet, and tails with fins for sculling. That makes sense only if you know little of zoology. A great number of aquatic creatures do not follow the predictable pattern. The most fearsome aquatic reptiles today are gigantic crocodiles in Australia that can reach 7 metres (23 feet) long and weigh over a ton (up to

2,600 pounds or 1,200 kg). In many ways they are close to the way I envisage dinosaurs, with a semi-aquatic lifestyle. Are crocodiles' extremities paddle-shaped, and their digits webbed? Certainly not – crocodiles have claws. Are their tails finned for swimming? Indeed, they are not. They have tails just like those of dinosaurs. Indeed, many of them have projecting dorsal scales very like those of stegosaurs. One particularly revealing picture was taken by Kevin Ebi of Living Wilderness Nature Photography. It shows caudal projections on an alligator, which, if you inspected them in close-up, could easily be taken to be those of a *Stegosaurus*. The eyes and nostrils of crocodiles are 'high on the head' much like the typical dinosaurs. Dinosaurs and crocodiles are distant cousins, so it is not surprising that they share the genes that give them uncanny similarities.

You may think my suggestion of a lack of education in zoological matters among many palæontologists is high-handed and condescending, but it's true. One example is the way they have interpreted the holes that are often found in a dinosaur skull. All dinosaurs have large openings (*fenestra*, from the Latin for window) in their skulls. At the back of the skull is the post-temporal fenestra; then there are infratemporal and supratemporal fenestræ that accommodate well-developed jaw muscles, and in front of the eye socket is the antorbital fenestra. Nobody has ever put forward a convincing reason for its development (apart from the vague possibility that it helps to reduce the weight of the skull). In some theropod dinosaurs there is also a small maxillary fenestra just behind the opening for the nostril (the naris) and a mandibular fenestra in the lower jaw. And there are other holes too: dinosaur skulls are often scarred by holes the size of *T. rex* teeth. These are rounded holes that penetrate the bone, and, as we have seen, the standard explanation in palæontology is that these dinosaurs were attacked by predators. The evidence they provide seems to fit: these are holes in the skulls of dinosaurs which a *T. rex* tooth fits perfectly. Thus, these wounded dinosaur skulls show that *T. rex* was a voracious predator, and the evidence is revealed through these bite marks.

Except for one important fact – this explanation does not work. A crushing bite by a predator doesn't drill a single hole in the skull; it cracks it open. If a dinosaur were to be attacked by a predator, its skull might be broken apart, but it wouldn't be perforated by a tooth. Picture the mechanism: the jaws of a *T. rex* open wide, they seize the head of an opponent, and then they start to close with colossal force. How is this going to leave a single hole? The jaws each contain a row of teeth. If there were to be tooth-marks left on a bone, they would be in the form of rows of indentations, left by the array of the teeth (and they would be left on opposing sides of the skull). The holes that we find in dinosaur skulls are not spaced in a way that fits the dentition of predators, and the only way a single hole could be left is if a dinosaur had been attacked by a *T. rex* with a single, isolated peg-like tooth remaining in its upper jaw. Yet no skull like this has ever been found. Jaws with most teeth missing are unusual in *T. rex* fossils, and nobody has ever excavated a skull with just one tooth remaining in its jaw. Interestingly, the margins of these holes show that most of them have undergone healing since they were inflicted, so we know the 'attack' was not fatal. The fact that the holes fit a tooth is because teeth are tapered, so a huge range of holes will fit – you just have to slide the tooth in further if the hole is larger in diameter. Yet, although the standard explanation cannot make any sense, it is still widely promoted.

To me, there is a different way of looking at these holes – perhaps they were eroded through disease. It could be that an infection burrowed its way through the skull. The first time anyone provided evidence for the structure of these holes was in July 2017, when a scan was carried out to reveal their profile. The conclusions revealed that the holes tapered and were narrower inside the skull than they were at the surface. The researchers claimed that this made it impossible for an infection to have spread from inside the skull, so the idea was ruled out. They also dismissed the notion that the holes were made by a predator's tooth, though that was already obvious.[15]

The infection idea is an explanation that I had already developed. An infection in bone that began on the head and worked its way inwards could leave a tapered hole of exactly the type that we find, and – in an aquatic environment – it would be likely for microbes to revel in the moist surface tissues and cause abscesses to develop. Even if the infection was initiated within the head, it could leave a tapered gap as it eroded the bone. Actinomycosis is a condition of this sort that I have studied microscopically, and I have a number of microscope slides showing the erosion of bone. Yet, even though the idea that these sporadic holes could ever have been caused by biting is self-evidently ridiculous, it is widely accepted by palæontologists.

The rebuke that *Laboratory News* published also dismissed my method of volumetric analysis as 'dunking toy dinosaurs in water'. I do not think other scientists would consider it justifiable to dismiss this procedure as amateurish. This water-displacement method was derived specifically for my original *Laboratory News* article, though I find that it is an approach which is widely accepted by experienced scientists and indeed is also employed by palæontologists elsewhere in their discipline. The Lawrence Livermore National Laboratory (L.L.N.L.) in California teaches precisely this method for dinosaurs.[16] This is the method employed by Colbert in 1962, by Paul in 1997, and by Henderson in 1999, and it is widely referenced in the standard scientific papers that consider the volume and weight of dinosaurs. Pretending we were merely 'dunking toys' is condescending, in the way that schoolchildren like to tease a rival, but it turns out to be a tried and tested technique that has been relied upon for centuries. It is, indeed, fundamental to Archimedes' Principle.[17]

Many of those protesting about my *Laboratory News* article said that an outsider to any discipline should not be given credence without the support of their peers, and you will recall that the rebuttal published in *Laboratory News* questioned 'the abilities of a non-specialist to declare expertise in an area where there is no evidence of prior experience.'[18]

That seems to be a reasonable argument, but can it stack up in science? The role of the outsider in the progress of scientific research is a crucial concept to consider. We believe that scientists need to be trained in the ways of their discipline and, as we know, this lies behind the questionable concept of peer review in assessing new research. Unless you've been properly trained, they say, you cannot practise; and, unless your peers have agreed with your views, they cannot be published. The system arose to ensure consistency and reliability in science, though we have seen that it also has a secondary consequence: revolutionary ideas and innovative theories can be crushed by the weight of orthodox opinion. Once a topic is accepted, then peer review works wonderfully. But until then, it can be exceedingly hard to launch any new approach in science. Scientists are given their grants to further accepted lines of inquiry, but if a new approach is launched, then the currently conventional concepts are suddenly obsolete, so startling changes in direction are difficult to start. There is every likelihood of obtaining finance for a project that everybody already understands, but none whatsoever for innovative ideas.

In fact, many of the great leaps in scientific have been created by outsiders. Steady progress is made best in universities; but the radical new insights are the province of the devoted amateur. Think back to the amazing imagination of Leonardo da Vinci in the 1400s; he was a painter. A century later we had a self-taught aristocrat, John Napier, who revolutionized mathematics with his invention of logarithms. It was Robert Hooke in the 1600s who invented the spring balance and the spirit level and who gave us our first insights into the microscopic world – all without training. Antony van Leeuwenhoek, who discovered microbes with the world's first high-powered microscopes, was a draper. In the 1700s we had Benjamin Franklin, inventor of the Franklin stove and electrical pioneer, a printer who became a politician. We know of Antoine Laurent Lavoisier and his discovery of oxygen; but may not realize that Lavoisier was a tax collector (hence his execution by guillotine). His British counterpart Joseph Priestley, who made many pioneering chemical discoveries, was a cleric. William Herschel, who discovered Uranus and nebulae, was a church organist. Thomas Young, who gave us Young's Fringes in interferometry and Young's Modulus in elasticity, and was also the person who deciphered the Rosetta Stone, was qualified only as a physician. Michael Faraday gave us a host of electrical insights, including the anode, cathode, electrode and electrolyte, the dynamo and the transformer, yet he was an unqualified bookbinder's apprentice. Gregor Mendel, who pioneered the study of inheritance in the 1800s, was a monk. Louis Pasteur, who gave us the germ theory and who pioneered bacteriology and immunization, was a chemist who had never studied microbiology. John Boyd Dunlop, who invented the pneumatic tyre, was a veterinary surgeon. Joseph Swan, who perfected the light bulb and created the world's first electrically illuminated buildings, was really a chemist. Thomas Alva Edison was a telegraph operator when he went on to become America's greatest inventor, perfecting the phonograph and inventing an electrical voting machine. George Eastman, who patented the first photographic film and the Kodak camera, was a bank clerk, while Leopold Mannes and Leopold Godowsky, Jr., creators of colour photography, were both concert musicians who travelled to gigs with cases of chemicals for experimenting in their spare time. Most of the luminaries we have discussed in this book were amateurs, from Sir Charles Lyell (who was a lawyer) to Thomas Jefferson (who was studying fossils 40 years before the word 'dinosaur' was even coined).

Now, I know what you're thinking: this was all very well in the olden days, but in the modern world you must be academically qualified to pursue a career as a professional pioneering scientist. Not so. What has been the most significant development of our lifetimes? The computer, without doubt; it was pioneered by Charles Babbage, who dreamed up the idea to help him bet on the best racehorses, and the idea was then developed in Poland during the 1930s by Marian Rejewski, Jerzy Różycki and Henryk Zygalski into an electronic computer which they used to interpret Enigma messages sent by the Germans (work always claimed to be British). Their innovation re-emerged as the world's first programmable electronic computer in 1943 when a new version was built by Tommy Flowers, a brilliant British telephone engineer who used his own money to fund the project. Flowers built his version with electronic valves, and his vision became the world's first computer. Then there were the pioneering inventions by Victor Glushkov in the Soviet Union, who in the 1960s created the first personal computer, complete with a keyboard, a monitor and a light pen for manipulating text and creating images on screen. He created a high-level programming language that could handle fractions, polynomials and integrals, and named his invention MIR. In the Cyrillic alphabet, МИР stands for Машина для Инженерных Расчётов, meaning Machine for Engineering

Calculations (rather neatly, MIR translates from Russian as both 'peace' and 'world'). He was followed by my colleague Sir Clive Sinclair, who produced Britain's first mass-market personal computer, Bill Gates the Microsoft pioneer, Steve Wozniak who founded Apple and was joined by Steve Jobs, then Mark Zuckerberg who gave us Facebook.What unites them all? Not one of them was qualified. They are all rebels who dropped out of university. It was their personal insights that they pursued, not a formal academic career. The most far-reaching twentieth century development in the life sciences? Surely that is the discovery of the structure of DNA, published in 1953 by Francis Crick and James Watson. Both had been given strict instructions not to meddle with research into DNA. Neither of them carried out any research with DNA but relied on data from Rosalind Franklin and Maurice Wilkins; Crick and Watson weren't qualified in the field, yet their discovery revolutionized the whole of biology (although when it was first announced by Sir Lawrence Bragg at a major scientific conference in Belgium, the press took no notice). The greatest scientific theory of them all – relativity – made the name of Albert Einstein renowned throughout the world. Yet in 1895, when he was 16, Einstein failed the entrance examinations for the Swiss Federal Polytechnic in Zürich. After he eventually qualified he could not find a teaching position and instead became a clerk in the patents office at Zürich. It was here that he compiled his ground-breaking papers on relativity; it was his spare-time obsession. Like all the others, he was a hobbyist, and in the true sense of the word, an amateur.

Amateurs have always been at the forefront of innovating science, and the true path of scientific progress is very different from the way it is usually portrayed. New ideas in science take a long time to percolate through the consciousness of the community, and being an outsider is an advantage here. The spectator in the stand always sees more of the game than the players on the field. That is why most dinosaur discoveries are made by the spare-time hobbyists and enthusiastic amateurs for whom I have such admiration, and you can see why it is exceedingly difficult for new ideas to penetrate from outside.

Only the aquatic theory can explain how dinosaurs grew to be so fabulously large – yet the very idea challenges the basis of the fake news that modern palæontologists perpetuate. No wonder they don't like it. Although dinosaurs are huge, it is hard to comprehend their real size. Sauropods were the size of a jet airliner, and they weighed even more. The largest dinosaur was over 110 feet (34 metres) from nose to tail, the length of a Boeing 737-300 airliner. Those gigantic sauropods weighed more than 100 tons, while the Boeing weighs a mere 36 tons (72,100 pounds or 32,700 kg). Although the biggest blue whale (*Balænoptera musculus*, named from its baleen food filters and its muscular body) is 90 feet or nearly 30 metres long and weighs 150 tons, it is somewhat smaller that a substantial sauropod and some 50 per cent heavier – though it needs to be permanently supported by its marine habitat. Most of the other large whale species are similar in size and weight to the giant dinosaurs. The finback whale *B. physalus* measures 80 feet (24 metres), the right whale *Eubalæna spp.** and the sperm whale *Physeter macrocephalus* each reach 60 feet (18 metres) – they all grow to weigh about 80 tons.

From our studies of today's elephants (and their antecedents) a 'rule of thumb' begins to emerge. In order to hang together, the largest a land animal can weigh is about 10 tons. Allow an animal to evolve in an aquatic habitat, where buoyancy can help its skeleton to support its organs, and an animal can evolve to be ten times larger at 100 tons (perhaps 120 tons). We must always base our conclusions on scientific evidence, and the facts of science show that a large sauropod dinosaur is clearly comparable to other gigantic aquatic vertebrates, and it has nothing in common with animals that evolved to live on land.

* The right whale, known for centuries, has recently been found to comprise three genetically distinct species, the North Atlantic right whale *Eubalæna glacialis*, the North Pacific right whale, *E. japonica*, and the Southern right whale, *E. australis.* All are members of the family Balaenidae, which includes the smaller bowhead whale *Balæna mysticetus*, and they are all comparable in size to most giant dinosaurs.

9

COPULATING COLOSSUS

Dinosaurs had sex, of course; but how? And how can we tell the gender of a fossilized dinosaur? It is simple in placental mammals. The birth canal of the females exerts specific constraints upon the shape of the pelvis, and it is easy to make the distinction. There is a fossilized hippopotamus skeleton in the Sedgwick Museum of Cambridge University, for instance, which has been made up from several different fossilized skeletons. One half of the pelvis is female, the other side is male, and the structure of the pelvis can be easily seen to be markedly different on each side. Dinosaurs laid eggs that, for the size of the adult, were surprisingly small, and so the pelvis cannot be expected to show pronounced anatomical distinction between the sexes. There are other cues we can use, however. In 2004, Dr. Mary Higby Schweitzer of North Carolina State University discovered that some of the bones from a *T. rex* fossil were preserved with microscopical details still visible. She described seeing osteocytes (bone cells) preserved within the bone.[1]

One of her observations was of medullary bone, which is found only in female birds, a layer of mineral-rich tissue that is laid down within the bones to provide a source of calcium called upon to generate the shell of a newly forming egg. Here was a clue – medullary bone found within a fossilized bone could arguably be linked to the gender of the dinosaur.[2]

This needs well-preserved tissues, which are rarely encountered, and it also requires sections to be cut from the long-bones, which is hardly ever permitted, so this is not a handy test one could apply to any fossilized skeleton. However, Schweitzer's research suggests that this skeleton may have come from a female *T. rex*, and so anatomical disparities between this known specimen and others of the same species could perhaps point to sexual distinctions, and some similar skeletons from different sources also show slight disparity in structure. Of the 30 skeletons of *T. rex* that have been excavated – most of them incomplete – it does seem that they fall into two groups, or morphotypes, with slightly different skeletal proportions. There is a good display in the Palæontology Museum of Manchester University, with the pelvic structure of the two morphotypes being distinguished as 'gracile' and 'robust' forms. Because the pelvic structure of the robust form is slightly wider, it seems arguable that this may be an evolutionary adaptation for egg-laying; if so, then this morphotype would actually be female and the gracile form could be the male. It is a sound line of reasoning. As matters stand, we might be reasonably confident that we can tell the sex of a dinosaur.

The central problem remains, how did dinosaurs contrive to copulate? If there is one single field where palæontologists have resorted to rampant invention, dressed up as fact, it is in the way dinosaurs had sex. What do we know? Where is the evidence? Since dinosaurs were reptiles, we can assume that they would have possessed a cloaca for copulation – the term is Latin for 'sewer' – a system in which the sex organs are brought together for the exchange of spermatozoa. Lizards do this, and so do birds, and we also find a cloaca in the most primitive mammals. Those that (like dinosaurs) lay eggs, the echidnas, and the platypus, possess a cloaca. In some reptiles, like crocodiles, there is an organ that acts as a penis within the cloaca, and a similar feature is found in a few types of primitive present-day birds. It would be reasonable to suggest that male dinosaurs had developed a penis in this way for the intromission of sperm into the female.

We cannot know the method of erection, though some mammals have a penis bone (the baculum), which serves to make penetration more reliable. Some reptiles have a penis that is brought to erection by means of blood pressure, as happens in humans; and we know that the largest birds (like ostriches) have an erectile penis that is inflated by lymph, rather than blood. So there are several possibilities – though, since such soft tissues are unlikely to leave traces in fossils, we may never know for certain.

How could gigantic dinosaurs manage to copulate? It is impossible to imagine how these huge creatures copulated on dry land. This is a problem that has been faced for over a century, and nobody has ever come up with a satisfactory explanation. The first example I can find in which evidence is suggested for mating behaviour in dinosaurs was in 1906 when, in their joint paper, Henry Fairfield Osborn and Barnum Brown assured their followers that the greatly reduced forelimbs of *T. rex* were used 'for grasping during copulation'.[3] Similar suggestions were subsequently made over the thumb spikes of *Iguanodon*. None of these ideas gained any following, and all were eventually abandoned.

In spite of these claims, there is no scientific evidence to reveal how dinosaurs could copulate, though this has not prevented prominent palæontologists from transmuting their own idle speculation into hard facts. One of the first to deal with the subject in depth was

Beverly Halstead was one of the few who wrote of dinosaur sex. In 1988, he was interviewed by Sandy Fritz for a feature in *Omni* magazine, and Ron Embleton created some impossibly impracticable studies of how sauropods copulated.

an English investigator, Beverly Halstead, who also went under the curious name of Lambert Beverly Halstead Tarlo. He published several books expounding on his views about dinosaurs and became well known for settling with unusual directness the question of dinosaur sex. Without any knowledge of how dinosaurs could copulate, he confidently concluded that they mated as do present-day reptiles: 'All dinosaurs used the same basic position to mate. Mounting from the rear, he put his forelimbs on her shoulders, lifting one hind limb across her back and twisting his tail under hers.' This all confidently asserted, though stated without a shred of scientific evidence.[4]

Once the territorial tyranny had seduced the world of palæontology, writers decided there was no need for scientific evidence. So keen were editors to publish on dinosaurs, and so eager were the mass media to report new findings, that palæontologists simply took to inventing scenarios without anything to back them up. They created whatever took their fancy and published speculation as fact. Edwin Colbert in a 1977 book wrote that two male *Brontosaurus* would frequently face each other: 'nodding their heads up and down or weave them back and forth through the considerable arcs, and at times they would entwine their necks as they pushed against each other.'[5] There isn't a scrap of science behind any of this.

In *The Dinosaur Heresies*, a book of 1986 that capitalized on the fashion for regarding dinosaurs as dynamic and fleet-footed, author Robert Bakker baldly stated: 'sexual practices embrace not only the physical act of copulation, but all the pre-mating ritual, strutting, dancing, brawling, and the rest of it.' This is, of course, pure fantasy and has no place in any serious scientific discussion.[6]

In 2007, Phil Senter suggested that it was sex that had driven the development of the long necks of dinosaurs like *Mamenchisaurus* and *Diplodocus*. To Senter, these were secondary sexual characteristics, and adults which had especially long necks (and were particularly adept at intertwining them during the sex act) had an evolutionary advantage over less well-endowed dinosaurs.[7]

That long neck imposes demands upon the circulatory system of a dinosaur, which would have to pump blood up to the height of the head. What size of heart would they need to pump enough blood all the way up to the brain of *Brachiosaurus*, if its head was more than 25 feet (8 metres) above the heart? Based on the body mass of this dinosaur, it was calculated that the heart would have had to weigh about 440 pounds (200 kg). There is the notion that the blood moving down the neck would – like a siphon – pull blood upwards, but that only works if the blood vessels are rigid tubes, like those of the plastic siphon in a toilet cistern.[8]

In reality, we are faced with the fact that a huge dinosaur would not be able to provide enough impetus to force blood up to the top of its neck if it were standing upright.[9]

The elongated tails of dinosaurs were often implicated in sexual display. Modelling of the skeletons of dinosaur tails has led some to conclude that gigantic sauropods (like *Apatosaurus* and *Diplodocus*) whipped their tails like a stockman cracking a bullwhip. It was even claimed that the wave motion along the distal extremity would have the tail snapping at speeds faster than sound, emitting an extremely loud noise that would attract the opposite sex.[10] This whip-cracking works well for a tanned leather whip, no doubt; but the damage that would be inflicted upon an articulated skeleton and the tissues that cover it make such concepts incredible. And I am not certain that a loud, supersonic bang would be immediately appealing as a stimulus to sex. Even for a dinosaur.

Is there any sound scientific evidence of how a pair of gigantic mating dinosaurs could bring their sexual organs into proximity? There is no obvious means of finding fossil data, for only in a few cases have fossilized animals been preserved in pairs. About 320 million years ago a mating couple of sharks were caught in a flow of mud and became trapped in their embrace for eternity. Closer to our time, 47 million years ago, a pair of mating turtles were similarly fossilized. The trapping of insects in amber, as well as rocky strata, has given several examples of their mating. To date more than 30 examples of fossilized copulating insects have been discovered. The latest example is a couple of copulating froghopper insects found by Shu Li of the Capital Normal University in China. These date from the Jurassic and date back 165 million years.[11]

There is no fossilized evidence of dinosaurs copulating. The closest we come are a few scratch marks that might possibly have resulted from scraping materials together for nest-building. A more fanciful interpretation was reached by a team of 15 palæontologists based at the University of Colorado at Denver, who concluded that the dinosaurs had been performing a ritualized mating dance. They discovered dozens of scrape marks preserved in Cretaceous strata in Colorado, and in their 2016 study they claimed to recognize similarities between these marks and the scraping performed by male birds in their mating dances.[12] Some birds (including the sage grouse and the puffin) perform a scraping dance, possibly to demonstrate to a would-be mate how proficient were their nest-building skills. During these ritualized performances the birds strut about, fanning out their tail feathers and puffing up their breasts. To the Denver team, this led to the conclusion that dinosaurs did the same. Rather than hint at a possible relationship in their paper for *Nature Scientific Reports*, they spoke in the title of large-scale physical evidence that occurred in 'ceremony behaviour by dinosaurs' (I think they meant *ceremonial*). The editors clearly liked this; journals will publish anything, no matter how feeble the scientific basis, that perpetuates the terrestrial myth of those majestic monsters. Much like dying film stars and drug-addicted musicians in the popular press, dinosaurs help to sell scientific journals. Their editors know that, and so do the wily palæontologists.

Popular accounts of the mating habits of dinosaurs published in 2012 quoted Kristi Curry Rogers, Assistant Professor of Biology and Geology at Macalester College in Minnesota, who reportedly told the Discovery Channel: 'The most likely position to have intercourse is for the male behind the female, and on top of her, and from behind; any other position is unfathomable.' Yet in truth, nobody could propose a workable hypothesis. None of it has anything scientific from which to work.

An academic study of biomechanics was the life's preoccupation of Robert McNeill Alexander, Professor of Zoology at the University of Leeds, and he concluded that the physical dimensions of gigantic dinosaurs meant they must have mated in the same way as today's elephants. The problem he saw was one to which I have alluded: in conventional copulation, the weight of a male dinosaur is being supported not on four legs, but only on two, which doubles the load upon each limb. He wrote of the difficulties a female dinosaur would experience in bearing the weight of a mounted male upon her back. In some ways he erred – the weight would be mostly borne by the hindlimbs of the male as he mounted, 50 tons on each limb, and not shared equally with the female beneath. It would be an unbearable burden. We have already seen that for any dinosaur even walking would be a practical problem, and supporting additional mass would be difficult to square with reality. Alexander had argued that the weight of a walking dinosaur doubled the loading upon each of its limbs, so he thought that the act of mating would impose the same burden on the female. As he wrote in 1991: 'If dinosaurs were strong enough to walk, they were strong enough to copulate. They were presumably strong enough to do both.' There are two conditionals hidden in that statement – he said 'if' they could walk they were 'presumably' strong enough. He was wise to be cautious; I am now certain that dinosaurs did not evolve primarily to walk. Those were big 'ifs'.

The most recent book on the topic was published in 2012 and was entitled *The Dawn of the Deed: The Prehistoric Origins of Sex*. The author, John A. Long, is Strategic Professor of Palæontology at Flinders University in Adelaide and he was previously Vice-President of Research and Collections at the Natural History Museum of Los Angeles County. In a section concerning dinosaur copulation he concludes that mating was accomplished with the male mounting from behind, doggy-style, which seems to be the unanimous conclusion of the other palæontologists.[13] They're all wrong.

Copulation featured prominently in the BBC documentary series *Walking with Dinosaurs*, for which Tim Haines the producer had consulted the worldwide community of palæontologists to distil the essence of everything that was known, so that they could re-create a realistic representation of these copulating creatures. They showed a pair of wooden-looking *Diplodocus* adults trudging unconvincingly towards each other through the scrub. Says the script:

> Our female is approached by a young male. She responds to his calls first by stamping, and then by generating very low-frequency mating calls. This so-called infra-sound is too low for most animals to hear. However, he picks these signals up through the ground, and responds by walking close to her, rubbing his body down hers. She shows she is receptive. Mating is a dangerous activity for the female. She is going to have to carry at least an extra ten tons on her back. As she has grown older, the vertebræ over her hips have become fused, and reinforced, to help her cope with this ordeal.

There are lurid CGI dinosaurs to accompany this contrived scenario, and the commentary explains: 'The males have started to display – during this, they rock back on their tails to impress potential mates.' There is not a shred of scientific evidence of any kind presented for this description, of course, and in any normal scientific documentary that kind of wild invention would not be considered for an instant. The images showed sharp spines down the back of the *Diplodocus* which would have made the doggy-style mating procedure highly compromising for the male. Not only that, but the way the male dinosaur thuds back onto dry land after his aerial adventure shows how impossible the manœuvre would have been: the huge animal, with its extended neck, crashes down to the ground with its head held out at length, as though on the end of a reinforced and inflexible girder. Any animal landing with such force would have its neck flexing and its head landing on the ground too. The solid and rigid structure given to this computerized image is completely unrealistic. Even to the untrained eye, it looks wrong. The mechanism of mating simply could not work as portrayed.[14]

This mating process becomes most difficult to square with conventional concepts of copulation when we look at a dinosaur bristling with defences, like the stegosaur *Kentrosaurus*. It is adorned with sharp spikes, and a land-based method of copulation is impossible to devise without the male having his genitals shredded. Heinrich Mallison, a scientist at the Museum für Naturkunde, Berlin, created digital simulations to try to work out how dinosaurs could have mated, and his re-creation of *Kentrosaurus* immediately revealed the greatest problem that a mounting male would face – he confirmed that it would be castrated by the razor-sharp spines that adorned the female's back. 'These prickly dinosaurs must have had sex another way,' Mallison concluded in an interview with *The Times* on March 24, 2013. 'Perhaps the female lay down on her side and the male reared up to rest his torso over her. Other species would have used different positions, like backing up to each other.' To exemplify the difficulties these dinosaurs faced, my wife Jan and I went to examine the well-preserved *Stegosaurus* skeleton at the Naturmuseum Senckenberg in Germany and compared it with the fine specimen in the Natural History Museum in London. They gave me no easy answers, and nothing by way of evidence. Trying to imagine how a dinosaur like *Stegosaurus* could have mated is problematic. With the well-armoured body and those fearsome tail spines, a male mounting a female from the rear, like an elephant or rhinoceros, would have been impossible. Even if it were possible, it would be suicidal.

The answer is obvious the moment we imagine dinosaurs evolving in the aquatic habitat that I now propose. Crocodiles, snakes and turtles simply approach each other and copulate without difficulty, their mass being buoyant in their watery environment. Once we envisage dinosaurs as aquatic creatures, all those mechanical problems disappear. Could *Stegosaurus* have been aquatic? I am sure it was – but what is the evidence? To unravel the scientific basis for this assertion, we should go back to the naming of this dinosaur by Othniel Charles Marsh in 1877, based on a partial skeleton excavated from north of Morrison, Colorado. That was the first time that *Stegosaurus* was recognized.[15]

As we have seen (p. 155), the first interpretation made by Marsh was that the dorsal scales were actually a shell, and for this reason he concluded it might have been a gigantic form of turtle. His initial conclusion that *Stegosaurus* was an aquatic animal was grounded in those early interpretations of the evidence. Once it appeared that the scales were actually raised along the back, and not laid down like tiles, he abandoned the view and reverted to the terrestrial alternative. Everyone concurred that *Stegosaurus* was certainly a terrestrial dinosaur, and this quickly became the only acceptable view. Of all the

Stegosaurus was difficult to imagine as an aquatic species, until Jonathan Poulter at the University of Leeds prepared these studies, showing the view both above the water surface and below. His wading dinosaurs precisely fit the new aquatic theory.

dinosaurs, it was the stegosaurs that gave me the greatest problems. Trying to envisage this armour-plated dinosaur as aquatic was something that I had always found difficult to grasp. Having the notion nurtured in the brain was not enough; as always, one must have objective, scientific evidence.

It came from an ingenious reconstruction by Jonathan Poulter at the University of Leeds. Poulter created computer graphic simulations of a stegosaur, showing perfectly its appearance as an aquatic dinosaur. He prepared meticulous digital interpretations, both looking from above the water surface and from beneath. These made instant sense of my proposal. A later interpretation was prepared as this book was being compiled, when a dinosaur enthusiast Johan Nygren sent in a sketch of a stegosaur that was not wading, like Poulter's example, but swimming. Both interpretations show *Stegosaurus* functioning perfectly in water (and seeming to move far more comfortably than it could have done on dry land). In this watery environment, mating would not be a problem; the copulation conundrum was solved. The aquatic habitat similarly makes perfect sense for those mating *Diplodocus*. If they were immersed in shallow water, then buoyancy takes care of the weight-bearing problem. Their tails could be effortlessly floated out of the way. Similarly, the neck and head of the *Diplodocus* are supported, and the problems posed by the imponderable mechanics are solved.

Although palæontologists persist in their terrestrial tyranny, eminent scientists from other disciplines have sometimes looked more objectively at the realities of dinosaur mating. A biologist at the American Museum of Natural History in New York, Stuart O. Landry, became convinced that dinosaurs would have been incapable of mating on land. Landry was a biologist who taught at the University of Missouri and in 1963 became Professor of Biology at the State University of New York at Binghamton. He worked there for 30 years, studying rodents, and became known for support of free-thinking, sceptical science. Landry was a long-time proponent of the need to conserve the environment and nurtured many other spare-time interests. He became an authority on the writings of Shakespeare, the music of Bach, the history of his native city of New Orleans – and the mating of dinosaurs. A symposium of vertebrate morphologists was held at the University of Chicago in 1994, where Landry gave a short

In June 2017 dinosaur enthusiast Johan Nygren sent in this sketch of a stegosaur that was not wading in shallow water, like Poulter's example, but swimming. He had discussed aquatic dinosaurs with his artist colleague Zeljko Zsrdic who then prepared this study of an aquatic stegosaur.

presentation called *Love's Labors Lost: Mating in Large Dinosaurs*. He described a huge sauropod rearing up to copulate and said: 'It would have to support 10 to 20 tons in a precarious position two or three meters off the ground.' If a male *Apatosaurus* had attempted this manœuvre, he said, it would inevitably have toppled over and taken the female with him. Landry suggested that the largest dinosaurs must surely have searched for mudholes to buoy themselves up. A reporter from the *Chicago Tribune* wrote that someone from the audience asked Landry if he was actually claiming that all dinosaurs must have mated in water. Landry paused to think for a moment. 'I would say the very large ones must have,' he finally had to agree.[16]

Gregory Erickson, a palæobiologist at Florida State University, gave a typical response: 'It's going to be very touch and go. It's an awkward thing. I've heard speculation that they did it in the water, but they're not aquatic animals. Just because they're large animals doesn't mean they can't mate on land; after all, elephants do it.' Yes, elephants do; that's undeniable. But an elephant is one-tenth the size of a dinosaur. The only animals as big as dinosaurs are whales – and they have sex in the sea.[17]

All these findings were arriving on my desk and I was keen to publish an update in response to the worldwide rejection of my views. The editorial staff at *Laboratory News* had been upset by the barrage of complaints they had received, and felt they had to avoid any further controversy. Although they went on to publish some of my later articles on discoveries made in other areas of scientific research, they steadfastly declined to consider publishing anything more to do with aquatic dinosaurs. They had been frightened off by the angry response from the world of palæontology. I don't blame them. It was intense.

The Royal Society in London had often published my papers, and they were the obvious choice for a follow-up. I proposed a paper, and they seemed pleased at the idea. Knowing how busy people are these days, I thought it best to keep it short and succinct, and hoped it would cover the main points. I confined myself to showing the problematic points in the existing literature. My submission had fewer than 1,200 words and about a dozen references, yet I hoped that it covered all the main points. I could think of no reason why a referee could reject it. This is what I sent them:

EVIDENCE FOR DINOSAUR EVOLUTION IN AQUATIC ENVIRONMENTS

INTRODUCTION

Dinosaurs pose many problems, including their large mass and apparently constant body temperature, which current models do not satisfactorily address, and it has been proposed that they evolved under the constraints of an aquatic habitat. (1) The suggestion has received a hostile response by palæontologists. (2)

However, recent research is consonant with this view and a summary is here presented to encourage a reappraisal of the way dinosaurs are currently conceived. This evidence is clearly suggestive of an aquatic habitat for these dinosaurs. The thermal buffering of an aquatic environment immediately solves the controversy over whether giant dinosaurs were poikilothermic or homeothermic, since mean temperatures approximated to 34°C (93°F), close to the normal metabolic body temperature for present-day humans. (3) Evidence for poikilothermy has been based on the observation of lines of arrested growth (LAGs) in dinosaur bones, which was said to imply thermally-regulated seasonal growth. (4) However, LAGs are observed throughout the vertebrates and are equally evident in homeothermic creatures. (5)

HERBIVOROUS DINOSAURS

Current interpretations of gigantic herbivorous dinosaurs (Sauropoda) are now open to reassessment in a manner that reflects well upon the aquatic hypothesis. The sauropods have been portrayed as inherently unstable in water, through an incongruity between the centre of buoyancy and the centre of gravity. (6)

These are not fixed points, and a subtle readjustment would have shown these dinosaurs to be stable. Henderson concluded that large dinosaurs would 'tip over' in water, though evolutionary constraints do not ordinarily confer inherent instability on organisms. Many bones of large dinosaurs were partially buoyant, and the elongated neck and tail would float in dinosaurs evolving in a watery habitat. Martin and colleagues published evidence that for a 'ventral bracing hypothesis' (VBH) to explain the existence of elongated calcified bodies within the necks of herbivorous dinosaurs. They attempted to analogize these cervical structures as structural beams. This VBH would be predicated upon primary periosteal bony bodies whose fibres were oriented perpendicular to the long axis and originating from connective tissue that existed between the overlapping cervical ribs. (7)

Are these ossified objects there to make the neck rigid, or to confer flexibility? Nicole Klein and colleagues now show that the structure of these components is not what the VBH predicts. They show that the cervical ribs of the sauropods are composed of primary bone tissue consisting of longitudinal mineralized collagen fibres and are ossified tendons. There is no periosteal bone and the predominance of fibres oriented longitudinally is exactly what one would anticipate, not in a type of sesamoid bone intended to confer rigidity, but in calcified tendons. These would imply that tension forces acted along the length of the cervical structures, giving greater flexibility and reduced mass. The VBH would have led to compressive forces that would render the neck more solid and less amenable to flexion. A lighter, resilient neck is exactly what one would expect in a semi-aquatic creature. (8)

The largest carnivorous genera, grouped as theropod dinosaurs, are typified by *Tyrannosaurus rex*, though there are fish-eating theropods (including *Spinosaurus* and *Baryonyx*) which may have been larger and heavier. Recent research suggests that *Spinosaurus* was a theropod dinosaur that inhabited areas between North Africa and Australasia from the Albian to the lower Cenomanian stages of the Cretaceous, 112 to 97 million years ago. The related genus *Baryonyx* is believed to be fish-eating, since it has been shown to retain acid-etched fish scales (assumed to have been eroded by digestion) within the rib cage. (9)

A skull of *Spinosaurus* in the collections of the Museo Civico di Storia Naturale di Milano, catalogue no. MSNM V4047, was examined in some detail by Cristiano Dal Sasso and his colleagues and in 2005 they concluded, by extrapolation in comparison with other skulls, the creature would have been 18m in length. (10) It would thus have been larger than the accepted dimensions of *T. rex* with a calculated body mass of some 20 tonnes. we can now construe *Spinosaurus* as the heaviest of all theropods. (11) Supporting large masses of this order imposes a high metabolic demand for any creature in a terrestrial environment, and this implies that the evolution of the spinosaurids might have been in an aquatic habitat. No present-day terrestrial animals have a similar mass: *Hippopotamus amphibious* can exceed 3.5 tonnes and is regarded as semi-aquatic, whereas the elephant *Loxodonta africana* weighs some four tonnes, occasionally more, though is essentially a terrestrial creature. The suggestion that dinosaurs evolved in an aquatic environment addresses many of the paradoxes that remain attached to these formidable animals.

The well-preserved MSNM V4047 skull was further examined by Cristiano Dal Sasso and his colleagues, who in 2009 reported results from X-ray computed tomography of the snout. Their images revealed that external foramina were directly connected to spaces that lay within the snout, suggesting to the authors that *Spinosaurus* had developed pressure receptors within these spaces. Were the snout level with the water surface, the authors surmised, these would facilitate the creatures detecting swimming prey even if they could not be visualized. The dinosaurs, 'when positioned on the air-water interface, [would have] an unexpected tactile function, useful to catch swimming preys without relying on sight'. (12) This is a further pointer towards an aquatic life habit.

The following year isotopic analysis appeared which further supported the same hypothesis. Romain Amiot *et al* submitted spinosaurid teeth to analysis of the oxygen isotope ratios from tooth enamel. To this they incorporated data from the analysis of other predators such as *Carcharodontosaurus* and compared these with the composition of samples from contemporaneous theropods, and also with marine and aquatic types including turtles and crocodilians. *Spinosaurus* teeth from a variety of sites revealed oxygen isotope ratios that were similar to those of turtles and crocodilians, but different from teeth of disparate theropods retrieved from similar localities. (13)

Furthermore, the spinosaur theropods have a dorsal fin, a feature which, from the sailfish *Istiophorus* to *Triturus* the newt, characterizes aquatic organisms. Thus there are anatomical and physical constraints which would lend support to large dinosaurs being adapted for life in water; and now the isotopic analysis of teeth, inspection of presumptive gut contents, elucidation of skull anatomy and the conclusion that the snout would have required orientation 'positioned on the air-water interface' all substantiate the view that these large theropods also evolved under the constraints of an aquatic environment. To aver that the proposition is 'wrong-headed and contrary to evidence and research' cannot reasonably be sustained. (2)

Although the giant dinosaurs clearly moved to land to lay eggs, the constraints under which they evolved were not those of a terrestrial environment. Current research substantiates the view that such dinosaurs were aquatic.

REFERENCES

1: Ford, BJ, 2012, A Prehistoric Revolution, *Laboratory News*: 24–26, 3 April.

2: Naish, D, 2012, Palæontology bites back, *Laboratory News*: 24–26, 11 May.

3: Skinner, BJ, and Porter, SC, 1995, *The Dynamic Earth: An Introduction to Physical Geology*. 3rd ed. New York: John Wiley & Sons: 557.

4: Chinsamy, A and Hillenius, J, 2004, *The Dinosauria*, 2nd edition, (edited by Weishampel, D; Dodson BP; and Osmolska, H), University of California Press: 643–659.

5: Köhler, M; Marín-Moratalla, N; Jordana, X and Aanes, R, 2012, Seasonal bone growth and physiology in endotherms shed light on dinosaur physiology, *Nature* doi:10.1038/nature11264, 27 June.

6: Henderson, DM, 2004, Tipsy punters: sauropod dinosaur pneumaticity, buoyancy and aquatic habits, *Proceedings of the Royal Society, Biological Sciences*, 271 (Suppl 4): S180–S183.

7: Martin, J, Martin-Rolland, V and Frey, E, 1998, Not cranes or masts, but beams: the biomechanics of sauropod necks. *Oryctos,* 1: 113–120.

8: Klein, N, Christian, A and Sander, PM, 2012, Histology shows that elongated neck ribs in sauropod dinosaurs are ossified tendons, *Biology Letters* rsbl20120778; published ahead of print October 3, doi:10.1098/rsbl.2012.0778 1744-957X.

9: Charig, AJ and Milner AC, 1997, *Baryonyx walkeri*, a fish-eating dinosaur from the Wealden of Surrey, *Bulletin of the Natural History Museum, Geology Series*, 53: 11–70.

10: Dal Sasso, C, Maganuco, S, Buffetaut, E and Mendezm MA, 2005, New information on the skull of the enigmatic theropod *Spinosaurus*, with remarks on its size and affinities, *Journal of Vertebrate Paleontology*, 25: 888–896.

11: Therrien, F and Henderson, DM, 2007, My theropod is bigger than yours … or not: estimating body size from skull length in theropods, *Journal of Vertebrate Paleontology*, 27 (1): 108–115.

12: Dal Sasso, C, Maganuco, S and Cioffi, A, 2009, A neurovascular cavity within the snout of the predatory dinosaur Spinosaurus. *First International Congress on North African Vertebrate Palæontology*: 25–27. Muséum National d'Histoire Naturelle, Marrakech, Morocco.

13: Amiot, R; Buffetaut, E; Lécuyer, C; Wang, X; Boudad, L; Ding, Z; Fourel, F; Hutt, S; Martineau, F; Medeiros, A; Mo, J; Simon, L; Suteethorn, V; Sweetman, S; Tong, H; Zhang, F and Zhou, Z, 2010. Oxygen isotope evidence for semi-aquatic habits among spinosaurid theropods. *Geology* 38 (2): 139–142.

Digital online submission had recently been inaugurated on the Society's website, and I submitted the paper as soon as it was ready. The automated reply came on October 15, 2012.

'Dear Professor Ford,' it said, 'Your manuscript entitled "Evidence for dinosaur evolution in aquatic environments" has been successfully submitted.' At least I knew it was entering the academic production line. The message added that my submission was to be given 'full consideration' for publication in the Royal Society's journal *Biology Letters*.

Eagerly I awaited the referees' decision. But it never came. Instead, I was informed that the paper had undergone a procedure I'd never heard about before. The Society's editorial office wrote to day that my manuscript had been 'unsubmitted to *Biology Letters*.'

Unsubmitted? Curious word. Most online dictionaries say that word doesn't exist, though it does apparently have occasional use in the legal sense, when used reflexively, of yielding the power of attorney to another. However, I knew at once what they meant: my piece had been rejected before it was even considered. There was a note which offered advice on what to do. The paper could be considered as an Opinion Piece to *Biology Letters* and I was advised to write to the editorial office with my proposal so that they could have a board member assess the article for publication in *Biology Letters*.

So I wrote.

On 17 October 2012, exactly as requested, I sent my message to the editorial office saying: 'I hereby present a topic that would make an "opinion" contribution to *Biology Letters* that is of wide general interest: RSBL-2012-0980 withdrawn from submission,' and, two days later, had an acknowledgement: 'I will discuss your opinion piece with an expert palæontologist on the board and get back to you with a decision as soon as possible.' It was signed: 'Best wishes, Charlotte.' You may imagine my thoughts; I was sure that their resident palæontologist would hate it. Within hours I had a response. 'Dear Brian J Ford,' it ran, 'Thank you for considering *Biology Letters* but we do not feel that your manuscript would be suitable for our journal. I wish you the best of luck with your opinion piece elsewhere. Best wishes, Charlotte.'

This speedy rejection was no surprise, but I was hesitant about accepting it at face value. The Royal Society is an old-fashioned body in many ways, its Fellowship emerging from the conformist community of establishment scientists who travel steadily up the escalator of academia until they are sufficiently eminent to be chosen to join. The Society does not court nonconformity. On the other hand, it has always been open-minded, and it claims to be receptive to new ideas. I wondered whether a single rejection was enough to refuse to publish a paper. The advice had been to submit my submission as an opinion piece, which surely gave a certain latitude, and I hoped that they might provide comments on the reasons for the rejection. That is what I thought, but I was wrong. Their answer came on October 24, 2012, and Charlotte was becoming less formal as the exchange went on:

'Dear Brian,' said the message, the tone now seemingly friendly. 'Can I just clarify, as there seems to be some confusion, your paper has not been subjected to peer review.' This was curious; the process of peer review – checking by experts in the field – is, as we have seen, the conventional course for any scientific paper. My article hadn't managed to get that far.

The email explained that the Society sent all submissions along to an 'expert board member' who would decide whether a new paper was suitable for *Biology Letters*. 'It is upon the recommendation of this board member that we have decided that your paper is not suitable for *Biology Letters*' said the message. This was a surprising verdict. When a paper is read by referees, they naturally have the right to reject it, and the reasons for rejection are conveyed to the author. This helps to redesign a submission so that, next time, it might be acceptable.

But in this case the submission was submitted, and then unsubmitted. As the message made clear: 'As the paper has not been through peer review there are no comments to provide.'

So it had not been rejected by their resident palæontologist; it was not even going to be reviewed. My submitting an unconventional opinion on dinosaurs was, simply, unacceptable. There was a conciliatory note at the end, however: 'I do hope that this does not alter your relationship with the Royal Society.' Well, of course it wouldn't. It simply opened one's eyes a little to the power of the establishment in restricting the publication of ideas that clash with those of the established church.

I responded at once:

> Ah. That explains matters – thanks. I truly must assure you that this passing incident doesn't in the least affect my regard for the Society … People seem as interested by the politics of the matter, as much as by its consequences. It's proving to be a fascinating saga. So yes, no more correspondence on this one, merely a slightly regretful curve to my eyebrow this morning. Very best wishes, Brian J. Ford.

By this time, I had been presenting lectures on the topic, gaining enthusiastic support from audiences, and I had also been discussing it with scientists from other fields, all of whom were enthusiastic. Then I had a message from a distinguished American academic. Not only did he approve, but he had realized that my theory could solve a range of other problems that had dogged the heels of science. The suggestion came from William 'Bill' Hay, Professor Emeritus at the University of Colorado in Boulder, the climatologist who knew so much about the world of microscopic fossils. When Hay read my initial publication, he realized that it solved a series of major problems in the study of the Cretaceous era. 'I was astonished,' he has since said. 'For more than 50 years we had been faced with insurmountable problems – and this new theory solved them all. It was a tremendous breakthrough. The aquatic dinosaur was just what we needed.'

Hay said that, not only did the theory answer the problems with dinosaurs, but he believed I had found an answer to some of the leading outstanding problems with current models of the Cretaceous climate. He wrote:

> Your idea has caused me to rethink the whole way we have interpreted Cretaceous climate. If dinosaurs were aquatic, it means that there was a lot more water on land. There has always been a

> problem in assuming conditions like those of today because with higher temperatures both the precipitation-evaporation rates go up. But the amount of water available for precipitation is limited by available water surfaces, whereas evaporation simply increases. This means that a warmer Earth should be drier, but that is not what we see. It has been a mystery that we didn't like to discuss.

He said that he was compiling a paper at the time and would incorporate incorporate my ideas. This meant that my theory could be endorsed by an independent scientist. Hay suggested:

> What I have been thinking is this: Would you be willing to contribute to a multi-authored paper using the aquatic dinosaur hypothesis as the basis for a new suite of climate models? We can't do this right away, because we would need to get a new suite of paleogeographic maps with better topography, and outline on them the areas of meandering rivers. We would need to figure out the rules for the change from straight to meandering river (I'm sure this exists in the literature somewhere), and then run appropriate climate models. As possible co-authors I am thinking of Rob DeConto, Sascha Floegel, and perhaps Joao Trabucho-Alexandre and myself, and perhaps a river expert and a hypsographer like Chris Harrison. This paper would emphasize how important the aquatic dinosaur hypothesis is as a breakthrough to understanding Mesozoic climates. It will take some months to make new paleotopographic maps, get them digitized, run the climate models, etc. so I would expect it would be the end of the year before we could have a paper ready. I will be attending the European Geosciences Union in Vienna in April-May, and making a visit to Kiel before then, where I will see Sascha Floegel. In Kiel I'll check with Wolf-Christian Dullo, editor of the *International Journal of Earth Sciences*, to see if he would like this – and could he let us have color illustrations etc. Let me know what you think.

Hay was then writing his 1,000-page bible on the Earth's climate entitled *Experimenting on a Small Planet*. This is the definitive account of climate change since the world began, and writing it was a monumental exercise. Every senior scientist should write a book like this. It has chapters setting out the progress of his lifetime of research, interspersed with what Hay entitles *Intermezzi*, autobiographical sections printed in italics between the chapters that tell of his career, the people he encountered and those with whom he cooperated, and the anecdotes that gave his life its unique flavour. Readers can be engaged by the progress of his ideas and the startling revelations of his research, while being diverted by the reflections of a full and varied life. The result is not merely a summary of academic insights, but a personal story that shows how motivations can occur, how chance meetings can lead to long-term relationships of supreme significance, and how the meanderings we make through life can ultimately have long-term crucial consequences for the way in which research progresses. We often muse on how useful it would be to download someone's brain at the end of a long and successful career – well, this stout book is the next best thing. It has since been reprinted as a larger-format volume running to 800 pages, and is sold for a modest price, unlike most scientific volumes that are beyond the reach of most individual academics.[18]

This was a bolt from the blue – a totally unexpected offer of support and cooperation from a clutch of the world's greatest experts on the Cretaceous period, when dinosaurs held sway. None of them was a palæontologist, but all were acknowledged experts in their fields. I have rarely published anything under joint authorship. I like to take sole responsibility for my views and normally my research has been a personal endeavour. In this case, my knowledge of palæoclimatology and the tectonic disposition of the continents is slight – less even than my rudimentary knowledge of dinosaurs – and so the offer of cooperation was timely. We began to pool ideas, and a major paper slowly started to take shape.

Meanwhile, Hay was finalizing his paper that would lay the foundations for my proposed change in the way we would view the Cretaceous period. Research could perhaps provide reasons to reconsider the current models for the ocean floor, the topography of the continents, and the temperature range on land. Hay pointed out that, when the effects of global warming had been considered, some significant greenhouse gases, including methane (CH_4) and nitrous oxide (NO_2), had traditionally been omitted from the calculations, though they were known to play an important part. He argued that my aquatic dinosaur theory had profound implications for the study of palæoclimatology and set out to show that the theory could solve the outstanding problems. My new theory now had support in a major academic journal.[19]

Specialists working in the field were generally supportive. Hay reported that, when he went to speak on my ideas at a conference, those in the audience approved. Everywhere else, the sense of hostility from the dinosaur palæontologists was unanimous. I had offered the idea of a talk to the Linnean Society, where I have served for many years as a trustee, a member of council and an officer of the Society; and also to the Geological Society who were next door to the Linnean in Burlington House, Piccadilly. Neither would touch it. This posed a problem. I felt there should be a follow-up, but the ranks had closed. This was becoming a fascinating saga and provided first-hand evidence of how a new theory can be rejected wholesale. The journals were discouraging. A clear choice was *Nature*. I wrote to Philip Ball, who had reported my earlier work in that journal, and on October 25, 2013, he had replied: 'Dear Brian, You evidently have an uphill struggle ahead of you. I'd be happy to see a copy of the paper, although the degree of controversy around this would probably make it hard for me to do much with it, beyond educating myself about the issues. Good luck with it.'

Tom Whipple at *The Times* wrote: 'Thanks for this, Brian' but then explained that their science section was currently in a state of flux and he was waiting for things to settle down. He added: 'When I know what space we have I may come back to you.' He didn't. Simon Gaskell at *Trinity Mirror* noted that I was sticking to my stance in spite of 'the wave of popular opinion' but nothing could be reported.

I had first published articles in *New Scientist* 50 years earlier and had continued to contribute from time to time, but on this occasion their Liz Else wrote to say: 'I must say it does look most interesting. My problem however is that we are fully stocked until mid-February – largely because my boss wants to take some of the pages and give them to special issues which he has planned. So I really have no room at the moment.' I raised it with their reporter Andy Coghlan, who wrote to say: 'Brian, I'm not surprised palæontologists poo-pooed it … anything that challenges the status quo in science gets a rough ride!' He also said: 'This is very interesting. Several of my colleagues here already have copies and have been discussing it. It's more their subject than mine, so they'll decide whether to proceed with anything. Is it being made available to everyone? I assume so?' And then: 'If only we had a time machine! That would settle it! It is a real problem when scientific cages are rattled. Have you tried *Scientific American*? You'd get more space than in *New Scientist* to lay out your case?' Looking for an avenue for publishing was suddenly elusive. One of my writer friends, Mavis Nicholson, thought it might be interesting to publish an illuminating account in *The Oldie* magazine. It was edited by Richard Ingrams. When he was editor of *Private Eye* in London I sometimes used to write for the magazine and he had always liked controversial ideas, so I wrote a light-hearted article without hesitation. This was a curious choice of magazine, but it could possibly provide an outlet – and I kept the text close to 1,000 words:

DIE-HARD DINOSAURS

People sometimes keep cuttings in a file. Keep this one. It may not amount to much now, but in ten years' time it'll be a hot topic. First, let me check that there are no palæontologists around, especially dinosaur experts. I am being roasted by palæontologists everywhere in the world because I have published a theory that blows their work out of the water. Needless to say, I have been careful not to stand close to an unfenced mine-shaft ever since the theory appeared.

Many of my scientific theories have been published around the world, in magazines from *Scientific American* and *New Scientist* to journals including the *British Medical Journal* and *Nature*. They have been graciously received, and some have brought awards in their wake. An international yearbook said my research was a 'scientific highlight of the year'; in the journal *Nature* I was recently described as being the 'world's greatest authority' on some of my microscope research, which is close to embarrassing. My dinosaur theory, though, brought an entire scientific discipline up in arms. A year ago I proposed that dinosaurs had evolved, not as terrestrial creatures pounding about an arid landscape in clouds of desert dust (as we regularly see them on TV), but in water. They had developed, I concluded, in an aquatic environment and not on dry land. You might think (as did I) that this would trigger a timely debate, but no. As we have seen, the theory was dismissed as a 'rotting corpse' and 'bad science', said the Smithsonian blog. 'Who the hell' was I to say this, shouted another. 'Dinosaurs were not aquatic!' they all proclaimed. Not a single supportive word was heard in the length and breadth of palæontology. A colleague from Cambridge wrote to warn me of the risk to my reputation by even postulating this absurd idea. 'Stop publishing!' he warned. An entire scientific discipline, without exception, bellowed that I was wrong. The international campaign was unremittingly hostile.

The science, however, shows I'm right. Dinosaurs seem to have been warm-blooded, though no reptile has evolved to control its own body temperature. If they had evolved under the constraints of an aquatic environment, the water would have provided the buffering effect and maintained their body temperature – and the shallow lakes that abounded when dinosaurs were at their peak were around our body temperature. Dinosaur footprints left in mud are shallow. Only if their body mass were supported in water would this be possible. The fossilized trackways sometimes show the prints of only the forelimbs. This is related to the fact that the giant sauropod dinosaurs had front limbs that were longer than their hind legs and so – when buoyant in water – only the forelimbs would reach the bottom. The skulls show nostrils on the top of the snout and they possessed conical teeth (exactly as in aquatic reptiles today); indeed, some fossil dinosaurs show partially digested fish scales in the abdomen, proof of their aquatic diet.

Most curious of all is that there are no tail drag marks among these fossilized footprints. Today's large reptiles certainly leave such tracks. Palæontologists now say that this is because dinosaurs must have held their tails aloft – but this would expend vast amounts of energy for the muscle cells, which (for a plant-eating creature like an herbivorous dinosaur) would not have encouraged the evolution of a huge tail. A giraffe (similar in configuration to a dinosaur) has a fly-whisk tail, for example. The *Diplodocus* in the Natural History Museum, who originally had her tail resting on the ground, was remounted to show it held high in the air when the lack of tail marks proved that this could not fit the facts. There is no living creature that is so wasteful of its energy. Dinosaurs could not have run as we see in reconstructions, because a massive body on two legs would be difficult to turn to face in a new direction. This, the effect of yaw, has perplexed palæontologists for decades. Meanwhile, recent measurements of the oxygen isotopes in fossilized dinosaur bones show ratios like those of aquatic creatures, which are very different from those of animals that lived on land.

A few old reconstructions of dinosaurs showed them lolling about in swamps, as though taking the weight off their feet, but this is not my argument – my view is that they evolved in water, and subject to the constraints of an aquatic environment. They did not retreat to swamps to rest, but emerged from lakes to breed. It is the vast bodies of shallow water with which the world was covered at the time, I contend, that made dinosaurs possible. We know that dinosaurs could certainly move about on dry land. They must have, since they laid eggs in nests and some may have cared for their young. But they clearly cannot have been primarily terrestrial. All the science shows that they must have evolved in water. The article appeared in April 2012. The media reports were detailed and exuberant. The BBC *Today* programme presented the theory as an interesting new light on dinosaur evolution though the presenter was unwise enough to compare my work with that of Galileo (not unlike comparing an ant to a lobster). The expansive story in the *Daily Telegraph* initially said I was proposing that dinosaurs evolved 'under water' rather than in it, but we advised them to correct the article in later editions. The story was reported around the world. Scientists liked it (one noted American biologist said this was the 'greatest wake-up call in the history of paleontology') – but every palæontologist who responded did so with venom. These were exciting and timely developments, so we sent a note along to chums in the scientific press – but nobody would say a word. The Royal Society (which has often published my work in the past) said they wouldn't even consider a paper for publication. One or two magazines said they were keen to publish, but feared they'd risk the wrath of their readers for reporting a view that had offended an entire field of science.

And so, can a single individual challenge the orthodoxy that drives a discipline? It has happened in the past, of course; indeed, this is how science often makes its changes in direction. In this case, the convention is so deep-rooted that nobody dares to challenge the establishment, even when the accumulating research reiterates the need for revolution.

I guess it will take a decade to percolate through the system. Then people will look back at the time when we envisaged dinosaurs thumping across the deserts as laughable. In the future, people will say 'It was obvious!' – even though they don't like to admit it now.

So keep this article in a safe place. In 10 years' time it will remind you that scientific revolutions are sometimes hidden from view because it is so convenient to cling to current conventions – but, eventually, discussions of new scientific theories do find themselves in print. Even if only in *The Oldie*.

That was a reasonably comprehensive summary, I thought, and Nicholson did too; but not Richard. After several weeks of prevarication (it was probably the time taken for an objective appraisal) he decided he would not publish. Nicholson was surprised. I was too. There was no haste for this popular version to appear in print, of course, because I knew that Bill Hay was working on his academic paper, but I was becoming frustrated by the fact that I could find nobody who would allow me to publish anything of my own. However, I did have one possibility to release an update. I write a regular 'Critical Focus' column for *The Microscope*, a journal published in Chicago by the McCrone Research Institute, who organize the Inter/Micro conference where I lecture each year. My theme is always microscopical, ranging from microbes making food to microbes influencing the climate. The expenditure of energy by the microscopical muscle cells in the tail of a dinosaur was one of the assumptions that had always made me realize that their tails could not be held aloft all day long. It could fit my brief. The journal's editor-in-chief Gary Laughlin agreed that a review of the response to my aquatic dinosaur theory would make for a timely topic, and their wise managing editor Dean Golemis said he'd love me to write on the research, so I retrieved the article I had written for the Royal Society (which they had rejected; I beg your pardon, *unsubmitted*) and I based my column on an extended version of that. Although the Royal Society submission had kept below 1,400 words in length, by the time I fully spelled out the story, its version for *The Microscope* ran to more than 5,300 words. Most recapitulated what we have already seen, so we shall not revisit that here; in my column I reminded readers that:

> Paleontologists get dinosaurs wrong. They look at them as gigantic terrestrial monsters, but there are other ways of contemplating them: I prefer to envisage them as communities of microscopic cells. This understanding of life at the cellular level leads me to one great truth – dinosaurs must have developed for life in water and not on land. I am not simply suggesting that they retreated to swamps to rest; in my view, dinosaurs evolved under the constraints of an existence in shallow water and everything about them points to an aquatic habitat. They were certainly not the terrestrial monsters we see in the films and books, perpetually pounding across desert dunes with the energy of an express train and the speed of a tank.
>
> A dinosaur's cells are very like yours; it is just that there are more of them. For example, I know the size of a dinosaur's leukocyte and the dimensions of their muscle cells. Paleontologists

don't. I am aware of these things because I can study the cells of present-day reptiles, and somatic cells alter little with time. There's something Freudian about the machismo of a monster which the paleontologists prefer to perpetuate, but to me dinosaurs behaved more like a hippopotamus or an alligator. Paleontologists suffer from terrestrial hysteria, and it is misplaced. For decades I have wanted to say that dinosaurs were aquatic, and I am glad I saved this theory for later, for it has dropped me into enough hot water to take the peel off a pepper …

Not a single paleontologist, anywhere in the world, spoke in support of the need for an open discussion on this new approach. Their blogs poured scorn on my ideas and there were many abusive messages on Twitter. Detractors insisted that dinosaurs would have needed webbed feet were they aquatic – heedless of the fact that crocodiles do not have webbed feet. One referred to my technique of volumetric analysis, saying that I had merely 'dunked models in water.' That disparaging term does, indeed, describe volumetric analysis – though he spelled it 'volumentric analysls' which did little to inspire confidence. We used the technique to good effect, and what's more, we spelled it right. Nobody produced any evidence that disproved my theory. We all knew that dinosaurs could exist on land – there had never been any mystery about that. But I had assembled evidence to suggest that they evolved in water – and nobody produced any evidence to disprove that view. Scorn was poured on my concept of *Tyrannosaurus rex* as an aquatic dinosaur. Yet the most exquisite reconstructions made it look exactly like an aquatic creature, a super-crocodile. I could see it gliding through shallow lakes, scavenging on dead and dying herbivores and sometimes using its massive hindlimbs to leap upon unsuspecting prey with its sharp teeth and gaping jaws. The forelimbs of *T. rex* have become so rudimentary as to be functionally useless, which was to me a clear sign of their redundancy in an aquatic environment. One person who wrote to offer a function for these dwarf forelimbs was Rob Reyes of Los Angeles, California. Rob proposed that the small front legs acted as a vacuum breaker, should Tyrannosaurus have been resting in a mud bank and become embedded. Like the few others who spoke in favor of my views, Rob is a central heating engineer and not a professional paleontologist.

Morphology, as we saw, was important in my redrawing of *Spinosaurus* and it is the morphology of the dinosaurs that provides another tranche of evidence in my favor. Giant dinosaurs all have legs about the same length – between 10 and 15 feet (3 to 4.5 metres). In animals the length of leg varies with the overall height of the species. Longest neck? Giraffe. Longest legs? Giraffe. Groups of animals – deer, for example, or lizards – show a variety of limb length, with the larger species evolving the longer limbs. That isn't an absolute rule, but a guiding principle of morphology. Uniquely, it does not apply to the great dinosaurs. If their necks had evolved to reach high plants and they were terrestrial species, then their legs would similarly have lengthened. The fact that they didn't is predicated upon their evolution for life in shallow water. Deeper seas were inhabited by swimming species (spinosaurs and plesiosaurs among them). The aquatic environment is the only sensible explanation of this curious consistency.

I am certain that all those dinosaur books on your shelves at home, every TV program on a DVD nestling in your study, each account of dinosaurs in every encyclopedia, is wrong. The study of cells made me realize how misleading is the current convention – and the adventure of publishing this heresy has been one of the most illuminating experiences of my life in science. Paleontologists always insist that their dinosaurs evolved to be terrestrial. Having reviewed the evidence, I am certain that they are misguided. Dinosaurs were creatures of lakes and vast, shallow seas. But in paleontology, like so many areas of science, reputations rest on religious adherence to convention, and you challenge fashionable faith at your peril. The facts don't matter as much as preserving the comfortable security of the status quo. Galileo found that out to his cost, and the lesson he learned is with us today.[20]

The article was well received and now I felt a sense of relief. I had managed to publish an updated review and was fortunate to have my regular column in *The Microscope* where it could appear. Although nobody else would allow these ideas to be published, the truth was irresistible, and reactions now began to emerge in the press. Meanwhile, the repercussions of my new approach were beginning to spread, and the New Year was marked by the first reports of palæontologists conceding that some of the fossilized footprints left by dinosaurs were actually made in water and not on land. At the Lark Quarry Park in Queensland, Australia, a series of dinosaur trackways 95–98 million years old was found in thin strata of siltstone and sandstone from the bed of a prehistoric floodplain, and they had been perplexing palæontologists at the University of Queensland. Their spokesman, Anthony Romilio, now had this to say: 'Some of the more unusual tracks include "tippy-toe" traces, where fully buoyed dinosaurs made deep, near vertical scratch marks with their toes as they propelled themselves through the water. It's difficult to see how tracks such as these could have been made by running or walking animals.'[21] Perfect.

Within a few weeks, the same conclusion was announced from the University of Alberta, Canada. Their researcher Scott Persons had found similar trackways in the Szechuan Valley in central China. His

A year after the aquatic theory appeared in *Laboratory News*, Anthony Romilio at the University of Queensland revised the interpretation of trackways left by 'stampeding' dinosaurs. His drawing showed that they must have been wading.

conclusions fitted well with the findings previously announced from Australia: 'What we have are scratches left by the tips of a two-legged dinosaur's feet. The dinosaur's claw-marks show it was swimming along in this river and just its tippy toes were touching the bottom.'[22]

At last the message was getting through. No palæontologist wanted to suggest that these creatures were aquatic, or even that they were wading; but the truth was peeking out. It was not long before Bill Hay made contact again. He had found time to digest my column.

He wrote to say that he had finally read my article 'Aquatic Dinosaurs Under the Lens'. Much to my relief, he said how he 'especially enjoyed the style of writing' that I had decided to adopt. Hay wrote that my re-telling experiences and interlacing them with personal anecdotes was far more fun to digest than the crisp terseness of standard scientific accounts. Then he made the crucial comment that meant so much to me: 'I think you have got it exactly right; everything now makes sense to me. Dinosaurs must have lived in water.' When I had formally proposed the idea it had seemed daring; but with Bill Hay supporting my views so heartily I felt far more confident about the way ahead. He mentioned that when he had been the Director of the University of Colorado Museum he had hosted Bob Bakker shortly after Bakker's book *The Dinosaur Heresies*[23] had been published in America. Much interest had been shown by the Engineering Department at the University of Colorado in the mechanics of dinosaurs – how they could have supported their weight and moved about – and the engineers concluded that dinosaurs could certainly never have 'danced around or skipped.' Bakker's book is liberally decorated with cartoons of animated dinosaurs (p. 255) but the engineering specialists had poured cold water on the idea. Hay had also been concerned about the mechanisms that dinosaurs needed to regulate their body temperature, and now said: 'Your solution is perfect.' This was highly encouraging. Hay added that my theories also solved another important anomaly. He said that everyone assumed that, not only were dinosaurs unable to swim, but they rigorously avoided salt water. He now felt my theories changed all that: 'Do you think dinosaurs could have tolerated a bit of seawater?' he asked.

Seawater was no problem. We automatically believe that drinking seawater is fatal, simply because we cannot do it. But vast families of creatures obtain all their water by drinking from the oceans. Seabirds survive by drinking salt water, excreting the unwanted salts through glands near the base of the beak. The cetaceans – whales, porpoises, dolphins – are all content to obtain their internal water supply from the sea, and so are reptiles from salt-water crocodiles to sea-snakes, and ocean-going birds from penguins to the wandering albatross. Living permanently in and around the sea would have posed no problem for dinosaurs. When I explained this to Bill Hay, he concurred: 'Yes, I like the salt-gland idea,' he replied. He also said:

> I don't think I sent you a copy of my presentation at the Cretaceous Conference in Ankara. It attracted a lot of attention, most of it favorable. The aquatic dinosaur hypothesis raises the possibility that the continents were much wetter than has been suggested previously, and some climate model runs based on that assumption found that the meridional temperature gradients are significantly decreased. The editor of the *International Journal of Geological Sciences* (formerly the *Geologische Rundschau*), Christian Dullo, is a close friend. I will ask him if he would entertain an article by you. It has published a number of 'controversial' papers in the past, such as those of Alfred Wegener, and has a very wide international audience.

Needless to say, being published in the journal that had given space to Alfred Wegener and his controversial concept of continental drift was an enticing prospect. By now our team was comprised of Bill Hay and myself, with Robert M. DeConto, Professor of Geosciences of the University of Massachusetts at Amherst; Ying Song from the Department of Geology, China University of Petroleum, Qingdao, China; Andrei Stepashko at the Kosygin Institute of Tectonics and Geophysics, Far East Division, Russian Academy of Sciences, Khabarovsk, Russia; Poppe de Boer at the University of Utrecht, Netherlands; and Sascha Flögel at the GEOMAR Helmholtz Centre for Ocean Research, Kiel, Germany. They were all consumed with their own research but said that they would willingly lend their energies to our joint investigations. Hay was eager to respond to the appearance of my 'Aquatic Dinosaurs Under the Lens' article, and he published this letter in support:

> I read Brian J. Ford's 'Critical Focus: Aquatic Dinosaurs Under the Lens' (*The Microscope*, 60 (3) pp. 123– 131, 2012) with great interest. It is always useful when someone from one field of science takes a look at another area. The idea that large dinosaurs were aquatic, popular in the first half of the last century, was replaced 40 years ago by the idea that they were almost exclusively terrestrial. Films like 'Jurassic Park' (with its Cretaceous dinosaurs) showed large dinosaurs galloping across savanna landscapes and captured our imagination. I started my career as a micropaleontologist, studying fossils of ultra-small oceanic plankton, but over the last 30 years I have concentrated on trying to understand conditions on a warm Earth, particularly the Cretaceous. Ford's ideas need to be explored, because if many dinosaurs were lake or swamp dwellers, it means that we have left out a very important aspect of the boundary conditions for our numerical climate models – water on land. Our models have assumed dry land with the major atmospheric water source there being evapotranspiration. If there were extensive wet surface areas on the continents hosting dinosaurs, this might well explain some vexing problems in model data comparisons.
>
> William W. Hay, Professor Emeritus, Department of Geological Sciences, University of Colorado at Boulder.[24]

It was encouraging to have this in print, but Bill Hay was not the only person to respond; it also brought a stern rebuke from Donald Henderson, author of that curious 'tipsy punters' article. In his response, also published in *The Microscope*, he was unabashed, and made a strong point that *Spinosaurus*, whatever else it might have been, was no aquatic dinosaur:

> Ford's article is full of unsupported assertions and errors of interpretation, but I will only highlight one. The most laughable claim is that the set of elongate neural spines forming a 'sail' on the back of *Spinosaurus* are the equivalent of a dorsal fin and a clear signal of an aquatic mode of life. Genuine, secondarily aquatic tetrapods such as living whales (both toothed and baleen) or the extinct *Ichthyosaurus* that possess a dorsal fin also have a suite of additional characters to complement the function of the dorsal fin. These include a caudal fin for axial propulsion; hydrodynamic fins/flippers for attitude and roll control; a smooth, fusiform body to minimize drag; and a vertebral column flexible enough to permit lateral or dorso-ventral undulations of the body for propulsion. Spinosaurid dinosaurs have none of these associated features to go with their 'dorsal fin.' They would not function very well as aquatic animals, if at all.[25]

Henderson's letter reiterated the familiar generalizations that have led palæontologists astray in the era of terrestrial tyranny: the notion that present-day whales and the prehistoric *Icthyosaurus* had a streamlined body and fins is because they swim in the sea. Reptiles that live in and around the sea (but do not swim beneath its surface) lack these extreme features, including crocodiles and turtles, alligators and caimans. Trying to relate crocodiles to whales is not a helpful comparison. Even so, Henderson left one abiding impression from his letter: *Spinosaurus* could definitely *not* have been aquatic. That was impossible, and he reiterated this conclusion in an email:

> I should also say that your article in *The Microscope* is full of unsupported assertions, and a highly selective use of evidence to support your claim. The most laughable part of the article is the claim that the elongate neural spines on Spinosaurus are the equivalent of a dorsal fin and a clear signal of an aquatic mode of life. Genuine, secondarily aquatic tetrapods possessing a dorsal fin have a suite of additional characters to complement the function of the dorsal fin. These include a lunate caudal fin for axial propulsion, hydrodynamic fins/flippers, and smooth, fusiform

> body to minimize drag, and a vertebral column that permits lateral or dorso-ventral undulations of the body for propulsion. Have a look at ichthyosaurs and whales. Spinosaurs have none of these associated features. I think you should go back to your microscope and not dabble in other branches of science where you do not have sufficient knowledge.[26]

Although some of his diagnostic criteria apply to swimming creatures (like dolphins), they do not apply to reptiles that evolved to inhabit water (like crocodiles). They do not have a caudal fin, nor hydrodynamic flippers; most crocodiles and their allies do not even have webbed feet, but clawed fingers and toes. In explaining his views, Henderson was reiterating what all palæontologists have been saying. Yet I was becoming certain that all large dinosaurs had evolved in an aquatic habitat and that the spinosaurs were among the most aquatic of them all. Surely palæontologists must concede this obvious conclusion?

I continued to experience difficulty in finding anyone to publish an update in a popular publication. My *Oldie* manuscript was sitting in the computer, and this was an article that needed a home. I mentioned the fact to the editor of the *Mensa Magazine*, Brian Page, who said they'd certainly like to see it. That was encouraging – so I looked again at the 'Die-Hard Dinosaurs' article I'd submitted to Ingrams and rewrote it. Magazines take a long time to process articles, and it eventually appeared in March 2014:

> We are used to seeing dinosaurs portrayed as monsters pounding about an arid landscape in clouds of dust or thrashing through the undergrowth. They are everywhere on TV and in the cinema. I recently proposed a very different view: that they actually evolved for a sedentary life in shallow water. They had developed, I concluded, in an aquatic environment and not on dry land. My purpose was to trigger a timely debate. Dinosaurs are conventionally portrayed as creatures of the plains, but my view was that they had specifically evolved for an aquatic habitat and were essentially creatures of the shallow seas so prevalent before

today's continents began to emerge. This theory solved most of the outstanding problems facing palæontologists.

To begin with, dinosaurs seem to have been warm-blooded, though no reptile has evolved a metabolic mechanism to regulate its body temperature. This has long posed a puzzle. The evolution of large dinosaurs under the constraints of an aquatic environment would have water buffering their body temperature – and the shallow lakes that abounded when dinosaurs were at their peak were typically around 34° Celsius (93° Fahrenheit). So, in water, they could have maintained a constant body temperature without a physiological mechanism. That's one major problem solved. Dinosaur footprints left in mud are shallow. I argued that only if their body mass were buoyant in water would this be possible. Otherwise, they'd sink in up to their thighs. Equally paradoxical is the fact that there are never any tail dragging marks among the fossilized footprints, although today's large reptiles certainly leave such traces as they walk. Palæontologists now say that this is because dinosaurs must have held their tails up in the air – and the skeletons of dinosaurs in museums around the world have had their tails repositioned in recent decades. You may have noticed that the *Diplodocus* in the Natural History Museum in London once had its tail resting on the ground but – when the lack of tail marks proved that this could not fit the emerging discoveries – the tail was remounted to show it held high. I argued that this is impracticable, since supporting such a massive tail aloft would expend vast amounts of metabolic energy. For a plant-eating creature like an herbivorous dinosaur, the work performed every second by the caudal muscles would not have led to the evolution of a huge tail. A giraffe (similar in configuration to an herbivorous dinosaur) has a small tail like a fly-whisk, for example. The 'counterbalance' idea makes no sense at all.

Sometimes preserved dinosaur trackways show the prints only of the forefeet. As it happens, the giant sauropod dinosaurs had front limbs that were longer than their hindlegs and so – when partly buoyant in water – I have argued that only the forelimbs could reach the bottom. The aquatic hypothesis is the only answer

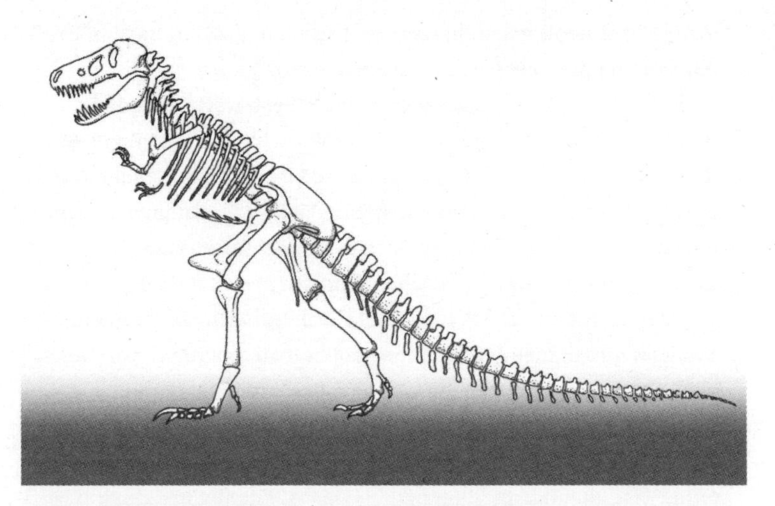

The traditional view of *Tyrannosaurus rex* showed it as a colossal monster with its tail on the ground for stability. No traces of tail-marks have been found, among thousands of footprints; so *T. rex* is now portrayed holding its tail aloft.

to that recalcitrant riddle. Many dinosaur skulls show nostrils on the top of the snout, which is exactly what we see in aquatic reptiles today. These giant creatures were endowed with conical teeth like those of alligators. Recent research has shown sense organs around the snout of many fish-eating dinosaurs, just like those in today's crocodiles, and some fossil dinosaurs show partially digested fish scales in the abdomen, proof of their aquatic diet. So much recent evidence points to an aquatic habitat for the giant dinosaurs, yet still they are portrayed as living on land, snatching fish by dipping their heads into the water as the prey glides past.

Dinosaurs could not have run as we see in reconstructions, because a massive body on two legs would be difficult to turn to face in a new direction. This is the effect of yaw, familiar to anyone who has tried to turn a supermarket trolley on castors at the end of an aisle. For years, palæontologists have been perplexed by the problem of yaw. Life in water, steering with a tail, solves this problem too. Furthermore, recent measurements of the oxygen isotopes in fossilized dinosaur bones show ratios like those of aquatic creatures, and are very different from those of animals that

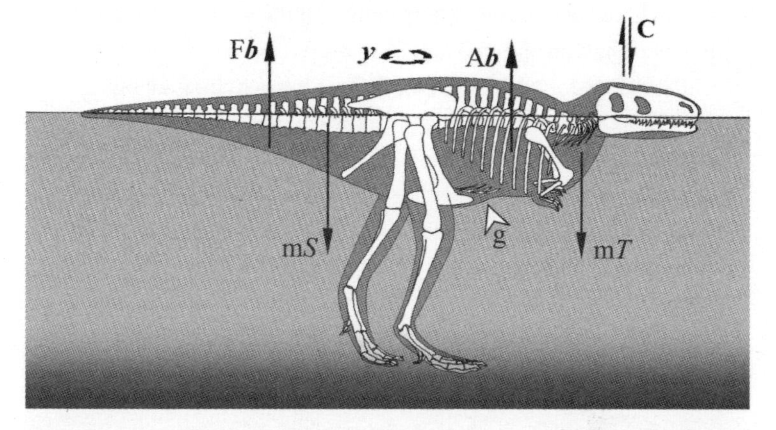

A tyrannosaur on land would exert downward pressure of some 5 tons (≈50 kN) in each limb, and to change direction of motion would need to produce torque in the limbs sufficient to twist the body mass. In water the situation is different. The mass of the axial skeleton and body mass (m*S*) plus that of the thorax and head (m*T*) are balanced by the flotation of the entire body (F*b*) and the low density of the air-containing thorax (A*b*) resulting in near-neutral buoyancy. Hydrostatic pressures are low and are controlled by the bones of the gastralium (*g*). As the head is raised or lowered (*C*) greater or lesser pressure is transferred to the substrate. Rotational forces (*y*) require no generation of torque by the limbs, since lateral caudal thrust – a simple flick of the tail – produces an instant and governable change in orientation. Each constraint exerted by a terrestrial habitat is thus mediated by the aquatic environment.

ever lived on land. An environment of shallow water solves these problems too. A few old reconstructions of dinosaurs showed them lolling about in swamps, as though taking the weight off their feet, but this is not the argument – my view is not that they merely wallowed in water, but that they evolved subject to the constraints of the aquatic habitat. Rather than retreating to swamps to rest, I believe they emerged from lakes to breed and it is the vast bodies of shallow water with which the world was covered at the time, I contend, that made dinosaurs possible.

Objectors shouted that physical analysis showed dinosaurs could move on dry land, but there is no mystery there. Obviously dinosaurs could certainly move about on land, since they laid eggs in nests and some may have cared for their young. But they clearly cannot have been primarily terrestrial. All the science shows that they must have evolved in water.

The debate for which I hoped did not happen. Instead, my theory was drowned in hostility. In the length and breadth of palæontology, not a single supportive word was heard. A palæontologist from Cambridge wrote to warn me of the risk to my reputation even by postulating this absurd idea. 'Stop publishing!' he warned. An entire scientific discipline, without exception, bellowed that I was wrong though nobody produced a single shred of scientific evidence to disprove my hypothesis. New scientific theories are often greeted with antagonism. In several of my books (and often in my lectures) I have explained how often this occurs, for the notion that revolutionary ideas in science are always welcomed is far from convention. Academics derive their grants from convention and conformity, and new ideas can be unwelcome. Discussing the instinct to reject new notions was something I have done all my adult life – and now I was experiencing it first-hand.

It was the BBC interview with me on the *Today* programme that brought about the most hostile response of all. Palæontologists rose up as one and called for the BBC to 'retract' the interview. The BBC's science reporter, Tom Feilden, mentioned how Galileo had also been vilified for offering his new theory (in my view, comparing my theory with Galileo's is like comparing a gnat to a lobster). Not a single palæontologist spoke out in support. The article appeared in April 2012 and the press reports were detailed and exuberant. The *Daily Telegraph*'s expansive story initially said I was proposing that dinosaurs evolved 'under water' rather than in it, but corrected the article in later editions. The story was reported around the world. Although palæontologists were incensed at the idea, scientists approved of it and one noted American biologist said this was the 'greatest wake-up call in the history of palæontology'. Every lecture I have given on the subject has engendered universal approval.

But what about the latest research? That is where the theory will stand or fail. In fact, the key research currently being published is substantiating my theories. Anthony Romilio of the University of Queensland is studying what were thought to be tracks of 'stampeding' dinosaurs, and now concludes that the facts don't add up. 'These dinosaurs,' he now concludes, 'were swimming.' When I re-examined my theories in an American article, I produced a picture of *T. rex* in the shallow lakes that abounded at the time, and now Scott Persons of Alberta has re-examined footprints of a fossilized *T. rex* and he too concludes that, after all, it was 'swimming along'. These palæontologists are wrong in one respect. Our parents used to tell us, when we kept our feet on the bottom of a children's swimming-pool, that you aren't swimming if you are still touching the bottom. The existence of the footprints shows the dinosaurs were not swimming at all – they were wading, precisely as I had proposed over a year earlier.

Yet palæontologists could never bring themselves to accept that dinosaurs were aquatic. Martin Lockley at the University of Colorado at Denver claimed in 2008 that the existence of the tracks of only one set of limbs was due to a failure of the second set to leave footprints at all. He and his colleagues were determined that none of this provided evidence of aquatic dinosaurs – their conclusions (to quote from the paper) 'lends support to the theory that brontosaurs were terrestrial and not aquatic in their adaptations.'[27]

Other researchers have joined Lockley in their belief, and they dismiss many of the earlier reports as simply mistaken. Curiously, they repeatedly doubt the evidence that dinosaurs could swim – though they say nothing about wading. The term does not appear anywhere in their paper of 2016. They conclude that, when only the hind feet left impressions, it must have been because weight borne on the forelimbs was insufficient to leave a mark. That's absurd – of course they would leave an impression on mud; it is just that the impression would be shallower.[28]

Indeed, it is the fossilized trackways of *Diplodocus* footprints that have led to changes in Dippy, the Natural History Museum's famous dinosaur. She was originally erected with her tail on the floor, remember, and later had her tail raised; and now the curators have conceded that everything we know about Dippy makes even that view incorrect. Restorers in Toronto have been asked to supply new casts of the forefeet, smaller in size and more like hands; the Museum is having to admit that she did not leave prominent tracks from her forelimbs. This, her third iteration, reminds us all how little palæontologists know.

Can I, as a single individual, challenge the orthodoxy that drives an entire science? It has happened in the past, of course; indeed, this is how science often makes its changes in direction. In this case, the establishment convention is so deep-rooted that nobody dares to challenge the current theories, even when the accumulating research reiterates the need for revolution. I don't doubt that some smaller dinosaurs lived on land, though even these can betray their aquatic heritage – indeed, a recently discovered microraptor no bigger than the sole of your foot fed exclusively on fish, even though it was based on land.

It may take a decade for this view to percolate through the system. People may well look back at the present time when we envisaged dinosaurs thumping across the deserts as laughable. In the future, palæontologists may well say, 'It was obvious!' – even though they won't admit it now. Our present-day indicators of what dinosaurs were like are misplaced. No, you cannot see them in birds; no, they are not like iguanas; no, you will not see them reflected in the Komodo dragon. If you want to see a present-day creature rich in resonances of *T. rex*, then it's a fat crocodile that provides the example you seek. The legs are different in proportion – this is a crocodilian, after all, not a dinosaur – but everything else: the teeth, the eyes, the snout, the claws, the tail, the scaly skin, the method of feeding by tearing (and never chewing), the egg-laying … today's crocodile takes you as close as can be to a long-extinct dinosaur. That's the model you need. This is the most fascinating conclusion, though another is the way that the palæontology establishment has prevented anything further being published on my theory. The original article appeared in *Laboratory News*, where several of my major articles have been published. They are 'considering' whether to allow me space for an update. I expanded the idea in my 'Critical Focus' magazine column for *The Microscope* in America.

Other bodies that have published my work have been warned off by their palæontology advisers – *Nature*, *New Scientist*, the *Biologist*, have all said that, with regret, they cannot consider anything on the theory. The Royal Society (who have often published my views) and the Linnean Society (of which I'm a former officer, and where the 'yaw' debate has centred) won't touch it; nor will the Geological Society. My previous publications with bodies like these have been graciously received and some have brought awards in their wake. An international yearbook said my research was a 'scientific highlight of the year' and in the journal *Nature* I was described as being the 'world's greatest authority' on some aspect of research, which is close to embarrassing. But this time it was very different, for my views were the subject of a blanket ban. Thank heavens for *Mensa Magazine*, or these words would have nowhere else to emerge in print.

This is not the fault of the organizations or their journal editors. It is their referees and advisers, conventional palæontologists, who have firmly rejected my idea. It is for innovative proposals like this that 'peer review' can sometimes become a death trap. So keep this article in a safe place. One day it will remind you that scientific revolutions are sometimes hidden from view because it is so convenient to cling to current conventions – but, eventually, new scientific theories do find themselves in print. Confining the study of dinosaurs to palæontologists has given a blinkered view. It has taken an outsider's perspective to look again at their problems with dinosaurs and offer them the answer they seek.

It's the fat crocodile they need, not the chicken.[29]

At last the article had found a home in print. The original draft for *The Oldie* had been 1,200 words long; this new version ran to almost 2,000 words. It brought positive responses, including this one from Michael S. Potter published the following month in the same magazine:

The accepted view of a tyrannosaur shows it teetering on two legs. This *Gorgosaurus* seems ungainly when we imagine it balancing on its hindlimbs. Rotating its axis to move in a different direction would be virtually impossible.

> From time to time along comes a feature in our excellent magazine that rings a clarion bell of authenticity, even to those readers who would not claim to be experts in the field described. So it was with myself enjoying the article by Brian J. Ford about the environment in which most of the dinosaurs likely existed (*Mensa Magazine*, March). Ever erudite yet always readable, Brian once again has shown how an entire scientific discipline could have got it wrong. The negative response from said establishment is so sad in its predictability as to bring into question the academic willingness to consider alternatives, surely the foundation of good science. Brian himself of course lays no claim to being a palæontologist but his sharp mind and ever questioning curiosity give rise, as always, to some very persuasive arguments … Long live Brian and others among us willing to challenge the establishment over their reflex condemnation of good sense above academic dogma.[30]

Once we add the water, the dinosaur's movement and manœuvrability problems are solved. Its aquatic habitat provides buoyancy to support its weight, and a slight movement of its tail allows it easily to turn to face in a different direction.

My papers had all been made freely available online, they were being discussed around the world, and I was enormously encouraged by the fact that the idea of dinosaurs in a watery environment was at last starting to raise its head.

By this time I had been studying the many published reports of dinosaur footprints and needed to check them out on site. In August 2014 I was invited to give a plenary address at the Microscopy and Microanalysis conference in Hartford, Connecticut. It was a grand affair, with an audience for my speech of well over 1,000. Just 12 miles (20 km) south of Hartford lies a major museum of dinosaur trackways – Dinosaur State Park in Rocky Hill. I was keen to study the footprints and see how they fitted with my theories. Photographs of dinosaur trackways were plentiful, but they could not reveal the three-dimensional contours of the impressions in the rock. My scientist friend Rich Brown proposed that we should make a day of it, so we headed down on Highway I-91 and I had the chance to take a close look. This is a remarkable site. On August 23, 1966, clear dinosaur tracks were dramatically uncovered during excavation work for a proposed state building. Further building work stopped while the rocks were cleared, and eventually over 2,000 clear footprints were revealed. They dated back 200 million years and were made by several dinosaur species. The decision was taken to build a geodesic dome to cover all the tracks and present them to the public, but funding could not be found for such a large construction project. Eventually, a display centre was built to show 500 of the footprints, and earth was shovelled back to cover the remainder. The other 1,500 footprints are currently buried once more and will remain hidden for scholars in the future to examine if the money is ever raised to expose them again.[31]

Looking at the footprints and the way they were made confirmed in my mind that they had been left by dinosaurs wading through shallow lakes. Their spacing, their consistent depth, and the way that the impressions of the toes had been made, convinced me that these were the tracks of aquatic creatures wading along through shallow water. If dinosaurs lived on dry land or around muddy swamps, fighting and brawling, we would have a range of tracks that told the tale – those on soft mud would be extremely deep. Those on dry land would be slight impressions. We would see huge claw marks as they leaped to fight, deep footprints where they were forcing themselves on their prey, huge skid-marks as they turned to resist, pounding trackways from their thunderous raids … but we don't. Dinosaur footprints are of fairly constant spacing, and of reasonably consistent depth – and they are always shallow. These trackways can have been left only by creatures moving at a steady pace, and largely buoyant in water. There is no other interpretation. Some of the tracks preserved so crisply at the Rocky Hill site suddenly stop; that further substantiates my view. These were tracks left by dinosaurs who had been propelling themselves on the lake-bed but had suddenly found the water becoming deeper and had therefore taken to swimming instead. The aquatic dinosaur was the only explanation.

As I rounded a corner in the museum I suddenly saw a small display – it was a revelation. The plaque included the small sketch of a dinosaur up to its neck in water. It was remarkably reminiscent of the reconstructions I had published, yet this picture had been prepared by

Walter Coombs, a vertebrate palæontologist, writing in 1980. He had published his findings in *Science* and his conclusions were interesting. 'Dinosaur tracks from Lower Jurassic rocks at Rocky Hill, Connecticut, were apparently made by a floating or half-submerged animal that was pushing along the bottom with the tips of its toes,' he had written. 'These tracks were probably made by large carnivorous dinosaurs (Theropoda) and are apparently the first evidence of swimming by such animals.'[32] This was astonishing, and precisely mirrored the conclusions that I had been reaching for myself.

When I looked into his publications, I came across an earlier paper. He had looked at many aspects of the dinosaurs and had considered the published research into the fossils. He had even considered whether dinosaurs ever ventured into the water. Even so, his final conclusions had been that dinosaurs were essentially terrestrial creatures.[33]

Indeed, Coombs's 1975 paper has been cited as one of the key publications that triggered the move in favour of the terrestrial imperative that was to seize palæontologists' attention. He had actually been a founder of the revolutionary notion of the spritely dinosaur with its feet firmly planted on dry land. Yet, within five years, he had been suggesting that dinosaurs had, after all, lived in shallow water.[34]

This alternative view was immediately discounted, but the fact that Coombs had reached conclusions that so strongly fitted the theory I

During a research visit to Dinosaur State Park, Connecticut, in 2015, I chanced across this remarkable drawing made by Walter Coombs, a vertebrate palæontologist, in 1980. It is uncannily like Romilio's aquatic dinosaur study of 2013. Coombs' proposal was rejected by all other palæontologists.

was now advancing proved to be a great encouragement. And my ideas were percolating through academia. Occasional blogs were beginning to soften their opposition. Five months after my article in *Laboratory News* had been published, the Smithsonian blog dramatically changed its tune. In a curious reversal of its previous stentorian condemnation of the idea, it featured a post headed *Did Dinosaurs Swim?* They conceded that perhaps, after all, they did; and the blog cited work in the 1930s by a palæontologist named R.T. Bird, who had found dinosaur trackways in Early Cretaceous rock in the vicinity of the Paluxy River, Texas. He had thought they indicated that dinosaurs went into the water. The Smithsonian blog quoted Bird's findings for sauropod tracks found at the Mayan Ranch, near Bandera, Texas: 'The big fellow had been peacefully dog-paddling along, with his great body afloat, kicking himself forward by walking on the bottom here in the shallows with his front feet,' it said. What a wonderful description. But the blog went on to dismiss the idea. There was 'no indication that the sauropod that made the trackway was swimming,' it concluded. The trackways Bird had seen was produced by a dinosaur that shifted its weight as it moved. Could he have been correct in assuming that some dinosaurs might have had an aquatic lifestyle? 'Not at all,' said the blog, resolutely.[35]

Yet the idea kept cropping up. In 2014 a feature published in London by Dave Hone in the *Guardian* newspaper was headed 'Were Dinosaurs all at Sea?' and posed the new possibility that some could have been able to swim, but he repeated the standard dogma that they would float so high that they would be unstable if they ventured into the water. Hone did concede that 'while many would have likely been capable swimmers, they didn't primarily eat water plants (as is seen by their strong teeth and stomach contents), and it is unlikely they used the water as a safe haven from predators.' After that promising headline, the article concluded conventionally: 'dinosaurs were very much creatures of dry land.'[36]

With so much evidence appearing, surely the notion of the aquatic dinosaur must break through? It was now so obvious to me, and all the newly published research seemed to confirm what I claimed. I wondered how long palæontology could hold out.

10

TRUTH WILL OUT

On Thursday September 11, 2014 there was a dramatic announcement. The news media resounded with a curious claim that gave us all a sense of déjà vu. After all the strenuous denials, it was suddenly proclaimed that there was, after all, a dinosaur that lived in water. It was (as the perceptive mind will already have deduced) our familiar friend *Spinosaurus*. Nizar Ibrahim, who purchased that *Spinosaur* fossil in Morocco, had joined Sereno's team in Chicago and his few spinosaur bones had been intensively scrutinized. They realized, as I had previously pointed out, that the interpreting of these animals as aquatic dinosaurs was the only version that fitted the facts. The announcement of their conclusions was well rehearsed. A formal paper describing their investigations was published in the leading American journal *Science*.[1]

This was interesting; their artwork showing the skeleton was carefully contrived to divert attention from the fact that they had so few bones with which to work. They combined their studies with all the others that had been published, including Romer's fossils which had been destroyed by the Allied bombs. The bones in their diagram were colour-coded, with the bones Ibrahim had purchased shown red and those previously published by Romer in orange. To the casual reader, the orange and red images look very similar. It was easy to gain the impression that they had more bones than they did. At the same time, news of their findings was released by the Public Affairs department of the University of Chicago, and a simulation of the new *Spinosaurus*

– hobbling along like an inverted sloth with incurved claws on its stubby feet, was put online by the university.[2]

The team meanwhile agreed a deal to become a partner with *National Geographic*, who made expansive announcements of their own.[3] They saw this as a headline-grabber, and a life-size model was made from plastic and unveiled to an eager audience.[4]

The Chicago team made a serious mistake: my reconstruction of *Spinosaurus* clearly showed that it was bipedal and propelled itself along by its two hindlegs. The new model from Sereno and Ibrahim envisioned it as a quadruped, with front- and hindlegs of roughly the same size. To me, this made no sense. It still doesn't.

Even though this was three years after my announcement of the crucial conclusion that spinosaurs lived in water, the Chicago team were claiming the revelation as theirs. Their article in *Science* pointed out that the positioning of the nostrils would have guarded against 'the intake of water' and they concluded that *Spinosaurus* was 'primarily a piscivore, subsisting on sharks, sawfish, coelacanths, lungfish, and actinopterygians that were common …' Their prime conclusion was a reprise of my words from *Laboratory News*: 'We describe adaptations for a semiaquatic lifestyle in the dinosaur *Spinosaurus aegyptiacus*', and they argued in favour of 'aquatic foot-propelled locomotion.' It was a perfect fit with what I had proposed, though no mention was made of my previous paper in their list of references. The announcement was made by Sereno, their star palæontologist and the son of a mailman and an art teacher from Naperville, Illinois. He gained his master's in palæontology at Columbia University in 1981. In 1997 he had been named one of the '50 most beautiful people' by *People* magazine, a scheme described as one of the most well-known and popular franchises in the magazine's 40-year history and which also claimed to be known as 'World's Most Beautiful'. This notion of the 'world' could be taken with a pinch of salt. Most of these nominations were Americans, and there might still be a few beautiful people in distant countries the *People* editors had overlooked.[5]

This has a parallel in the 'World Series' which is a baseball competition with the teams confined to North America. The rest of the 192 countries in the world who are not involved, or even interested, might be bemused to think they are implicitly included in the title.

Sereno was one of the few people publishing on those crocodile-like spinosaurs from the Sahara, and I was not surprised that he wanted a piece of the action. The caption to my first illustration in *Laboratory News* showed *Spinosaurus* as conventionally described, and said: 'This interpretation of *Spinosaurus ægyptiacus*, posted on the jurassicpark web forum, is a typical reconstruction of a large carnivorous dinosaur. Each limb would have had to support up to 15 tonnes. It seems incongruous on dry land.' In the next picture I had added some water using Photoshop (don't get excited; it was an amateurish job) and I described it thus: 'In this revised image, I have shown it immersed in water to the shoulders. The tail becomes buoyant and the mass of the body is supported by the water. The loading per limb would be less than 1 ton – and would be reduced to zero if the dinosaur floated in water like a present-day crocodile.' I then included the image of a spinosaur from the BBC series *Planet Dinosaur*, the dinosaur standing ankle-deep at the edge of a stream, ready to grab a passing *Onchopristis* fish. 'Current interpretations of *Spinosaurus* confirm it lived near water, plunging in its head to find food. In my view it was an aquatic dinosaur and lived largely immersed in water where its fish diet was abundantly available,' I wrote. It was that conclusion – that *Spinosaurus* was an aquatic dinosaur – which was radical and new. This was a revolutionary interpretation, though the terrestrial tyranny kept insisting on the opposite viewpoint. I had brought together a range of scientific evidence in support of my thesis: the sail fin along its back, the X-ray tomography in Milan revealing the crocodile-like sense organs on the snout, and the Chinese research proving that its isotope ratios were like those of other aquatic creatures. So many of the current conclusions reached by palæontologists had been based on idle speculation, but I had assembled sound scientific data that confirmed my view that spinosaurs were aquatic – and the case had now been independently proved.

In 2004, Paul Sereno had published a paper entitled 'Birds as Dinosaurs', which, though presenting conclusions as if they were new, reiterated the pioneering proposals dating from the 1860s, when

As an example of the conventional view of *Spinosaurus* I took this image captioned (in Indonesian) 'Spinosaurus berjalan dengan dua kaki' (Spinosaurus walked on two legs). Like so many dinosaur images, it was shown walking by the water.

Archæopteryx had been discovered. Soon after that, Sereno was on a roll. In 2008 he was listed as author of eight published palæontological papers and by 2010 had added 11 more to his name. Then his output simmered down. In 2011 and 2012 he published just one paper each year. And that was when my article suddenly arrived on the scene. In 2013, five new papers appeared on routine discoveries like that of a long-necked turtle, and the osteology of a raptor, and Sereno also appeared as one author among a dozen in a paper on 'the semantic integration of biodiversity data.'

The appearance of Sereno's *Spinosaurus* paper in *Science* was a substantial surprise. I began to recieve messages from academics who were concerned that my work had been plagiarized, but it caused me no problems – and for two reasons. First, they had it wrong. A number of CGI simulations appeared on a television documentary by NOVA and *National Geographic* to demonstrate this 'new' interpretation, and they are truly bizarre. Their version of *Spinosaurus* is seen stumbling along on four feet through the undergrowth, or swimming like a dog undergoing aqua-therapy and threshing about wildly in the water. In another simulation, it is hobbling along up to its ankles at the edge of a river. You see? That pervasive terrestrial tyranny was still controlling

When the water level is crudely added to the image using Adobe Photoshop, we gain an impression of how water makes the massive creature more plausible. I concluded in my article: 'In my view *Spinosaurus* was an aquatic dinosaur.' Published by the author in *Laboratory News*.

their thoughts. There was not a single view that showed *Spinosaurus* as I had published it in 2011: living in shallow lakes, up to its shoulders in water, poised and partially buoyant and gliding gracefully along, propelled by its hind limbs. My concept was of a biped; their incorrect view of *Spinosaurus* was as a quadruped – and still they had it essentially walking on land. These people lacked the vision to make the conceptual break with the prevailing orthodoxy. Their recreations provided the most heaven-sent material for my lectures.

And the second reason why I was unperturbed is even more prosaic. In science, being published is what counts, and the date of an article appearing in print defines its chronology. Publications prove precedence. Everyone with whom I discussed this latest development was concerned by the fact that Sereno and his team had misappropriated my findings, and people were becoming so agitated that eventually I thought it best to raise the matter formally with the University of Chicago. They put me through to Steve Koppes, their associate director in charge of the university's research news. Koppes was originally a daily newspaper reporter, earning his master's in journalism at the University of Kansas and later a BSc in anthropology from Kansas State. I told him that I was planning a lengthy account of my research which Sereno's conclusions now duplicated, and I added that it would be helpful for Sereno to 'explain his position'. Koppes was quick to reply: 'Paul considers spinosaurs to be subaquatic, not aquatic – to him a significant distinction, and many researchers have mused about the abilities of Spinosaurus and other spinosaurs, including Baryonix (*sic*) and they actually published their work.'

Hold on, there was nothing anywhere in those publications to make the great leap that these dinosaurs evolved in a watery environment – and anyway, mine had also been published. Koppes gave their reasoning: 'So far as I can tell, Brian, you have nothing published on Spinosaurus in any peer-reviewed journals.' So that was the answer. Because *Laboratory News* was a scientific magazine, but not a peer-reviewed journal, the University of Chicago group felt they could disregard it. This is often a stratagem used in science, though it comes into play only when you want to supplant somebody else's research. The structure of DNA by Watson and Crick was not formally peer reviewed, but just given quick approval and was rapidly published by *Nature* as an item of correspondence. Similarly, the famous paper by Robert Brown announcing his discovery of the cell nucleus was privately published by Brown himself, and informal letters written by pioneers ranging from Darwin to Einstein are regularly cited, though nobody reviewed those before they were referenced. Articles in *New Scientist* are scrutinized, not necessarily peer reviewed, yet you would never hesitate to cite one of those if it was helpful to your cause. In truth, anything can become a reference in a scientific paper if you choose to acknowledge it, while any publication can be flagrantly plagiarized if the writer thinks they can get away with it. An article published by *Laboratory News* can certainly be cited – in many cases, that magazine's more rapid publication of articles gives scientists a head start in publishing a new idea.

Even though Sereno's team and *National Geographic* had created their life-size plastic model of the dinosaur, complete with scaly skin and bright, piercing eyes, they still couldn't understand its aquatic lifestyle. Whereas I knew it would have existed in shallow water, seizing fish that it sensed with those glands around the snout, their plastic model still had it standing on dry land. It was holding its prey on the gravelled ground beneath its clawed front leg. The model was initially put on display in Washington D.C.

The paper published by Sereno and his colleagues in *Science* featured one curious illustration that showed the dinosaur looking decidedly out of scale. The hind limbs were too stubby, and its bones seemed too slight. They had reported that the hind limbs were short, and the pelvis was narrow, concluding that its light-weight posterior meant it would have been happier in the water. My reaction was different; I was certain that the reason the hind limbs were so reduced in size was because they were the wrong legs. Several people have since insisted that the hind limbs are real, but it still seems to me more likely that portions of the skeleton came from a smaller animal altogether. Sereno seemed to have made the mistake of mixing bones from different dinosaurs and creating a chimera.

The bones they had retrieved were only a very small part of an entire skeleton.[6] The blog on the *National Geographic* website admitted that the images that had been created were 'based upon a hodgepodge reconstruction that draws from many different dinosaurs. There's the new subadult skeleton, digital representations of the original and long-lost *Spinosaurus* bones, vertebræ and hands that may or may not belong to *Spinosaurus*, as well as replacement parts from an assortment of spinosaurs, all scaled to fit together.' They added: 'The dramatic departure of *Spinosaurus* from the previous release is a hypothesis that will be tweaked as additional specimens are discovered.' And that was the essential problem – *Spinosaurus* was being represented as a quadrupedal dinosaur that could walk on land. I am certain that is wrong; like *Baryonyx* and the other fish-eating dinosaurs, I was sure that its hind limbs in life were substantially larger than its front legs. I knew that it had evolved as a wading dinosaur, and (like other species in which the weight was borne by the buoyancy of water, including *T. rex*) its hindlegs would have been large and muscular, with its forelimbs reduced in size and significance.

The NOVA/*National Geographic* computer simulation depicting a living *Spinosaurus* may be the most ridiculous CGI creation that you will ever see. Their video does not reveal it as the semi-aquatic dinosaur that the Sereno team were belatedly claiming; instead, their

The most unlikely version of *Spinosaurus* was produced for Paul Sereno at Chicago when, in 2014, he reprised my demonstration that this must have been an aquatic species (see also p. 406). Their digital dinosaur hobbled along the water's edge on incurved fingers.

white, waxy creature is a terrestrial dinosaur. It is portrayed limping along on clenched toes like the victim of a severe dietary disorder, walking on its knuckles, hobbling through a forest, until it eventually steps up to its knees into a stream, and stumbles along with its body exposed. The overwhelming influence of the terrestrial lobby led the team to show *Spinosaurus* as essentially land-based, even when they were setting out to make it aquatic. They might have borrowed my conclusions, but they couldn't fully understand what they meant. In the event, terrestrial tyranny won the day.

The media were following this with eagerness. They like dinosaurs almost as much as assassinations and royal weddings and this was an ideal subject for television. One example was screened in October 2016 by Fox News. Nizar Ibrahim was on screen, blinking smilingly at the cameras and throwing in hyperbole as a chef uses garlic. 'We are talking about probably the most mysterious dinosaur out there,' he claimed. 'It was found in southeastern Morocco and is bigger than *T. rex* by about nine feet but more importantly it had adaptations for a life spent largely in water. So it's a river monster. Its feet looked a little like paddle-like structures, so this is an animal that was really at home in the water.' Where could it be seen, he was asked. 'We actually mounted a life-size *Spinosaurus* at the National Geographic Museum in D.C.; it is also the cover story of the *National Geographic* magazine, so very exciting times for *Spinosaurus*,' he chuckled gleefully.[7] The comment about the 'paddle-like' limbs was counterbalanced by that ludicrous simulation of the beast in CGI, hobbling along as if in urgent need of remedial orthopædic surgery. 'River monster' was clearly their buzz phrase. '*Spinosaurus* is probably the most bizarre dinosaur out there,' said Ibrahim in a later television interview. 'It has a long slender snout like a crocodile, it has short hindlimbs …' (a voice at this point should shout: 'Oh no it hasn't, you have the wrong legs!') and then, with boyish excitement, he went on: 'Wow, this thing could just grab me and pull me into three pieces.' Hmm – only three? 'This was the first dinosaur that was actually spending a substantial amount of time in the water,' he concluded: 'It was – a river monster.' I think everyone got the point. Paul Sereno was also on screen, and for the *Daily Telegraph* website he gave a typical interview. He proclaimed:

> *Spinosaurus* is the largest predator to date, fifty feet long. That makes it special in the category of Olympic records for dinosaurs. But for science, it broke another barrier that I think is scientifically more important for us. It is an animal that is adapted to living in water. We call it a semi-aquatic animal, an animal that could go on land, like all of its predecessors, the predatory dinosaurs we call theropods, but from head to tail it is showing signs of adaptation for a water-loving life.[8]

And there you have it. Not perhaps set down quite as succinctly, or possibly as clearly explained, as my original paper, but the sentiment is the same. In a later video interview for the University of Chicago, Sereno explained: 'Dinosaurs, unlike many other groups, played with the water but never really adapted themselves for a life in the water as predators of fish, for example – until *Spinosaurus*.' He explained that the bones of this dinosaur were denser than usual, lacking a marrow cavity, and said: 'That resembles animals that actually are spending a lot of their time in the water. They want themselves to be a little bit heavier, so they don't float all the time …' which is precisely what my paper had proposed. This was all excellent news.

Chicago is a city I know and love. I first visited it 1970, and for 35 years I have lectured at the Inter/Micro conference hosted by the McCrone Research Institute. There is even an annual Monday night presentation which they have named 'An Evening with Brian', and in 2017 I celebrated the 100th lecture I had presented to that conference. Since I was close to the university I contacted Sereno and he invited me to come to see his department. We needed to clear the air. The date and time were agreed, and my car was booked. But – just before I left – I asked my assistant to call the university to double-check. The voice on the phone was apologetic. 'I am so sorry,' she explained, 'but Paul has had a family emergency and is not here right now. He is out of town for the next couple of days. But he says that, if Professor Ford would like to visit with us, that's fine, and we can show him everything he wants.' When they told me the news I decided against the visit. Subsequently, Sereno told me that they had only casts of the skeleton. The original specimens had all been returned to Morocco, as part of their heritage. Naturally, they were what I wanted to see. Good fortune smiled; I was shortly due on a lecture tour that included Barcelona, and at the city's famous Blue Museum they had the reconstructed skeleton on loan. I spent half a day with it, studying its structure and building an understanding of how it functioned. There was now no doubt in my mind; this was an aquatic creature, similar in many ways to a crocodile, but using longer hind limbs to propel itself along by wading in shallow water, rather than (as crocodiles do) swimming at the surface. The CGI reconstruction on which Sereno had worked, which showed it stumbling through the trees and stepping unsteadily into the water, was far from reality. My interpretation is that *Spinosaurus* was elegant and sleek, streamlined and quick to react, using its well-built hindlegs to wade through shallow water and those sense organs to follow the fish. Its forelimbs would be somewhat reduced (think of *T. rex*, where the same evolutionary tendency is taken to extremes) since it did not customarily need them to support its weight on land, and it would use them only to hold its prey.

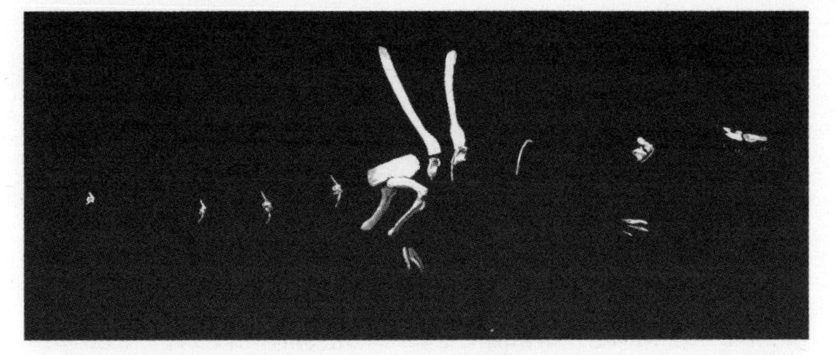

Nizar Ibrahim purchased some bones of a spinosaur in Morocco in 2008. The number of fossils retrieved was relatively few (they could be stood in a bucket) and there is no guarantee that they represented the same animal.

In spite of the lack of specimen material, and consulting the photographs of a few more spinosaur bones that had been destroyed in World War II, a guessed version of the entire skeleton was invented. The hind limbs seem out of proportion.

Here the pelvis and hind limbs have been adjusted by the author to a more feasible appearance. The tail and head have been raised, along with the abdominal gastralia, and we may now be closer to a real *Spinosaurus* skeleton.

The picture of *Spinosaurus* at the water's edge which the BBC had created (p. 318) had been shown to be incorrect by my revised interpretation, and the Chicago team had since agreed that spinosaurs were aquatic. Yet they were unable to abandon their terrestrial preconceptions, and their TV image still showed *Spinosaurus* as the BBC had done – paddling in the shallows.

It was Don Henderson[9] who had demolished the idea that larger dinosaurs could enter water, so they could never have swum; and he had condemned my original proposal that *Spinosaurus* must have evolved in an aquatic habitat as ridiculous. Henderson had told me: 'I think you should go back to your microscope and not dabble in other branches of science where you do not have sufficient knowledge.' Now he was about to publish an account of buoyancy and balance of dinosaurs in water; I could hardly wait to see what he was proposing.[10]

The aquatic habitat of *Spinosaurus* often came into my lectures at the time. Members of the public were always enthusiastic, though sometimes there were advocates of the terrestrial dinosaur model and they reacted with astonishing anger. One of the *Scientific American* bloggers named Darren Naish turned up at a lecture I presented in Singapore Science Centre. He picked up a ticket and came to the presentation with a friend. By the time he wrote it up it seemed very different, somewhat like an episode from James Bond: 'Via bizarre and unexpected circumstances I recently found myself secretly and furtively attending a lecture by Brian J. Ford.' It wasn't 'secretive' at all, he simply booked a ticket and sat in the audience, like everybody else, and it wasn't 'furtive' except that he didn't ask any questions. He had his male friend do so, though it was an incoherent question. It ran like this:

> I appreciate this isn't a technical lecture, and there's a lot of detail missing, but, the, the, roughly, dinosaurs were on there for about 150 million years, the scientific papers mention for the Triassic Period, which finished about 200 million years ago, dinosaurs went extinct roughly 65 million years ago on the basis of timeline, I, I'd like to understand more about how you actually see this in that, the evolutionary scheme; you talk about dinosaurs being, living in an environment which is all with shallow seas, you know, 160 million years of planetary activity was not unchanged, clearly, you know, the seas drying up; the scientists are saying yes, in the Triassic period 200 million years ago, there were, you know, semi-aquatic dinosaurs, you just said that the, the seas dried up 65 million years ago.

The style of this inquiry is interesting. You can work out approximately what is being asked, but the words are rambling and the sense incoherent. They were filmed, by chance.[11]

All if this reveals muddled thinking – which is precisely the way so many of these dinosaur experts look at their world. At the core of his question was the proposition that I'd claimed the seas had 'dried up'.

I interjected: 'No, no, obviously they didn't. My dear chap, the seas were everywhere.'

He replied: 'But you talk about the shallow seas drying up, that was not a meteorite, that is a separate issue, you don't need to get into the constituent doubters, but can you try to put this into perspective of that very, very long timeline?'

In answer, I reminded the audience of the extensive areas of sedimentary rocks in which dinosaur fossils were found – right across America, in England, across Continental Europe – and I explained that the seas certainly didn't dry up; rather, they became modified so that the dinosaur's habitat slowly vanished. To me, the chronology was clear.

The next questioner began by saying what an interesting topic this was, but he asked – as people often do – how dinosaurs living in a wet muddy habitat could be squared with finds in the arid sandy stretches of the Gobi Desert. I have also been asked how the skeletons of aquatic dinosaurs have been found up in the hills. The answer, of course, is that conditions weren't the same when the dinosaurs were alive. The Gobi Desert then would have been the 'Gobi Lake'. The question was answered, and – after a pause – another questioner raised their inquiry. On his blog post, it seemed very different. The questioner wrote that he had wanted to ask more about the Gobi Desert fossils, but he had been stopped as the organizer moved on. He hadn't.

When it went online, the blog attracted a torrent of abuse about my lecture, running to some fifty pages. 'Great article!' began one supporter, happy to see my theories attacked. 'Glad you put this naysayer in his place.' Said another: 'Ford obviously suffers from water on the brain and is clearly out of his depth. However, I suspect that articles such as this will simply be water off a dick's back.' Another comment asked: 'He's specifically claiming wading, right? I think he's suggesting a mode of life completely different from what we see in the world today. So he may reject modern analogues. ''Cause he's nuts,' while another remarked: 'We can thank Ford for this thread … cuz although his theory is a non-starter it certainly provided a lot of fodder for conversation.' One was an uncompromising summary: 'Well I attended Brian J. Ford's talk, which turned out to be 40 minutes of never-ending rubbish.'

Just a few, however, were less negative. One suggested: 'The "terrestrial dinosaurs" theory has been worked on for decades and got very elaborate, but what if the same amount of time and effort was spent on [an] "aquatic dinosaurs" alternative? All apparent contradictions would have been explained by now.' Further positive thoughts were these: 'I'm sure dinosaurs were very comfortable moving through and foraging in the water. Theropods could have capitalized on aquatic prey while herbivores may have also found abundant and nutritious aquatic foliage … I don't think dinosaurs that occasionally or seasonally spent a lot of time in the water need show any particular anatomical adaptations to do so.' There was also this remark: 'The long neck might have provided an advantage for the reaching of aquatic plants. After all, even if sauropods were mostly wholly terrestrial, they might still take advantage of aquatic plants on occasion, just as elephants in some populations often eat water-plants,' while another correspondent mused: 'Perhaps in our haste to distance ourselves from the patently ridiculous snorkelling duckbill dinos and bottom of the lake walking sauropods we pushed the pendulum too far away from the water?'[12]

Even though those few brave souls suggested that there might perhaps be some merit in the aquatic dinosaur theory, the responses in general were a tirade of invective. My proposal that dinosaurs evolved in water was ridiculous, crazy, impertinent. No right-thinking dinosaur enthusiast could be permitted to think otherwise. Other blogs had appeared that said the same thing, including one which was headlined 'Why the Aquatic Dinosaur Theory is Damaging to Science'.[13]

The comments following this posting were predictably caustic. 'This is the kind of shit that drives me up the god damn walls,' wrote one. 'Are people this fucking dumb?' asked another (and received the prompt response: 'Yes, they are.').

My theories depended fundamentally on the question of the climate, and the conventional view made the aquatic dinosaur seem impractical. The Cretaceous period was considered to be covered with large areas of arid landscape that were cold in winter. This enduring problem is summed up in the Cold Continental Interior Paradox, a concept first coined by Robert DeConto at the University of Massachusetts–Amherst.[14]

It centred on conclusions that the plants of Northeastern Siberia indicated that they flourished in an average summer temperature of about 14°C, falling in winter to 5°C. The climatic modelling by the palæoclimatologists argued against this. All their models pointed to a winter temperature that fell considerably below freezing, and this clashed with the evidence from the fossil plants.[15]

The levels of atmospheric CO_2 would have been crucial. Present-day values are about 406 parts per million. The levels in the Cretaceous have varied wildly. While DeConto's group relied on a concentration of about 1,230 parts per million, Bice and Norris published a paper in which they calculated that the level could have been as high as 4,500 ppm – more than 10 times present-day levels.[16]

Not only were the data so unreliable, but the calculations were incomplete. The standard models were using levels of CO_2 as an indicator of global temperature – and yet they had ignored methane, CH_4. This gas holds on to about 23 times as much atmospheric heat as carbon dioxide.[17]

Just when my paper was published, David Wilkinson of the University of Lincoln, England, released some intriguing findings. Not only did we have well-recognized sources of methane with which to contend (including release from swamps, metabolism by termites and venting from volcanoes), but substantial amounts would be produced by giant dinosaurs consuming vegetation. Wilkinson's team calculated that the amount could be more than 500 Tg (1 Tg $=10^9$ kg) of methane each year – about 500 billion tons. This amounts to 700,000,000,000,000 cubic metres of methane … though they were basing these figures on the conventional Cretaceous model, with its arid continents and lower temperatures.[18] My proposed revision, with hordes of dinosaurs breeding across endless verdant lakes and swamps, would add greatly to the total.

In today's world, even a diminutive creature like a termite produces a total of some 50 billion tons of methane per year,[19] and dinosaurs metabolizing enormous amounts of vegetation would (as cows do today) have poured out masses of methane into the air.

Methane is a serious matter, but this is not the only greenhouse gas that is widely ignored in the conventional calculations. Water vapour is also a powerful contender; 60 per cent of today's global warmth is due to atmospheric H_2O, not CO_2. The moist environment that I was now postulating would mean that the atmosphere contained far more water vapour than the standard models had considered – humidity is related to heat retention. With the endless watery landscape that I was proposing for the Cretaceous, the heat-retaining capacity of the atmosphere was far greater than we currently claim.

The environment in which aquatic dinosaurs would have flourished had to be consistently warmer, and consistently wetter, than any of the previous models had described. This had once seemed to be a problem, but Bill Hay had become a fervent supporter of my revolutionary new view and he encouraged me to continue my investigations. The contributions that were now coming together for our joint paper started to show that my theory provided an answer to all his outstanding problems with modelling the Cretaceous period. Hay had been busily compiling a scientific paper to support my theories. He pointed out that the ramifications of this revolutionary new proposal were widespread, and he concurred with me that it was time for a radical reappraisal of the environment in which the dinosaurs had lived. Throughout 2015 and 2016 Hay worked on a paper setting out the implications, and it was published early in 2017. Not only was the climate much warmer, but he now agreed with me that the landscape was very different – rather than those deserts in which dinosaurs were traditionally portrayed as pounding across the landscape, he envisaged expansive oxbow lakes, river deltas and shallow seas. Hay set down the problems with the accepted views and then explained: 'Impetus for completely rethinking the nature of the land surfaces in the Cretaceous came from an unexpected source. Brian J. Ford is a cell biologist who became interested in dinosaurs and their tracks.' Then he added: 'Although many vertebrate paleontologists working on dinosaurs reject Ford's ideas out of hand, they make good sense to many geologists. General aspects of the biology of the great sauropod dinosaurs have been reviewed [and] much of the biology fits well with Ford's "aquatic dinosaur" hypothesis.' This was a crucial scholarly publication. After five years of unending hostility, here was one powerful voice of support of my theories published in the academic literature.[20]

Hay said that, with my proposed revision, the Cretaceous period was very different from current thinking. The mean temperature was higher, the continents were wetter, shallow water was abundant, and dinosaurs flourished in vastly greater numbers than anyone had previously envisaged. Clearly, the effects of greenhouse warming would have to be increased over the previous calculations. Tropical warmth would have predominated, and revised climatic models would show that the cold zones of the present-day theories were a massive mistake.

Science has misunderstood the climate of the Cretaceous; once again, we will now have to revise our textbooks. The most recently published papers on dinosaurs continue to provide further evidence for the aquatic habitat that I had proposed. In Canada, research into their dinosaur fossils now concluded that the late Cretaceous landscape was one in which 'lush tropical forests covered the lowlands, incised by creeks and streams running from the very young Rocky Mountains, and filled with fresh water.'[21]

When a new dinosaur was discovered in Utah, it was named *Moabosaurus utahensis*, and the researchers at Brigham Young University reported: 'Utah is rich in dinosaur fossils, but it looked much different 125 million years ago. There was no desert. Instead, lakes, streams and trees were abundant.'[22]

Those trees were certainly swamp species; palæontologists insist that sauropods ate land plants, but the gymnosperms and pteridophytes among which they lived evolved in swamps during the Carboniferous Period. Their remains formed coal. Although today's representatives are typically terrestrial plants, in earlier eras they lived (like the dinosaurs) in shallow water.

Further evidence comes from sauropod teeth, which show scratches.[23] The usual explanation is that they attest to the diet being rich in spiny plants. Many plants contain tiny needles of silica which abrade the teeth – but the marks left on sauropod teeth are deep, and broad. I have worked with remains of humans from the twelfth century who had fed on seafood, and their teeth show broader scratches caused by sand in the diet. I think the marks on dinosaur teeth result from their aquatic habitat: by eating swamp plants, they inevitably chewed on sand. We can find out much of how dinosaurs chewed their food by studying these marks.[24]

Dinosaur fossils have been found at the Cleveland-Lloyd Quarry in Utah for over 90 years, and in 2017 Josh Peterson, geology professor at the University of Wisconsin–Oshkosh, produced evidence for the watery habitat for dinosaurs that I have postulated. He claimed: 'Our hypothesis now is that this area was a body of water that was seasonal, it was an ephemeral pond. During the wet season when the rains would come all these dinosaur carcasses would wash in …'[25]

It was also in 2017 that Africa revealed its first large theropod dinosaur, a forerunner of *Tyrannosaurus rex* named *Kayentapus ambrokholohali*, measuring 30 feet (9 metres) long and weighing about 6 tons. Its habitat, said the researchers, was clearly a 'prehistoric watering hole or river bank.'[26]

Not only were the surroundings in which dinosaurs were found increasingly said to substantiate a shallow-water environment, but the anatomical details of the dinosaurs themselves provided further confirmation. The remains of a small feathered dinosaur named *Sinosauropteryx* were discovered in China during the 1990s, and a postgraduate student at Bristol University, Fiann Smithwick, has been examining the details of the proto-feathers which the fossils have retained. It seems that there was a band of brown proto-feathers surrounding the eyes. The interpretation is that the band of feathers acted like a sunshield, to help reduce glare – sunlight reflected from the surface of water. Water was everywhere.[27]

It is those tiny forearms of *T. rex* that provide some of the most obvious explanations for my conclusion that dinosaurs were aquatic. All large animals need four legs on which to stand, which is why a gigantic elephant weighs 10 tons, while a kangaroo is 100 times lighter. The biggest-ever kangaroo was *Procoptodon*, which lived in Australia from about 1 million years ago, during the Pleistocene epoch. They stood almost 7 feet (2 metres) tall and weighed 500 pounds (about 225 kg). An elephant, because it has four sturdy limbs, can grow to weigh almost 50 times as much as that fossilized giant kangaroo. For a tyrannosaur to lose its forelimbs, there must have been an evolutionary benefit, and that could accrue only if the dinosaur lived in water. The front legs of a tyrannosaur are ridiculously tiny. I had said in my *Laboratory News* article that these small forelimbs might have been useful for it to hold its food, or to inspect what it was about to consume; but that could not be the whole story. As we have seen, research shows that those forelimbs are so reduced in size that they could not even reach the dinosaur's mouth. Many other meat-eating dinosaurs are now known to have had much reduced forelimbs, and this feature seems to have evolved separately on several occasions. There must be an evolutionary imperative that drives the trend – and a life in water is the only answer I could see.[28]

When *T. rex* was first investigated by Henry Fairfield Osborn, researchers didn't believe that the tiny front legs were possible. It was thought that those minute arms had come from a different dinosaur fossil. It did not seem feasible that such a huge beast could have such small front limbs.[29]

Although later finds seemed consistent with Osborn's, it was not until the near-complete skeleton of *T. rex* was discovered that the matter was finally resolved. This is the specimen nicknamed Sue, now on display in Chicago.[30]

Because palæontologists have been determined, against all odds, to portray tyrannosaurs as terrestrial animals, a whole list of purposes of these little limbs has been proposed. All are questionable; many are plain ridiculous. One theory is that they were there for sex; the male *T. rex* could hold onto its partner. It was even proposed that the small arms were useful for the male sexually to stimulate the female's underparts in acts of passion. It was later claimed that the food of this mighty monster took the form of creatures that lived in deep burrows in the mud, so the front limbs had to be small so that they did not get in the way when it inserted its vast head into burrows. Another theory was that they would have been useful in helping the dinosaur to right itself, if knocked to the ground, and there was also the idea that the tiny legs could have prevented a fallen *T. rex* from sliding forwards, if it was trying to rise again by pushing forward on its hindlegs.[31] Then there was the reasonable notion of Rob Reyes in Los Angeles, that they acted like vacuum breakers, if the *Tyrannosaurus* become embedded in a muddy river bank.

The only benefit in reducing the forelimbs to such an extreme extent was if they evolved in water. Then all the problems disappear, and *T. rex* emerges as a huge version of a crocodile, except that it waded rather than swam. Once again, the aquatic theory solves the puzzle. That ridiculous chase of a car by *T. rex* in *Jurassic Park*, the result of terrestrial tyranny (Bob Bakker was their advisor), was always impossible. Computer modelling of the limbs of *T. rex* proves that it could never have run. Calculations show it could only walk, and – in the watery environment in which I envisage it spending its life – it would move far slower even than that.[32]

A detailed report in *Scientific American* concluded that *T. rex* 'may not have been able to run at all' and emphasized that running was 'biologically unfeasible.'[33] Hannah Osborne in *Newsweek* featured recent scientific research proving that *T. rex* could never have run.[34] William Sellers of Manchester University concluded that *T. rex* would have been limited to walking. His group's research 'contradicts arguments of high-speed predation for the largest bipedal dinosaurs …'[35]

These conclusions appeared five years after my original paper was published, and the new findings were completely consistent with my own conclusions. All the scientific evidence shows that dinosaurs were not the massive monsters about which we have been taught, pounding about on an arid landscape and burning huge amounts of metabolic energy to support their extremities. That is all fake news. As we have seen, once they had held up their heads and their tails there would be no energy left to fuel their bodies. They evolved when the world was largely covered in vast shallow lakes and had clearly used the water to support their mass, buoy up their tails, regulate their temperature and provide a habitat for their food.

Fitting the evolution of dinosaurs into their family tree is continuing to unravel how they emerged, and the quest for 'missing links' has been a continuing preoccupation. One of the earliest of all ornithischian dinosaurs was excavated in Lesotho, that small nation entirely surrounded by South Africa, and it was named *Lesothosaurus* by palæontologist Peter Galton in 1978 to commemorate the country of its discovery. It was a bipedal dinosaur some 6 feet 6 inches (2 metres) long, with a long tail, slender limbs and jaws that were horny and shaped like a beak. This might have been identified as a primitive herbivorous dinosaur, but for the presence of the sharp teeth that were shaped like fangs. Clearly this was a creature that could live on flesh as well as plant matter. Galton claimed this as a crucial missing link – it seemed to bring together the disparate groups of dinosaurs into a common line of descent. In 2017 another gap in the fossil record was filled by a 245-million-year-old fossilized carnivore named *Teleocrater*. The fossil had originally been discovered in the Middle Triassic Manda Formation of Tanzania in 1933 by F. Rex Parrington and was named by Alan Charig in 1956 but was not formally described until 2017 by Sterling Nesbitt and the team at Virginia Polytechnic Institute and State University. This 10-foot

(3-metre) monster existed before dinosaurs and walked on four legs rather like a crocodile.[36]

Another missing link was proudly announced in August 2017 when there was a new flurry of excitement as a 'Frankenstein dinosaur' suddenly claimed the headlines. The fossilized remains of this curious beast date from 150 million years ago and they represent a small dinosaur about 10 feet (some 3 metres) long which presented a puzzle because it seemed 'to consist of body parts from unrelated species,' reported the news. It had limbs like those of the herbivorous lizard-hipped dinosaurs, yet with the pelvis like that of a bird-hipped stegosaur and a body that had features reminiscent of a carnivore like *Tyrannosaurus*. The obvious interpretation would be that this represents an ancestor of the later, more specialized, dinosaurs. This was indeed the conclusion that grabbed the news media. It seemed that a Cambridge student of palæontology, Matthew Baron, had proposed this answer in a joint paper co-authored with Paul M. Barrett and published by the Royal Society.[37]

This made global headlines. Sarah Gabbott at Leicester University was bubbling with enthusiasm. 'This is one of those rare fossil discoveries that provide much more evidence to unravel dinosaur relationships than your average skeleton,' she told reporters.[38]

It is always interesting when taxonomists make adjustments to evolutionary family trees. That happens all the time in biology, and ordinarily the public don't hear a word. This, however, concerned dinosaurs; and as we have seen, they are always big news. The title of Baron's paper was not terse, academic, abstruse, as are most such publications; it began with the tease, 'a dinosaur missing link?', to ensure it caught maximum attention, and added a question mark in case it was wrong.

The background is much more interesting. Remember, it had been in 1887 when Harry Seeley had proposed his distinction between the two main dinosaur groups, the lizard-hipped Saurischia and the bird-hipped Ornithischia. This conclusion had underpinned dinosaur studies ever since.[39]

Baron had already published a revision of the family tree of early dinosaurs in a joint paper published many weeks earlier. He now suggested that the theropod dinosaurs (like *T. rex*) might have evolved from the ornithischian group, which is an interesting possibility. He also emphasized that the earliest dinosaurs might have evolved in Britain.[40] 'The northern continents certainly played a much bigger role in dinosaur evolution than we previously thought, and dinosaurs may have originated in the UK,' he told the BBC.[41] This was revolutionary. David Norman, one of the senior palæontologists at Cambridge, added: 'All the major textbooks covering the topic of the evolution of the vertebrates will now need to be re-written if this suggestion survives academic scrutiny and becomes accepted more widely.' Praise indeed for the importance of the paper – though you should note that Norman was a co-author of the paper that he was so enthusiastically espousing. All this helped promote the cause.

The conclusion now was completely different. It was being claimed that *Chilesaurus* was the 'missing link', yet this one came from Patagonia some 8,000 miles (almost 13,000 km) south, at the opposite end of the world. In politics, or sport, the media would have ridiculed such a switch in polarity – but, in dinosaur science, newspapers just print what the experts say.

What of *Chilesaurus* itself? It was found in southern Chile in the Aysén region of Patagonia, an astonishing and largely unexplored land of verdant valleys and arid deserts. In 2004, Manuel Suarez and Rita de la Cruz, two geologists, were surveying the area of Black Hill, near General Carrera Lake, when Manuel's 7-year-old son Diego found the fossil. Indeed, it was later named *Chilesaurus diegosuarezi* in honour of the little boy who discovered it. The scientist who pointed out its curious features was Fernando Novas, at the Bernardino Rivadavia Natural Sciences Museum in Buenos Aires. 'It was so shocking,' he said, 'because this new animal combined several features belonging to three different lineages.' It was originally nick-named the platypus dinosaur, because of its curious blend of anatom-ical features (but 'Frankenstein dinosaur' proved to be more newsworthy). The conclusion that it might be a common ancestor neatly settled the confusion. However, this did not resolve the issue of where they evolved. The same team in Cambridge had first claimed that the area was up in the far north and, later, that it was down in the far south – which remains a puzzle to us all. Such science values headlines far more than consistency.

These 'missing link' preoccupations regularly provide startling new examples of dinosaurs that seem to exhibit a curious mosaic of features. One of the latest is being claimed as definitely amphibious. It is a small, bird-like creature named *Halszkaraptor escuilliei*, which lived some 75 million years ago in the Ukhaa Tolgod area of the Gobi Desert. Reptiles like these were originally included by Othniel Charles Marsh in the family Ornithomimidae as long ago as 1890, but as more was discovered about their anatomy they were later moved to the family Coelurosauria. This particular example lies in a block of stone that was smuggled out of Mongolia to Japan and then to Great Britain, and was eventually purchased by François Escuillié, a fossil dealer. In 2015, he travelled to Brussels with the specimen to ask for the opinion of two palæontologists, Pascal Godefroit and Andrea Cau, who first concluded that it was comprised of several skeletons of unrelated species. To prove the point, it was taken to the particle accelerator at Grenoble, France, for scanning. This confirmed that it was actually a single dinosaur – though a highly unusual one. This result was published by Andrea Cau and a team of 10 experts in the scientific journal *Nature* in December 2017. They emphasized the curious hybrid assortment of characteristics in this new creature – it was a predator with an extendible neck, it was bipedal (like other theropods, including *T. rex*), it had a posture like a bird, and it had forelimbs that seemed to be adapted for swimming.[42]

The news that *Halszkaraptor* was amphibious was encouraging, though there was no aquatic evidence when the discovery was announced in the spring of 2019 of *Moros intrepidus* (from the Greek μωρό, meaning stupid), a Cretaceous dinosaur from Utah. This may be the ancestor of *T. rex*, but was 5 feet tall (1.5 m) and weighed about 170 pounds (77 kg). Though it may have been terrestrial it would need an aquatic habitat to evolve further.

The origins of dinosaurs will long be debated, and so too will their abilities. What can we discover of the senses that dinosaurs possessed? Present-day crocodiles and lizards have colour vision, a fact we can demonstrate by studying sections of the retina microscopically. There are two types of light-receptor cells in the retina: rods (which detect light intensity) and cones (which respond to wavelength, and thus reveal colour). Since both reptiles and birds have cones in the retina, we know they can all perceive colour and we can reasonably assume that this was probably true of dinosaurs. A study of likely vision in fossil animals was published by M.P. Rowe, a research psychologist at the University of California, Santa Barbara.[43] He drew attention to an important paper by Lawrence Witmer, a palæontologist at the Department of Biomedical Sciences at the Heritage College of Osteopathic Medicine in Ohio University. Witmer pointed out that most reptiles (and some birds) have a curious feature in their eyes, a bony reinforcing ring of flat bones known as scleral ossicles.[44] These are easily recognized and, yes, they are a well-known feature in the dinosaurs. This ring of scleral ossicles consists of small overlapping bony plates between the margin of the retinal margin and the conjunctiva, and at the College of Medicine at Wayne University, Detroit, their Research Associate in Ophthalmology, Gordon Walls, decided that the function of the ossicles was to reinforce the place where the cornea and sclera meet (the rim of the 'white of the eye') and where pressure could cause the eye to become distorted. In the view of Walls, the pressure inside the eye would increase as the animal tried to focus on near and then on far objects, and these ossicles would prevent harm.[45] They seem to have evolved first in dinosaurs; in my view they were an adaptation to the pressure exerted on the eyes by a life in water.

Many of the theropod dinosaurs, including *Allosaurus* and *Carcharodontosaurus*, had binocular vision similar to present-day alligators, restricted to 20° in width, though *Carcharodontosaurus* had a conspicuously large optic nerve, and some others – the deinonychosaurs like *Deinonychus* and *Troodon* – had a field of binocular vision up to 60° wide. The tyrannosaurs seem to have had the best sense of sight, as you would expect from a carnivore.[46]

Present-day plant-eating animals (like antelope, deer and cattle) have eyes on the side of the skull, so that they have a wide field of view that will warn them against approaching predators, compared with the predatory meat-eaters (like lions, tigers and owls), which have eyes that point forward, giving precise three-dimensional acuity. This was also true in the realm of dinosaurs. The sauropods typically had eyes on each side of the head, which would have given them a huge field of vision, and their sense of sight seems to have been complemented by a well-developed sense of smell. The predatory theropods had forward-facing eyes, precisely as we would anticipate. Trying to compare dinosaurs with their living descendants provides some clues, but the most objective means of assessment comes from examining their skulls. The development of computed X-ray tomography, which gave us the C.T. scan, has allowed anatomists to study the volume of the olfactory bulbs of the brain – the centres where smell and taste were detected. The larger the cavity in the skull, the larger the olfactory bulb. Darla Zelenitsky at the University of Calgary in Alberta, Canada, led a project to examine a range of 21 different dinosaur skulls, and he and his colleagues concluded that theropod dinosaurs all had well-developed olfactory senses. The types with the largest olfactory bulbs, they found, were *Velociraptor* and *T. rex*, which apparently rivalled bloodhounds in their ability to smell their prey.[47]

The structure of the olfactory lobes of the brain gives part of the answer, but this tells us only what the brain could do with the information it received – it must be supplemented by a study of the olfactory organs that detected the signals in the first place. We have already seen that the spinosaurs seem to have possessed sense organs in the snout suitable for detecting their aquatic prey, just like the sense organs of crocodiles and alligators. However, finding traces of sense organs depends upon the state of the skull, and such fine details are not always preserved. This time, they were. A revealing report was published online by palæontologists from μVIS X-ray Imaging Centre at the University of Southampton. The Greek character in this perplexingly named facility gives it the pronunciation 'mu-vis', which few outsiders know, perpetuating the long-standing tradition in science of creating names that are impossible for outsiders to understand. Palæontologist Chris Barker led a team that investigated a dinosaur named *Neovenator salerii* from the Isle of Wight, using advanced X-ray and 3D computerized imaging technology. This was a meat-eating creature measuring 25 feet (7.5 metres) long and weighing about 5 tons. The preservation of detail within the skull was remarkable, and the results clearly revealed fine nerve channels and the sites where sense organs existed, showing that *Neovenator* had a highly sensitive snout. Like those seen in the *Spinosaurus* scans, these nerve pathways are reminiscent of those of present-day crocodiles and suggest that these dinosaurs had well-developed senses that were exquisitely well adapted for an aquatic lifestyle. Of course, the terrestrial tyranny militates against anybody saying so, and Barker resolutely insisted that there was no evidence to suggest the dinosaur was aquatic. Instead, he maintained that their facial sense organs existed for erotic pleasure.[48] Barker said:

> The 3D picture we built up of the inside of the *Neovenator* skull was more detailed than any of us could have hoped for, revealing the most complete dinosaur neurovascular canal that we know of. The canal is highly branched nearest the tip of the snout. This would have housed branches of the large trigeminal nerve – which is responsible for sensation in the face – and associated blood vessels. This suggests that *Neovenator* had an extremely sensitive snout – a very useful adaptation. Some modern-day species, such as crocodilians and megapode birds, use their snouts to measure nest temperature, and in the case of crocodiles even pick up their young with extreme care, despite their huge mouths. *Neovenator* might well have done the same. There is plenty of evidence that carnivorous dinosaurs engaged in face-biting among themselves, perhaps targeting the sensitivity of the face to make a point. As well as being sensitive to touch, *Neovenator* might also have been able to receive information relating to stimuli such as pressure and temperature, which would have come in useful for many activities, from stroking each other's faces during courtship rituals to precision feeding.

To ram the point home, Barker helpfully told the reporters that: 'Dinosaurs used their heads for most activities …' The mind boggles.[49]

Senses are all very well, but what did dinosaurs do with them? In short – were dinosaurs intelligent? There could be a clue from their nearest descendants, the parallel family of crocodilians. In the Adelaide River in the Northern Territory of Australia lives an elderly saltwater crocodile, about 80 years old, that the locals have named Brutus. He has only one forelimb, his other arm having been torn off years ago in a shark attack, and he measures 18 feet (5.5 metres) long. The largest of these saltwater crocodiles can reach 23 feet (7 metres) and weigh 2 tons (1,800 kg). The Bowman Brothers company run their Adelaide river cruises for visitors and hang hunks of meat over the side of their tourist boats to attract Brutus. He still swims vigorously, and then leaps from the water to a height of some 7 feet (2 metres). In September 2013, Peter Jones from Cambridge, England, was on the aptly named East Alligator River in the Kakadu National Park when he recorded an extraordinary event on his camera. Brutus seized a shark. The events that unfolded seem to suggest an unexpected level of intelligence on the part of the crocodile, for Brutus dragged the threshing bull shark out of the water in its jaws and carried it up the beach. There he threw it to the ground.[50]

This is revealing – a crocodile conventionally captures prey by seizing a creature as it comes to the water's edge; it then holds the hapless animal underwater to drown. In this case, the crocodile had captured a creature that lives in the water, so attempting to drown it would have been counterproductive. Brutus worked that out for himself and so, instead of dragging the shark underwater, as instinct might cause him to do, Brutus decided to come out of the river and threw the shark on the bank. This would prove fatal to a fish, just as immersion will kill a land-based animal. Rather than relying (as we imagine they do) on habit, this giant crocodile used its wiles to reverse its normal behaviour in order to incapacitate its prey. Only later did he drag it off to the comfort and shelter of a mangrove thicket. This was a novel situation, and the crocodile adapted to it. That is a sure sign of intelligence. If crocodiles show this level of mental ability, then we could assume that dinosaurs may have done so too. Dinosaurs were crude, thuggish and idle – but they were almost certainly smart.

11

THE LIFE AND DEATH OF DINOSAURS

Now we can see how dinosaurs lived their lives. They owe their vast size to their aquatic habitat, and they always relied on the support provided by that aqueous environment. The water supported their weight, gave buoyancy to their extremities and regulated their body temperature. These creatures became vast because they moved slowly through the water that buoyed them up. The gigantic sauropods were the biggest plant-eating creatures that have ever existed, and they cruised gracefully through those huge shallow lakes and wide rivers, leaving shallow footprints in the sandy mud beneath and sweeping up colossal amounts of vegetation. They lived in water, relaxed in water and copulated in water. Only when the time came to lay eggs might they have hauled their massive bodies onto shallow sandbanks to deposit their eggs where they sensed they might be safe.

Preying on the dead and dying sauropods were the carnivorous dinosaurs – the gigantic theropods like *Spinosaurus* and *Tyrannosaurus rex*. They also moved slowly about in the water. As we have seen, these were not aggressive hunters so much as solitary scavengers. In the lush, hot, verdant forests sauropods fell in their millions, dead of old age after (perhaps) 20 or 30 years of life. The flesh on those leviathans soon softened in the heat, and the theropods would pull off massive mouthfuls of meat, tossing their heads up and repositioning their food, until it could be safely swallowed. Crocodiles do the same to this day. Although we always think of them seizing animals that come to the water's edge to drink, crocodiles prefer to consume the decaying corpse of a buffalo that drowned days before and is softly rotting in the tropical temperatures. Crocodiles do not chew, but they pull off meat by seizing it, and dragging it. If the corpse is fresh and the meat is still tough (they do this also with a creature they've caught) then the crocodile will seize a portion with the tip of the snout and rotate rapidly in the water, twirling and threshing around until a huge hunk of flesh twists off and becomes detached. Meat-eating dinosaurs could not have performed this athletic twisting movement; and we can find a hint of this in the sauropods' skeletons, for most are found with no skull. This suggests that their theropod counterparts simply chewed off the nourishing head and swallowed it whole. That alone can explain why sauropod skulls are rare, whereas larger theropod skulls are found far more frequently.

There are other strange signs in dinosaur skeletons that help us see how they evolved and where they lived their lives. Those of the giant sauropods contained many bones that were spongy, and which in life seem to have contained air. This notion of the *pneumaticity* of the bones of dinosaurs was first postulated in 1870, so it was not a new discovery, and more recent research confirmed that these lightened bones occur widely in all the giant sauropods. These light and spongy bones would be essential for a wading dinosaur wishing to stay high in the water.[1]

Another obvious feature of dinosaur skeletons which nobody has explained is a strange group of bones, rather like fingers, that extend across the abdomen. They look like auxiliary ribs, but they are unattached to the rest of the skeleton, and they have been given the name *gastralia*. Nobody has worked out what function they perform, and they are unusual features, found most prominently in dinosaurs, but they are also present in crocodilians and in the curious tuatara lizard found only in New Zealand, a living fossil that dates back 200 million years.[2]

Why would we find these curious reinforcements in the abdominal walls of creatures like crocodiles and dinosaurs? What would be the evolutionary imperative that would cause them to develop? A perfect answer is that the gastralium was an adaptation for creatures living partially submerged in water. Water pressure would cause the abdomen to collapse inwards, making it difficult to inhale. The gastralia alone would militate against that, supporting the abdominal wall against hydrostatic pressure. Even in the skeletons of dinosaurs I was beginning to find new scientific evidence that supported my aquatic theory.

We can now see that the environment in which these creatures flourished was the ideal ecosystem, a realm in balmy equilibrium. The hot Sun beaming on those wet, luscious forests nurtured a hugely diverse plant population including the biggest trees the world ever knew. Many of those plants are with us today, though we rarely stop to consider them. A radio programme was recently burbling in the background, and somebody was asking horticultural experts how they could create a garden of plants from the age of the dinosaurs. The answer showed how we've lost sight of these remarkable plants. The question came from a member of the public, Ruth Herbert, who asked: 'My son would like to create a dinosaur garden. What plants should he use that would have been around at the time?' The panellists cooed with pleasure, and one asked: 'How old is your son?' Mrs Herbert replied: 'He's in his twenties,' which made them chuckle. I didn't chuckle. I was delighted to think of a young gardener showing such commendable interest. The team's answers were unadventurous: they suggested *Gunnera*, a huge waterside plant with spiny stems; another suggested rhubarb. They also proposed he should grow tree ferns like *Dicksonia*, some of the *Equisetum* horsetails and the rice-paper tree *Tetrapanax papyrifer*.[3] I hope the young man's garden is big enough for planting *Gunnera*. This is a huge and spectacular plant, with stems up to 8 feet (2.5 metres) long bearing leaves that can be up to 11 feet (3.3 metres) across. The genus is known from Cretaceous fossils dating back 95 million years, so it was contemporaneous with dinosaurs. Another genus of flowering plants that were around from the same period but which they didn't mention is *Magnolia*, although those striking plants would make a fine feature in any garden. My wise friend Sir Ghillean Prance, formerly Director of the Royal Botanic Gardens at Kew, recommends the laurel family, for they were also prominent at the time. However, these flowering plants were relatively late arrivals; it was the non-flowering species – ferns, conifers, etc. – that typified the era of the dinosaurs. Tree ferns were a prominent feature of the Cretaceous landscape, and *Dicksonia* is widely available in garden centres. These spectacular ferns are cut down in huge amounts in Australia during building work or when new highways are being constructed. The trunks (rhizomes, actually) are topped and tailed, labelled to show that they were officially cut down, and bundled up like logs to be shipped overseas. Once they are planted upright in soil, new roots grow downwards, dormant fronds unfurl, and within two years you have a happy and healthy tree fern. *Dicksonia antarctica* – as the name implies – is the hardiest of tree ferns, they can survive hard frosts down to 19°F (-7°C) though some specimens are hardier than others. They grow in U.S.D.A. Plant Hardiness Zone 6, which approximates to the climate of southern England, but it is wise to give their crowns protection in mid-winter. The recommendation of *Equisetum* should be taken with a pinch of salt. These are the horsetails and many of them are invasive weeds that, once planted, are almost impossible to eradicate. During the Carboniferous period, around 300 million years ago, they evolved into colossal trees reaching up to about 100 feet (roughly 30 metres) and they were abundant at the time of the dinosaurs. The most common *Equisetum* plants we see in the present day are spreading weeds less than knee high, and they are the bane of a gardener's life. The largest still surviving is *Equisetum myriochaetum*, the Mexican giant horsetail, which is native to South America and Mexico. In tropical climates it regularly grows to 15 feet (about 4.5 metres), though the tallest on record reached 25 feet (7.5 metres). For temperate climates, the giant horsetail *Equisetum telmateia* will grow up to 48 inches (120 cm) tall and is comfortable in Zones 6–10, which makes it at home in most of Britain.

Evergreen trees similar to present-day sequoias grew at the time of the dinosaurs, and conifers with ancient relatives could certainly feature in our dinosaur forest. However, there are some other plants you might be able to cultivate that are unchanged since the dinosaurs were alive, and these are true living fossils. Most spectacular would be the monkey-puzzle tree *Araucaria araucana*, which dates from the Triassic period some 250 million years ago. It is an evergreen reaching up to 130 feet (40 metres) tall. This is the national tree of Chile, and it is also found growing wild in Argentina, where it often has snow around its roots – so it is hardy and comfortable in temperate climates. It is also vast, so don't think about growing it in a small garden, or near a house. Even taller is *Wollemia nobilis*, long known for its fossilized impressions found alongside dinosaur remains in Cretaceous rocks. This ancient species was believed to be extinct until September 10, 1994, when a forest of these gigantic trees was discovered alive and well in a valley, just a few hours' drive from Sydney. The trees are clustered in a narrow sandstone gorge in the Wollemi National Park in New South Wales, about 95 miles (150 km) northwest of Sydney Harbour. You could easily make a day-trip there from Sydney, yet until 1994 nobody knew these monstrous trees existed. The specific epithet *nobilis* is customarily used in botanical nomenclature to indicate the most noble species of a genus, but not in this case. This tree is named after the person who found it, David Noble, a field officer of the Wollemi National Park in the Blue Mountains of Australia. Curious to think that until recently nobody knew it had survived, yet you can now pick up modestly priced specimens in garden centres.

Another fairly recent discovery that the dinosaurs used to eat is the graceful conifer rejoicing in the name *Metasequoia glyptostroboides*. This was originally described in 1941 by Shigeru Miki, a Japanese palæobotanist, as a fossil in rocks dating from the Cenozoic era about 60 million years ago. In an unrelated piece of research in 1944, a small, graceful tree was discovered in Hubei, China, by an ecologist and plant taxonomist, Zhan Wang. In July 1944 he was travelling to Shennongjia when he went down with malaria, so he decided to rest up for a while in Wanxian County and made contact with Yang Lung Tsing, principal of the local agricultural school, whom he had known as a student. Yang told him about a curious local tree growing in a temple and asked if it could be identified. So on July 21, 1944, Zhan collected some branches and cones. He knew they looked similar to specimens of *Glyptostrobus pensilis*, the swamp cypress that grew extensively in southern China, but these were different – and Zhan realized he had a new species on his hands. World War II was raging at the time and nothing more could then be done, but in 1948 two botanists, Wan Chun Cheng and Hu Hsen Hsu, realized that this was the same tree as the prehistoric species already discovered as fossils. That same year, an expedition from the Arnold Arboretum at Harvard University collected some young trees and distributed them to universities and arboretum collections around the world for further study. Another living fossil had come to light. Like *Wollemia*, pot-grown saplings of *Metasequoia* are now widespread in garden centres around the globe. The most recent discovery of a living fossil plant was of a stonewort, an alga with giant cells, that was discovered in North America for the first time in July 2017. This is a pond weed, *Lychnothamnus barbatus*. It may not feature in our reconstructed dinosaur-era forest, but it is a plant that is surviving today from the dinosaur era. It reminds us that such discoveries are still being made, and other living fossils are certain to come to light.[4]

One of the most spectacular of all the dinosaur-era trees is the magnificent maidenhair tree or ginkgo that can grow up to 105 feet (32 metres) and live for 2,000 years. This is *Ginkgo biloba*, which was growing 270 million years ago in the Permian period. Its distribution slowly declined, until by the end of the Pliocene epoch it grew only in central China and it survived in the gardens of temples from where seeds were brought to Europe centuries ago. This is an important medicinal tree in Buddhism and Confucianism, and it has been widely planted in the temples of Korea and parts of Japan. The ginkgo is the symbol of the city of Tokyo, and also of the Urasenke version of the Japanese tea ceremony. There are also the cycads. Growing like stunted palm trees, these curious, bristly plants are the source of sago, and some species have the nickname 'bread palm' because of the high content of starch they contain. Cycads live long (up to 1,000 years) and have stiff evergreen leaves. They exist in a state of cooperation with symbiotic cyanobacteria that colonize their roots. However, these bacteria secrete a toxin that can cause neurological disease in humans and animals. This is a large family of plants, with 300 species so far discovered; they grow across the tropics and yet they are relatively hardy. Many can be cultivated as garden plants in temperate zones, and the sago palm *Cycas revoluta* can be raised out of doors in southern Britain.

So there you have it – a range of plants that the dinosaurs knew, and which have survived to this day. *Magnolia* would be a fine way to start a dinosaur garden, and *Gunnera* certainly adds spectacle (if your site is big enough to contain it). You could plant *Equisetum*, but be sure you do not let the smaller species spread out of control. The tree fern *Dicksonia* and the sago palm *Cycas* would both work well in today's temperate garden (if you can afford them some protection in winter), and so would those other gigantic trees, *Araucaria* the monkey-puzzle, and the slender, graceful *Metasequoia*. Then there is the statuesque *Ginkgo*. A few of these truly spectacular trees (the cycads and tree ferns) are still abundant in the wild, but others are miraculous survivors. *Metasequoia* was recognized only in 1944, and was found across the wet valleys of Hubei's Lichuan County and the neighbouring Hunan Province but only in small clusters amounting to a dozen or two trees. There is just one place where they remain in significant numbers: the Xiaohe Valley, where there are perhaps 5,000 trees still surviving. The monkey-puzzle tree hung on in isolated parts of South America, and other species survived in small territories on the other side of the southern hemisphere, like the well-known Moreton Bay Pine *Araucaria cunninghamii* and the popular *A. heterophylla* that lived naturally only on tiny Norfolk Island, both of which are now planted as ornamental species all around the tropics. And then there is the Wollemi pine, that remarkable survivor and a species that was discovered within living memory and had clung on in just a single valley. Were you to grow a selection of these trees you would have a plantation that a dinosaur would recognize.

There is only one place where a truly prehistoric environment still persists, and that is the Karori Wildlife Sanctuary of Wellington, New Zealand, where the tuatara *Sphenodon punctatus* can still be found. Many of the plants that are surviving there date back to the era of the dinosaurs, and the tuatara is testimony to how things once were. It is a cousin of the dinosaurs, and their closest living relative today.

Just as the modern-day imaginings of dinosaurs place them in the wrong setting, they also show them under heavens that might be the wrong colour. Light is scattered through fine particulates in the air (which is why sunset and sunrise are often under a red sky) and is also changed by the chemical composition of the atmosphere. Raised levels of carbon dioxide (CO_2) can cause the blue to shift towards yellow. The sky of the early Earth was not necessarily blue but could have been yellow or orange. We all know that it is the green chlorophyll in vascular plants that captures sunlight and converts it to metabolic energy, as the Sun shines out from a clear blue sky. The standard explanation for this is that, as it reaches the Earth's atmosphere, white light from the Sun is scattered by the particles in the air and by the gases of which it is composed. Because blue light is the shorter wavelength than the red end of the spectrum, it is more easily scattered and thus the sky seems blue. I believe that there may be another reason. The blue may be due to the excitation of oxygen molecules by the impact of photons from the Sun. Red, yellow and green peaks in the absorption spectrum of sunlight would give rise to a blue colour. Liquid oxygen is blue in colour, and the sky is blue when seen from space, as well as from the ground, so the colour may primarily be due to the oxygen in the air.

In the time of the early Earth, back in the Precambrian era, there were much higher concentrations of carbon dioxide and methane in the atmosphere. Sulphur dioxide from volcanic activity was also more abundant. Scattering from these molecules would have coloured the sky orange. One of the most succinct summaries of Earth's history was written by Cesare Emiliani, a brilliant Italian scientist who in the 1950s founded the science of palæoceanography. Born in Bologna, where he obtained his doctorate in geology, in 1948 he moved to the University of Chicago and earned another PhD, this time in palæoclimatology. His calculations of prehistoric worlds make fascinating reading; and he wrote that, during the unravelling of the early Earth, 'CO_2 and SO_2 accumulated in the Earth's atmosphere. As more CO_2 accumulated, the temperature rose. The sky turned red and the Sun turned blueish.'[5] A red sky would make much sense for green plants. Red light is readily absorbed by green leaves and thus would encourage the evolution of today's photosynthetic pigments. Perhaps today's plants are green because the sky under which they first evolved was red.

Carbon dioxide has wavered between high and low levels throughout the Earth's history, and there were peaks of concentration in the late Jurassic period, with another in the late Cretaceous. Levels of atmospheric methane are also believed to have been higher during the age of the dinosaurs. So their sky could not have been the same as ours. When the Sun is near the horizon, it has to traverse such a greater depth of air that all the blue wavelengths are scattered and only the longer, red, portions of the visible spectrum remain (that's why sunrise and sunset are usually red). Thus, when you have created your prehistoric garden, it should be photographed only near sunset. Then, when you look around, you will see plants that the dinosaurs saw – and when you look upwards you may be seeing their orange sky.

Are today's birds dinosaurs? That would be like saying lizards are frogs, or that humans are marmosets. Most dinosaurs were scaly though some developed wispy hairs which developed into proto-feathers and eventually evolved into the feathers that birds now use for flight. I believe they first developed as insulation for reptiles faced with a cooling world. Although some tyrannosaurs developed proto-feathers, there has been a view (enthusiastically espoused by bloggers) that *T. rex* was a feathery giant.[6] There was no evidence for this – and, when the skin of *T. rex* was examined, the claim turned out to be false.[7] That is another discussion which we can now close down.

Popular accounts feature *Tyrannosaurus rex* as a feathery monster. In fact, no *T. rex* ever ran, no *T. rex* ever charged its prey, and no *T. rex* ever had a covering of feathers. This image is a myth.

The possibility of reviving dinosaurs from their fossilized DNA – the idea popularized by Michael Crichton in *Jurassic Park* – remains a topic of discussion. It may appear simple, but seems to me to be highly unlikely. Even so, recent discoveries continue to raise people's hopes. In December 2017 a Spanish team announced their investigations of a beautifully preserved tick in amber from 90 million years ago, its body apparently engorged with blood, and close to fragments of what seem to be dinosaur feathers.[8] It may well be that this truly was a parasite that fed on dinosaur blood. So we are coming closer – yet the idea that DNA could survive intact, even preserved in amber, seems to me inconceivable.

In 2015 what appears to be real dinosaur blood was identified inside a fossil bone. Susannah Maidment of Imperial College London said: 'We stumbled on these things completely by chance.' Her team were studying the processes of fossilization by examining small samples of dinosaur bone. They were surprised to find small round bodies that they think look like blood cells dating back 75 million years, more than 10 million years before *T. rex* appeared. None of these fossils contains DNA, though those extracted from 'better preserved fossils' using the same technique may later do so, claims Maidment.[9]

For centuries there continued to be claims that monstrous reptiles still survived in secret places; an engraving of a 'great sea serpent' was published by Hans Egede, a Danish explorer, in 1734, and on October 28, 1848, the *Illustrated London News* published an engraving of a gigantic monster, similarly identified as a 'great sea serpent', which all served to perpetuate the notion that colossal creatures were still to be found. Some still believe that a Scottish monster survives to this day in Loch Ness. The first book devoted to the possible existence of the Loch Ness monster was written by Rupert Gould, an officer in the Royal Navy who became an expert on horology – indeed, he was responsible for the restoration of the famous marine chronometers of John Harrison, the eighteenth-century craftsman whose intricate design of watches solved the longitude problem for long-distance navigation.[10] After Gould's book appeared, he remained the principal authority on the alleged monster, which soon acquired the nickname 'Nessie'. Some distinguished scientists have been greatly in favour of believing in Nessie; Henry Hermann Bauer is Emeritus Professor of Chemistry and Science Studies at Virginia Polytechnic Institute and State University, and has written extensively on Loch Ness. He feels the evidence is strong. So does Robert Rines, an American lawyer, inventor and musician, who saw a 'hump in the water' in 1972 and devoted more than 30 years to searching for more evidence. He published a photograph that he claimed showed a plesiosaur's fin in the water in *Nature* magazine in 1975, and built himself a sumptuous house overlooking the loch. Another enthusiast was Tim Dinsdale, an aeronautical engineer with the Royal Air Force, who filmed a hump in the water of Loch Ness in 1960 and spent most of his life searching for more. Roy Mackal, Professor of Zoology at the University of Chicago, spent decades searching for Nessie and only later turned his attention to the search for Mokele-mbembe, a monster that was said to inhabit the Likouala Swamp in the Republic of Congo. Richard Greenwell, an

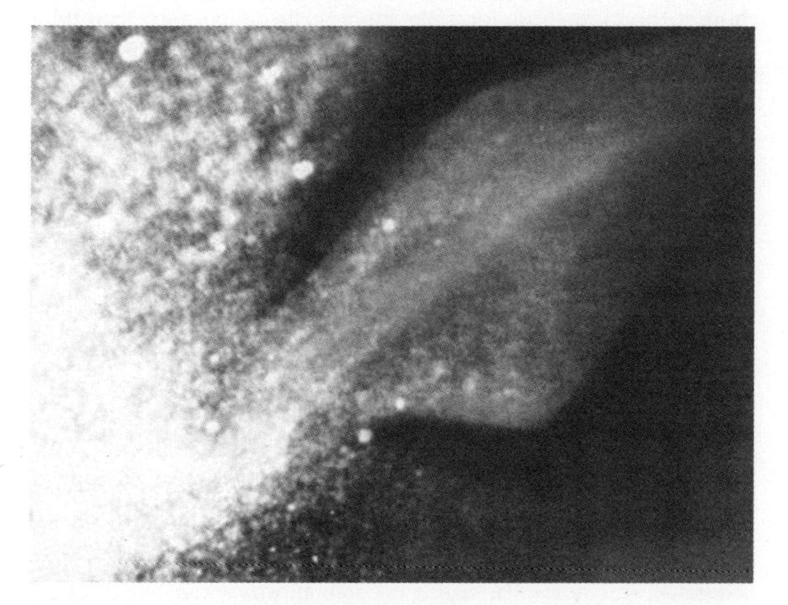

The leading journal *Nature* published this astonishing image on December 11, 1975. It seemingly shows a plesiosaur's fin, photographed in Loch Ness. The picture was later shown to have been extensively retouched by Dr Ronald Rines, the American lawyer who was determined to prove Nessie's existence.

ecologist from the University of Arizona, and Congolese biologist Marcellin Agnagna, accompanied him on several expeditions, which found no sign of the creature, but did record many first-hand reports from local villagers who consistently described a creature looking just like a dinosaur.

Adrian Shine is the greatest authority on this controversial topic, and I discussed this with him sitting in the balmy sunshine at the intriguing exhibition centre near the shore of Loch Ness. He points out that 1,000 people claim to have seen something, and he adds that it would be exciting if they were right; but he does not personally believe in the rumours. Regretfully, nor do I, though any plausible new evidence would obviously be of enormous interest. A collection of the findings has been assembled on site at the Loch Ness Centre and Exhibition, and their audio-visual presentation provides an excellent summary of our knowledge. On the opposite side of the parking lot by this building is a chin-high wooden fence. Most visitors never think to peek over the top. They should. Hidden on the other side, emerging from the rushes at the side of a swampy pool, is a concrete effigy of Nessie. It makes for a fine photograph.[11]

If Nessie is a myth, where else should we look for dinosaurs today? The tuatara is close; it is not a lizard, though neither is it a dinosaur. Another distant relative is the Komodo dragon (*Varanus komodoensis*), which survives on an isolated clutch of Indonesian islands: Rinca, Flores, Gili Motang, Padar and Komodo, where I have observed them in the wild. This is the largest of all the lizards, bigger than a person, reaching a mature body length of 10 feet (3 metres) and a weight of 150 pounds (70 kg).[12]

It has been suggested in the past that these giant lizards became so large only because of the lack of competition – there are no other carnivores that occupy the niche on those islands. This phenomenon has been dubbed island gigantism. But there is a better explanation. There were once many species of large vertebrates inhabiting Indonesia and Australia, which became extinct some 10,000 years ago, after the Pleistocene. Fossils of lizards very like the Komodo dragon have been discovered, the earliest dating back some 4 million years. These are not dinosaurs. *Varanus* is a typical lizard which sprawls on limbs that grow out from either side of the body, whereas dinosaurs' limbs were vertical. The two groups did have a common ancestor, however, and both lizards and dinosaurs belong in the diapsid group of reptiles.

Komodo dragons prefer decayed flesh to eat, and often consume carrion that has softened in the sunshine. Most of the time they sit still, basking in the heat. In that sense, their behaviour is just like that of crocodiles, and is similar to what I propose for dinosaurs. When I was observing these huge lizards on the island of Komodo, it was easy to see how hard it is for them to move about when fully grown. They have to haul themselves along; sometimes they can give a burst of speed, perhaps to seize a deer, but their primary method of attack is to bite an animal and inject it with venom. It is often said that Komodo dragons wait until their prey dies from a blood-borne infection caused by the bacteria between their teeth, but this is incorrect. Scientists studied glands taken from a terminally ill Komodo dragon at the Singapore Zoological Gardens and found that they produced a range of poisonous compounds that inhibited blood coagulation, caused a drop in blood pressure, muscle paralysis, loss of consciousness and death through shock. They argue that this would also have been true in the case of the extinct *Varanus priscus*, a 2-ton monstrous lizard identified by Owen in 1859.[13] If so, could it have been true of theropod dinosaurs?

There is just one primitive mammal that has a similarly venomous bite. The selenodon of the Dominican Republic is related to shrew-like mammals that existed before the demise of the dinosaurs. This is now extremely rare but is believed still to be capable of survival with the help of a conservation project. Until humans arrived, selenodons were the apex predators of Hispaniola. Their venomous bite reminds us that it is just possible theropod dinosaurs also had a bite that poisoned their prey.

When Michael Crichton published his epic novel *Jurassic Park* in 1990 he was criticized for giving some of his dinosaurs the fictitious ability to spit venom – but he may have been partly correct. Since the Komodo dragon has been shown to secrete venom in its bite, and so does the selenodon, then I think it possible that theropod dinosaurs may have done so too. What of today's crocodilians?* This group comprises the true crocodiles (the family Crocodylidae), the alligators and caimans (Alligatoridae) and the gharials (Gavialidae). The gharials are found in Pakistan and Myanmar, and sometimes grow to a length of 20 feet (6 metres). Alligators are found only in China and the southern U.S., while true crocodiles exist in the tropics right around the world. I have observed crocodiles in West Africa and Australia, and caimans in the Amazon – indeed, I once captured a juvenile caiman in Amazonia that seemed almost as apprehensive as I felt. It is to the crocodiles that we should look for a close comparison to dinosaurs. Just as dinosaurs once did, they live in shallow water, though they swim at the surface rather than wading on the bottom. They also spend almost all their lives gently cruising or basking in the sunshine. Instead of energetically hunting, crocodiles prefer to find decaying corpses and tear off pieces of flesh to eat. When they do attack their prey, it is by ambush rather than pursuit. They rarely feed in the cold season, when their blood has cooled, and once they have fed they can remain mostly inactive for months. This has many similarities with the way that I now believe theropod dinosaurs lived their lives. It is their sheer size that militates against too much frenetic activity.

If you want to understand dinosaurs, study those gigantic crocodiles. Even though the prehistoric genera were far larger than those we see today, crocodilians were always far smaller than the giant dinosaurs; *Deinosuchus*, a prehistoric alligator, grew to a maximum of 40 feet (about 12 metres) in length and weighed about 5 tons, while the prehistoric crocodile *Sarcosuchus* reached 42 feet (about 13 metres), weighed some 8 tons. The biggest individual crocodile alive today is believed to be about 20 feet (6 metres) in length. He has been

* Crocodiles and alligators can easily be distinguished. At first sight, alligators are black, while crocodiles are grey. Alligators have a blunt U-shaped snout, whereas those of crocodiles are more tapering and V-shaped. The upper and lower jaws of crocodiles are roughly the same width, but in alligators the upper jaw is wider. The upper teeth of an alligator are all that show; in crocodiles, both upper and lower teeth are visible. In this discussion the term 'crocodile' is used generically to cover both types.

named Dominator, he lives in the same river as Brutus (p. 422) and he must weigh over a ton. These saltwater crocodiles, *Crocodylus porosus*, are impressive beasts, but of course the biggest dinosaurs could be 100 times heavier and much longer. When we look at the behaviour of crocodiles, we must remember how much more difficult it would be to move as a dinosaur. Crocodiles are only distant cousins of dinosaurs, both belonging in the Archosauria group, but crocodiles can reveal to us what dinosaurs looked like. The scales of a crocodile are similar to those of dinosaurs, and I have tried (without success) closely to photograph swimming crocodiles in water.

A vivid impression of how a dinosaur would have appeared was caught on camera by crocodile tour guide Karen Letch in Adelaide. The tail of Brutus, an 18-foot (5.5-metre) saltwater crocodile, even shows parallel plates like a stegosaur.

My first attempts were in Gambia in 1979; the most recent were in Costa Rica in 2012; but, although there are plenty of instances where crocodiles sit sunning themselves contentedly on mudflats, there are fewer chances to catch them just at the surface. One excellent picture was captured by Phillip Lanoue, a specialist wildlife photographer; another came to me from Karen Letch in Adelaide, who guides visitors around the crocodile habitat with Adelaide River Cruises. Their photos illustrate a key point of comparison: not only is the ridged skin exactly like that of a dinosaur, but note the appearance towards the tail, where the projecting scales of the causal region form two V-shaped rows. This is precisely what we see in a stegosaur. Next, look at the feet; these are also dinosaur-like. Although the objectors to my theory insisted that an aquatic creature must have developed webbed extremities, this betrays unfamiliarity with zoological principles – as we have seen, none of these aquatic reptiles has webbed feet. Like dinosaurs, crocodiles and their relatives have claws. Look next at the head of a crocodile, and there is something extremely close to *Tyrannosaurus rex*. The eyes, those teeth, the moist scaly skin, the nostrils; the head of a crocodile gives you an immediate impression of what it would be like to meet *T. rex* face to face. It was like a fat, idle, gigantic crocodile that, instead of swimming in deep water, floated buoyantly through shallows, propelling itself along with its muscular hind limbs. *Tyrannosaurus rex* was no boisterous hunter. It scavenged.

My investigations had been so greatly encouraged by Bill Hay and the team he had recommended to join with me on a major paper to launch the whole theory. There is always so much rivalry in science, and such a sense of competition; but here were well-known academic scientists joining with me to establish a new approach to the study of dinosaurs and the environment in which they lived. We circulated several drafts around the group as the new information was assembled, and Hay invited me to work on my new theory with him in the Rocky Mountains of Colorado where he lives in a house perched above a steep-sided valley. Hay is an authority on the climate of the past and has long been expert at geology. He went through my manuscript meticulously and put me right on several important aspects.

Above all, he was excited at the way I believed the Cretaceous period could now be reconsidered. He was confident that this would revolutionize science.

Our other contributions were beginning to come in. We rang Bob DeConto from the house and he emailed over some of his latest revisions to the climate. With my new model for the Cretaceous, he had now calculated new maximum and minimum temperatures that were very different from those everyone had accepted for 50 years or more, and they would have provided an ideal environment for dinosaurs. Ying Song had meanwhile pointed out that the conventional view of the tectonics of Eastern Asia must be incorrect and he had provided his revisions. Andrei Stepashko had sent in amendments of his own, while Poppe de Boer (who invited me to visit him in the Netherlands) had contributions that he had often promised to send, but which had yet to arrive. Sascha Flögel had drawn up a range of original climate maps but was yet to provide anything new for us, and Hay began to wonder whether he should be left off the list of co-authors. However, Flögel had already made a number of important alterations that improved the paper, and I insisted that he should stay. At the core of the revision was my insistence on a wet and warm environment. Hay pointed out that all the research to date had been focused on dry continents, and now we had a major revolution under way. The reason for the conventional view was that an increase in temperature was always taken to be linked to a rise in evaporation rates, and precipitation of rainfall was always assumed to be reduced as one moved away from bodies of water. Hay explained: 'If you've got a continent, you can get water onto it. That's understood – evaporation keeps continuing, whereas the precipitation is intermittent, and so the continent gradually dries out. Everybody has always believed that. We have never considered the possibility that a continent might not actually be dry!' We had looked at DeConto's latest diagrams, which he had just sent by email, and agreed that these figures provided the final confirmation of my theory. Said Hay: 'It's *more* than that, because it shows that the dinosaurs lived in an environment very different from what we live in, where night and day temperatures range up and down. Their climate would have been balmy, just wonderful. The evening cools off just a little, the habitat warms up in the morning but is never too hot. We just have a wonderful even temperature.'

He added that my theory put an end to so many of the arguments about dinosaurs, including whether they were warm-blooded. The buffering effect of a warm-water habitat provided metabolic consistency. He laughed: 'You don't *need* any of that! If the temperature variation is only five degrees Celsius between night and day, and through the year, who cares whether you're hot-blooded or not.' This was now such an exciting project. I told Hay that, with the new evidence the group was providing, I thought we were bound to win the argument. He nodded enthusiastically: 'We are going to win. We are – there's no problem about that.' We drank a toast of Châteauneuf du Pape, and later went out to dine. Hay grinned. 'Next we have to start working on the next paper. It all makes so much sense,' he said.[14]

We must now turn to the final question – what happened to the dinosaurs? Exactly why did they become extinct? There have been many theories proposed, and I do not believe that the final answer has been generally agreed. Those enterprising palæontologists at Bristol University have assembled a page that claims to list 101 of them (it is actually less than half as many as they claim, though by now you will not be surprised at a certain tendency of palæontologists to exaggerate).[15]

The tiny brain of the dinosaur has been cited by many as evidence of stupidity, and some say that this must have led to their extinction. Palæontologists have pointed out that huge dinosaurs had very small brains in comparison to their total body size, possibly as small as a kitten's brain. As we have seen (p. 285), this works only if you consider kittens to be far more stupid than a dog. The size of the brain cannot be correlated with mental ability, or, as we have said, an elephant (brain weight 11 pounds or 5 kg) would be 1,000 times more intelligent than a shrew (brain weight 1 gram, 1/30th ounce). The size of the brain is more closely related to the weight of the body and the number of cells it comprises. It is the brain that does the thinking; the dinosaurs may well have had a 'second brain' that regulated the massive body. Although we think of the brain as the centre of all bodily control, large groups of nerves elsewhere in the body can also help control the body. Hidden in our bodies is a structure known as the solar (or cœliac) plexus which regulates many of the functions of the body. So well developed are these systems in birds that they can survive without a head. A headless chicken once toured the United States, living contentedly on liquid food from an eye-dropper. It was in Fruita, Colorado, on September 10, 1945, that a farmer named Lloyd Olsen was sent to bring a chicken for the pot. He beheaded one of his chickens with a glancing blow, chopping off its head. After being decapitated, much to Olsen's surprise, the bird shook itself and strutted away. Finding the headless chicken still alive next morning, Olsen began to feed it with a dropper, and soon was exhibiting the chicken in touring sideshows under the name of Mike. The press became interested, and Mike was featured in both *Life* and *Time* magazines. Press reports claimed that Mike was earning the Olsens the present-day equivalent of £40,000 (about $50,000) every month at the height of his fame – until Mike choked to death in Phoenix in March 1947. Now you know why people speak of someone running around 'like a headless chicken'.

Early on, when describing his newly discovered stegosaur, Othniel Charles Marsh had recorded a space in the hip region of the spinal cord, which he thought could have accommodated a structure up to 20 times bigger than the dinosaur's small brain, and from this came the notion that the dinosaurs might have possessed a 'second brain' near the tail, which could have been controlling responses in the rear of the body. In recent decades, the idea has been ridiculed – but it may have been correct. Fossilized remains do not provide firm scientific evidence, though the finding by Marsh of a modification of the skeleton is a hint in that direction, while the casual way in which sceptics have airily dismissed his proposal has no evidence at all in support. Even if the solar plexus in a dinosaur was not situated within the spine, something like it could still have existed. Perhaps the small brain nursed the intelligence, while a hidden nervous command centre ran the rest of the body.

Another reason for the extinction of dinosaurs that has been proposed is their sheer size, since slipped vertebral discs could have become common due to their inability to support themselves on land. This would lead to loss of mobility and the inability to compete for food. Another view was that changes in the atmosphere caused the demise of the dinosaurs. There were, as we have seen, raised levels of carbon dioxide during the dinosaur era, and it was postulated that this gas destroyed dinosaur embryos, or that the oceans became eutrophic and stagnant. The raised levels of CO_2 might have meant a lowered availability of oxygen in the oceans, it was claimed, so supplies of marine food would disappear. This theory was discounted, primarily because palæontologists insisted that dinosaurs never ventured near the sea and so it would have been irrelevant to their disappearance. Then there was the possibility of atmospheric change caused by volcanoes, resulting in raised levels of CO_2 and impenetrable clouds of ash that would have darkened the Earth and catastrophically compromised plant life. Or perhaps toxic elements like selenium and uranium would have been released into the environment, and the shells of dinosaur eggs would have become dangerously thin (much as happened to avian predators when D.D.T. was in widespread use).

Changes in climate have been advocated as the prime cause. The increase in atmospheric CO_2 would have caused the temperature to rise, adversely affecting life in the oceans and causing devastating changes in weather patterns and rainfall. This could also unbalance the ratio of males to females in the young dinosaurs, because the gender of the embryos is affected by the temperature at which they incubate. There could also have been an effect on sperm production in higher temperatures, resulting in lower reproduction rates, while adult dinosaurs could have been overcome by heat (especially if they were warm-blooded). On the other hand, the clouds of ash would probably have been the result of extensive volcanic activity, in which case incalculable amounts of dust may have blanketed the atmosphere and led to a drop in temperature, so the dinosaurs' body temperature would have fallen to a critically low level, their eggs would not have hatched, and they would have died during chilly winters. Others have suggested that metabolic disorders could have caused the shells of eggs to be too thin and fragile, so that the embryos would die through dehydration. Strange egg fossils discovered in the Pyrenees suggested that eggs with several shells may have been laid, causing the embryos within to suffocate and die. It was then argued that a supernova explosion would heat the upper layers of the atmosphere, disrupting the ozone layer and producing high-level clouds that would cool the environment. Similarly, the radiation from space could have produced blindness through cataracts forming in the dinosaurs' eyes, so they would have been unable to see where they were going, or what they were eating. This could also have resulted in changes to the DNA of the nuclei within the cells of dinosaurs, so they produced mutant offspring that were unable to survive. Recently it has also been postulated that dinosaurs died out because they were poisoned by blooms of toxic algae. Perhaps the climate became too wet, resulting in the environment becoming uninhabitable; or it was too dry, the arid surroundings leading to death from drought. It has been argued that dinosaurs died out as the mammals took over; however, mammals became prominent as fossils only in parts of Asia and North America, so that could not have applied worldwide. The emergence of pathogenic bacteria could have caused widespread epidemics that killed off all the dinosaurs through outbreaks of infectious disease. This was first put forward a century ago, and more recently Homer Simpson was shown sneezing on a dinosaur and causing them all to expire from the virus outbreak he triggered. Alternatively, outbreaks of parasitic infestations with worms, flukes, lice or flies could have overwhelmed the dinosaur population. The Bristol University group argue against this, saying that 'an epidemic sufficiently large to wipe out the dinosaurs would also affect other animal groups, but there is no evidence of such an effect.' That's unsound; bacteria are often host specific, and it is certainly possible that an epidemic of disease in one type of creature would not have affected other types of animals. The real reason against this is a lack of evidence. We do have those diseased jaws (p. 341), but nothing to suggest that the dinosaurs were wiped out through a pandemic. Perhaps their pituitary glands became overactive, leading to excessive growth of bones and cartilage which would hinder the movement and metabolic efficiency of dinosaurs; or the pituitary became underactive, which led to the growth of the frills, horns and spines seen in many later types and which impaired their ability to move about and feed.

Our Prehistoric Monsters Wiped Out by the Disease Germs of To-day!

Science Finds in Distorted Skeletons and Fossil Bacilli 30,000,000 Years Old Evidence That Microbes Destroyed the Huge Beasts and Birds That Roamed the World Before Man—and Thus Gave Man Room to Grow

By W. H. Ballou, D. Sc.

A Death Bed 20,000,000 Years Ago

Here the Noted Artist, H. Harder, Has Shown a Brontosaurus, a Gigantic Extinct Sea Monster, Weakened by Disease and Driven to Shore. Here It Becomes the Prey of Carnivorous Dinosaurs, Who, Becoming Infected With the Germs, Spread Them in Turn Among Their Own Kind and Those Who Devoured Them.

Infection as a possible cause of the dinosaurs' demise is not a new idea. This report appeared in the *South Bend News-times*, Indiana, on March 9, 1919, and described how palæontologists claimed to have discovered fossilized bacteria.

It has been claimed that many new types of fungi, including highly toxic types, evolved during the time of the dinosaurs. This is questionable; fungi must have evolved with earlier forms of life, though their diaphanous nature does not necessarily leave fossils behind. More significant was the evolution of the angiosperms, the flowering plants (like magnolias), which were appearing for the first time. They first began to emerge between 245 and 200 million years ago and we have fossils of flowering plants from 160 million years ago. By 120 million years ago they were replacing conifers and ferns as the dominant forest plants. The Cretaceous was not only the era of dinosaurs but also of new flowering plants. Their rapid metabolism would have raised the levels of oxygen, so that rates of dinosaur metabolism became too high and, in the words of one investigator: 'The dinosaurs may have well burnt themselves up, or out!' Another theory was that of overpopulation, causing severe competition and rivalry among the dinosaurs. Or there could have been excessive predation by meat-eating dinosaurs, leaving them no herbivorous

More recent is Homer Simpson causing the demise of the dinosaurs by sneezing into the face of a *T. rex* in 'Treehouse of Horror V', the 6th episode of the 6th series of The Simpsons. His 'flu virus spread like wildfire, and all the dinosaurs died. It was released on the Fox Network on Sunday, October 30, 1994.

species on which to feed. Some have suggested that atmospheric pressure could have altered as elevations of the land changed over time, making it impossible for the giant reptiles to survive. An additional complication resulting from the development of plant life could have caused a loss of open land on which dinosaurs could freely move, causing their communities to be under pressure. Perhaps the new angiosperms poisoned the dinosaurs, because flowering species produce alkaloids (we all know how poisonous these plants can be) and the dinosaurs died out through eating the wrong plants. As angiosperms became seasonal, many dropping their leaves in winter, sauropods that were used to feeding on them would die out through a lack of nourishment. The theropods would then lose out, having nothing on which to feed. Again, the evolution of butterflies might have caused their caterpillars to eat all the vegetation on which the sauropods subsisted, so they could have caused the starvation of them all. On the other hand, changes in the dinosaur diet may have led to a burden of bony changes like arthritis, bones fractures and tooth decay.

There have also been many celestial theories. Sunspots could have affected the climate and the atmosphere to the detriment of the dinosaurs, and changes in the orbits of the planets surrounding our Sun relative to the galactic plane, have also been cited. If the orbital plane varies, then perhaps meteorites and dust particles could be swept up by the solar system, causing more impact events than normal. It has been calculated that extinction events occur every 26 million years or so, and astronomers have found that the galactic plane changes on a similar time scale: perhaps they were causally related.

Then there is the periodical reversal of the magnetic poles, which is known to have happened in the past, on average every 500,000 years. This would cause a drop in the level of the magnetic shield, resulting in genetic damage and perhaps leading to eventual extinction. That's a neat explanation, but regrettably the occurrence of reversals does not coincide with the great extinction events that have been documented. The theories about newly evolved species of flowering plants causing damage to dinosaurs also work well at first sight, though the main diversification of the plant realm took place about 40 million years before the end of the Cretaceous. The slow but steady emergence of these plants was accompanied by a diversification in the herbivores, so they would have had time to evolve and to adapt to changes in the available food supply.

So now we come to the more scientifically respectable theories. The strike by an asteroid is the best known and is widely accepted. The notion that the Earth was hit by a meteorite first gained fresh momentum when surveyors from the Petróleos Mexicanos company discovered that huge Chicxulub crater in the edge of the Yucatán Peninsula in 1978.[16] This was supported by the discovery of iridium in the Earth's surface that was deposited some 66 million years ago. Surely this would fit the demise of the dinosaurs. We have seen how the theory of Alvarez, father and son, seemed to tie all this together. An asteroid or meteorite had hit the Earth with force a billion times greater than the combined power of the atom bombs dropped on Hiroshima and Nagasaki, heating the atmosphere, generating towering tsunamis and shock waves, and causing massive clouds of dust that darkened the Earth.[17] Walter Alvarez set all the findings down in a detailed book which (in this era of wanton exaggeration by the scientists and an insatiable thirst by the public for sensation) was given a title fit more for a sci-fi comic or a Hollywood thriller than a serious scientific work: *T. rex and the Crater of Doom*.[18] The result was that this theory has become accepted around the world. The swapping of the tongue-twisting term 'Chicxulub crater' with the far more eye-catching 'Crater of Doom' has even penetrated into the world of academic publication.[19]

People needed further evidence for all this. Between 2001 and 2002 the Chicxulub crater core was drilled by the D.O.S.E.C.C. project (standing for Drilling, Observations and Sampling of the Earth's Continental Crust), which was intended to resolve the issue, but it proved inconclusive. A team of scientists from the University of Texas, the National University of Mexico and the International Ocean Discovery Program launched a project to investigate the crater further. They planned to drill down 4,900 feet (1,500 metres) below the seabed, and eventually bored down 4,380 feet (1,335 metres), which was considered sufficient. The samples are currently being analyzed in Bremen, Germany. There was no sign of a solid body. The team found granite and thick beds of limestone covering a layer of graded sediment at least 330 feet (100 metres) in thickness.[20]

There can be no doubt that an asteroid (or meteorite) had struck the Earth, explosively evaporating as it vaporized into hot gas and causing catastrophic disruption. Remember that, as the conspicuously cratered surfaces of the Moon and planets like Mercury remind us, asteroids have bombarded the planets and satellites of the solar system ever since its birth. As I said earlier, the Earth was similarly struck, though most of the meteorites burned up in our atmosphere and almost all the innumerable craters that were once formed have long ago weathered away. Currently, NASA scientists are hunting down potentially destructive asteroids, and more than 12,000 near-Earth objects have been identified, of which some 10 per cent could cross the Earth's path in the future.

Yet are these findings certain? Can we be positive that the layers of sediment we see were due to a tsunami caused by that asteroid? People say so, but some researchers are not so sure. There are other possible triggers that are being ignored. At Princeton University, the subject was addressed by Gerta Keller and her team. Keller has led an interesting life. Born into poverty on a Swiss farm, she took low-paid manual jobs and travelled the world before she eventually qualified in geology and palæontology at Stanford University at the age of 33. An elegant speaker with fair hair and a refined manner, she gives elegantly constructed lectures with a trace of her Swiss accent. Her research into the K-T extinction event has been methodical and exhaustive, and her conclusions provide the perfect backdrop for my investigations. Keller decided to look more widely at the geological evidence and became increasingly certain that the Chicxulub explosion was not the reason dinosaurs had become extinct. Previous scientists had correlated the sedimentary deposits with sea-level changes and showed that their data fitted the known timescale. But the new figures that Keller was obtaining clearly showed that the dates did not match. The cataclysmic event occurred at least 300,000 years earlier than the mass extinction. The two events had occurred at very different times.[21]

Keller herself was still being pressured into supporting the generally accepted view. What science wanted was new evidence to support the asteroid idea, not theories that proved it to be wrong. Keller now says:

> In a perverse twist of science, new results came to be judged by how well they supported the impact hypothesis, rather than how well they tested it. An unhealthy 'us' versus 'them' culture developed where those who dared to question the impact hypothesis, regardless of the solidity of the empirical data, were derided, dismissed as poor scientists, blocked from publication and getting grant funding, or simply ignored. Under this assault, more and more scientists dropped out, leaving a nearly unopposed ruling majority claiming victory for the impact hypothesis. In this adverse high-stress environment just a small group of scientists doggedly pursued evidence to test the impact hypothesis.

It will not surprise readers by now that unconventional ideas are derided by the scientific establishment. For all its drama (and the immense appeal of the Crater of Doom) there was no firm scientific evidence that linked the K-T boundary with the timing of the impact crater. It was definitely not coincidental. Keller described the accepted conclusions as 'flimsy and unsubstantiated' even though the scientific world continues to accept the conventional view.

This is a contentious topic. The results of the earlier drilling had been unveiled at the EGU-AGU conference in Nice, France, in 2003 amid confusion caused by the security checks carried out by French authorities. There was apparently only one screening device for 10,000 people attending the meeting so very few people had managed to gain access to the hall when the opening session began. Queues of scientists reaching back 'several kilometres' meant that delegates could not attend their sessions in time, Keller now claims. The results of the drilling were presented as evidence that the Yucatán event had certainly caused the extinction event. It was followed by presentations of the opposite interpretation. Reported Keller: 'Pandemonium broke out with some scientists shouting obscenities – it was not a gentlemanly audience.'[22]

The acrimonious atmosphere of the conference was faithfully reported in the journal *Nature*, and it was clear that there was much disagreement. According to the journal's report, those attending were said to be 'shocked and stunned' by the sense of hostility.[23] They should not have been surprised; this is only to be expected when any new theory challenges the espoused orthodoxies of the time. Plenty of people were similarly shocked and stunned by the hostile reception that my dinosaur publications have been receiving.

After everything had been considered, Keller's team became convinced that there was no evidence of a mass extinction coinciding with the meteorite impact. They concluded that the effects on the environment of the Chicxulub impact were short-term and the catastrophic effects of the Chicxulub impact had been greatly overrated. What did they think might have caused the deposits of iridium and the evidence of disaster? They had one suggestion: Deccan volcanism – huge volcanic eruptions that took place in India. The Deccan Traps are a vast province of igneous rock on the Deccan Plateau of India. This is an inconceivably vast volcanic deposit – layer upon layer of solidified basalt over a mile deep (more than 2,000 metres) and covering almost a quarter of a million square miles (more than 500,000 km^2). Lava poured out of a great split in the sliding surface of the Earth for tens of thousands of years at the end of the Cretaceous, and the layers have weathered significantly since that time. It is calculated that the Deccan Traps originally covered more than half a million square miles (1,500,000 km^2). In some places in India, more than 2,000 years ago, local people excavated cave homes for themselves by digging into the lava. There are spectacular examples from about AD 600 in the Elephanta Caves not far from Mumbai Harbour.*

Deccan volcanism was itself a consequence of something bigger by far – the breakup of the continents and the destruction of the shallow lakes and seas on which the dinosaurs depended to survive. As the Cretaceous period came to an end, the dinosaurs' world began to vanish. The proto-continent of Pangea started to break apart. The crust of the Earth was moving, convection currents were causing continents to slide, and that is why those violent volcanoes were springing from the magma thus exposed. At the same time, the continents were raised up and the sea level fell. In a relatively short space of time, those huge shallow lakes and rivers vanished. This is why the dinosaurs died out – not because of the asteroid, not even because of the volcanoes. I now believe that they disappeared for the same reason that so many species are threatened today: habitat destruction. The dinosaurs needed extensive shallow water in which to live their lives. The cause cannot have been that fabled asteroid or meteorite. Larry Heaman at Alberta has used a novel dating method to show that dinosaurs survived at least 700,000 years after the asteroid impact: it did not cause their extinction. Had the atmosphere been so polluted, as the theorists claim, then all such life would have been doomed to die out. It would have spelled the end of all reptiles, or perhaps, the end

* The Deccan Traps did not trap anything, as is often assumed. 'Traps' is a geological term that derives from the Swedish word for stairs, *trappa*, and is an allusion to the stepped profile of successive layers of lava.

of all the large animals and plants, but this is certainly not what occurred. The reptiles that disappeared were the dinosaurs and their allies, and I am certain that they did so because of the loss of their habitat. Indeed, not all dinosaurs disappeared; one of the first dinosaurs to be discovered, *Lystrosaurus*, lived on. It lived at the edge of lakes and would not have been critically affected by the loss of shallow water. This genus appears in a BBC documentary, *The Day the Earth nearly Died*, which contains numerous errors. One of the most serious assertions is that *Lystrosaurus* became the ancestor to all mammals. This is not the case; no scientist believes that this family, the Dicynodontia, were the ancestors of mammals.[24]

The dinosaurs' close relatives, the crocodiles, alligators, caimans and turtles, lived on. Any global catastrophe would have killed them too – but they were not affected by the depth of the water, because they swam at the surface. Snakes and lizards, including giants like monitors and the Komodo dragon, inhabited the land and were indifferent to changes near the shore. But for the gigantic dinosaurs, the loss of their shallow seas was crucial. They had evolved specifically to inhabit a precisely determined environment that had endured for hundreds of millions of years, and now, as the Cretaceous period

Conventional interpretations of dinosaurs by present-day palæontologists are vividly exemplified by this inaccurate drawing Gregory S. Paul, who has spent over 30 years studying dinosaurs. This herd of 100-ton *Giraffatitan* are shown pounding through the landscape. In reality, they were far too large to walk.

ended, their world had disappeared, and so did they. The ichthyosaurs and plesiosaurs that also became extinct had similarly evolved for a life in shallower seas. This also marked the end of the environment in which the chalk-forming coccolithophores had multiplied in such huge numbers. The movement of the tectonic plates did not only kill off the largest of animals, but also severely restricted the proliferation of the smallest. It marked the end of the most remarkable era that life on Earth had ever seen.

Volcanoes have always indicated tectonic activity. The two are inevitably linked – as the Earth's continents crack and move, this is always going to release hot rocks and magma from beneath the Earth's crust. If we look for evidence of volcanism, there is a good association with mass extinctions throughout the geological history of our world, and Gerta Keller has demonstrated the connection. It seems that tectonic movements and volcanoes are the agency at work, not celestial bombardment.[25]

The antagonism Keller had experienced is typical of the way new theories are received in science. Established academics like to hold on to old theories, and, if an iconoclastic new view does eventually become popular, they are just as keen to claim it as their own.

This had gone far enough. Even though I had always sat on this theory and had reluctantly set out simply to publish it as a paper or two, it was becoming clear that a detailed account would have to be published. It was time for a book. I spoke to an old friend, Myles Archibald, the Publishing Director at Collins, and he was very positive about putting all my investigations into print. He wondered about a date for submission in late 2016, but I pushed it back by a year; it needed careful thought, and I remained hopeful that some palæontologist, somewhere, would publish it first. Eventually, over a sunny lunch in central London overlooking Tower Bridge, Myles introduced me to Julia Koppitz, one of their most experienced editors, and later – this time in Cambridge – Myles and I lunched together again to thrash out the way ahead. I remained reluctant, but everyone could see that a book was the only sensible response to the barrage of senseless criticism. I was determined to give full credit to everyone's views, though I soon discovered that dinosaur palæontologists would not

Gerta Keller at Princeton University pointed out the obvious – that the dinosaurs survived for 300,000 years after the asteroid struck Mexico. That cannot have caused their extinction. Challenging the accepted view caused her problems.

discuss it. One of my first ports of call was to Paul Barrett at the Natural History Museum. I thought it would be timely if we met to sort out our differences before I compiled a draft, but it was not to be. He wrote:

'Dear Prof. Ford,' he began. He went on to say that he could not think of any recent publications that would support my 'frankly incorrect' theories. A meeting could not be of benefit to either of us, because there was no common ground. He signed it: 'Sincerely, Prof. Paul M. Barrett, President, The Palaeontographical Society, Merit Researcher (Dinosaurs), Department of Earth Sciences.'[26]

Others were much more positive. At Gonville & Caius College of the University of Cambridge, my much-missed friend the late Sir Sam

Edwards – the college President, who was a former Chairman of the Science Research Council – relished the prospect of my revolutionizing an entire science; and his successor, Yao Liang, who came as my travelling companion on a lecture tour of Southeast Asia, found it all most intriguing. The eminent geneticist Peter Forster and I dined from time to time at Murray Edwards College, and he considered it a most exciting challenge. When Sir Tom Blundell invited me to the fellows' dinner over at Sidney Sussex College, and the dinosaur theory came up in conversation, it produced a buzz of interest, a number of helpful new thoughts, and a sense of incredulity – and this was most people's response. Scientists of every shade wanted to help and they all provided encouragement.

In the world of palæontology, however, the sense of hostility was unabated. Then, suddenly, there was a dramatic development over the joint paper I was writing with the American team. The proposed manuscript had continued to develop, and new findings were continually arriving; but (after months of warm correspondence) our cooperative venture took an unexpected turn. Bill Hay sent me the latest updated version of our paper, which now did not list me as the prime author. Hay's name was now first. His explanatory note explained the change: 'I am back in touch with Rob DeConto, off and on … I think things would go much more smoothly if I were first author, Rob second, and you third,' he wrote. 'I hope you can agree to being third author. I think it is really better to see your name embedded in a group of "well known" Cretaceous climate specialists.' The new version of the co-authored paper had all the names in a different sequence: my contribution had indeed been relegated to third position.[27]

As so often in the competitive world of science, the others now wanted to claim the limelight. In many ways, I had no objection to being placed anywhere in the list, but this was such a departure from what Hay had originally proposed. Because the new theory they were investigating was mine, it clearly made sense to have an academic expansion of it appearing under my name, with the team's input acknowledged as co-authors. Hay had already enthusiastically endorsed the fact that my new theory was the origin of our remodelling of the Cretaceous climate, and we could hardly now pretend that hadn't happened. But some saw how important this paper would be, and now others wanted to establish their claims to involvement. Naturally, I could understand their ambitions, but under the circumstances I clearly had to decline the proposal. The reply from America was courteous: 'In the next version of the paper I will list you as first author,' said Hay.[28]

This seemed correct. It was a civilized response and in line with what I expected. I had always held back on finalizing my introduction to the article, and so I wrote again to ask about the current contributions of the various authors, so that I could ensure everybody was given due credit. This time the reply I received was terse and uncompromising:

It warned me that the information in the draft paper on which we had been working was considered 'strictly confidential.' Hay said he hadn't specified that previously, because it was 'simply understood' in the field. He emphasised that he had published over 250 peer-reviewed scientific papers as sole (or joint) author, and that everybody always understood that a draft was confidential. I was, of course, aware of that. He added that I did not have permission to include any of the draft in the book, which is why I am not including here the findings we were jointly compiling.[29]

Yes, of course it was an overreaction, but this is what can happen when someone who is an authority in their field suddenly senses a chance to claim their own pre-eminence. The desire to make an individual mark had surfaced, and the original proposal for a jointly authored paper had collapsed. Instead of the wide-ranging and comprehensive new publication on the theory, with everyone's input combined into a well-documented and authoritative statement, Hay confirmed to me that the other authors in the group would be publishing separately. After so much support and cooperation, he wrote to say this:

> I have separated out part of it into a paper on Cretaceous climate, authored by myself, DeConto, de Boer, Flögel, Song and Stepashko. It will not include any mention or discussion of the aquatic dinosaur hypothesis. Because new information is coming to light,

> and because I do not expect the contributions of some of the authors to be completed before the end of February, I would expect it would be March 2018 at the earliest before we are able to make submission to a journal. Then the peer-review process, responses, etc. will take time, so it is usually about a year after submission that the paper appears in print.
>
> Section 7 of the draft, on the aquatic dinosaur hypothesis, will not be used. It was written exclusively by me, and I retain the rights to the text and figures. It might be useful in a later publication.
>
> The two paragraphs you submitted, one for the Introduction, the other for the Conclusions will not be used in any of the proposed work.[30]

And there we have it – I had now been banished from my own joint publication. It posed a problem: Bill Hay had repeatedly emphasized the prime importance of the new theory and had openly published the fact. As we know, scientific data appearing in print record the progress of academic endeavour. He could certainly now print something which circumvented his earlier conclusions, but it was not a dignified way to proceed. The conflict would be visible to everybody in the field. The revised model of the Cretaceous period, stemming from my new insights, was published in December 2018 and it offers a crucial contribution to the literature,[31] so you should keep an eye out for it. This promises to be an influential paper, for it will define our comprehensive new approach to the study of the Cretaceous landscape and climate. Of course, it will also present an object lesson in how rivalry in science and the pressure to publish can override the personal dimension. Scientific revolution progresses through priority, not friendly cooperation. This is an important lesson to learn.

In the long term, how these findings are published is inconsequential as far as our grasp of the Cretaceous period is concerned. Even though the published record reveals the contorted history of how an idea develops, the important consideration is that the new research should be available to everyone. I have no doubt that, within the next few years, the same palæontologists who attacked my original articles will be publishing findings that suggest they always knew it, anyway.

This is the way science progresses. It is not a quest for new truths, so much as a race by rivals. Personal animosity is a greater stimulus to science than the contented spirit of enquiry can ever be. I have sat on this new theory for half a century; and, now that it was out and available (just as you expect in science), others in the field want to claim it for themselves.

And so, I have developed three related theories.

First, those gigantic dinosaurs evolved only because they had developed under the constraints of a watery environment, and this alone can explain the vast dimensions they attained. It also explains how they could breed, and how their body temperature was regulated. The terrestrial tyranny has impelled palæontologists around the world to represent their fossils as being the remains of dynamic, lively, bouncy creatures that leaped about like lambs, but they were actually largely inert, the herbivorous dinosaurs munching continually at the luxurious vegetation in vast tracts around them, using their necks to penetrate through the verdant growth and their tails to steer them along. The meat-eating theropods were similarly slow moving, subsisting largely on the decaying carcasses of the gigantic sauropods.

Second, it was clear that we have persistently misunderstood the Cretaceous climate and the lie of the land 100 million years ago. The surface was flatter; mountain building was yet to start, the land was covered with huge lakes and vast shallow seas, with extensive ox-bow lakes and colossal deltas, and the temperatures were equable and ideal both for the proliferation of gigantic plants and the nurturing of enormous dinosaurs. Current research suggests that my new view is indeed correct.

My third conclusion is that we should abandon the asteroid impact theory for the demise of the dinosaurs. I contend instead that the dinosaurs dwindled away and disappeared when they lost their habitat as the continental blocks drifted across the Earth's surface and massive terrains began to collide, forming huge mountain ranges that towered above the land, with deep valleys and straightened river courses. During the Upper Cretaceous, about 70 million years ago, the Indian tectonic plate was travelling north at about 6 inches (15

Terrestrial dinosaurs are customarily seen prancing across the landscape (see Bob Bakker's sketch on page 255), and this view of *Brontomerus utahraptor* is typical. This was a 50-foot (15 metre) long, 10-ton monster bigger than a bull elephant; it did not skip up the beach.

cm) per year, and by 50 million years ago it had closed off the Tethys Ocean that once separated Europe and North Africa as India crashed headlong into Asia and the crumple-zone we call the Himalayas was thrown up. Mountaineers see the results, for the top of Mount Everest is composed of marine limestone that was once the bed of that ancient sea. Incidentally, we pronounce that mountain wrongly. It was named after a Welsh geographer, Sir George Everest, though he did not identify the mountain and indeed he objected to its being named after him. It was in 1852 that a young Indian mathematician, Radhanath Sickdhar, first identified this mountain, catalogued as Peak XV, the highest in the world at 29,002 feet (8,840 metres). Because Sir George Everest was the surveyor general of British-ruled India, the peak was eventually named after him – yet he did not pronounce his surname as 'ever-est' but as 'eve-rest'. This colossal mountain peak is an enduring reminder of the power of plate tectonics; how curious that everyone says its name incorrectly.

And there is an additional lesson for everyone to learn: science is not dignified and objective, but random and riddled with personal animosity and petty rivalries. Indeed, there was another radio programme on in the background only this morning. Speaking was Richard Henderson, who jointly won a Nobel prize for his pioneering work with electron microscopes. He was explaining how his new theories had been received: 'So I went along to this meeting,' he said, 'and I put up my data and they were really angry – really furious! They wouldn't agree with it, and I had to argue with them; I almost got lynched, actually.'[32]

When I was reluctantly planning this book, I thought back to a joke my grandfather in Brighton used to tell. 'An Irish woman was invited to see her son on parade in the army,' he said. 'As they came marching along she shouted out – "Oh look, he is the only one marching in step!"' When grandfather first told me, I retorted: 'Well, he could've been. Perhaps everyone else misheard the sergeant. Just because he was the only one who was different doesn't mean he must have been making a mistake. All the other soldiers might have been wrong.' Grandfather shook his head and insisted that I hadn't understood the point of the joke. But I knew that the outsider, once in a while, might be the only one to be correct. Now I have finished the project, I sympathize with the soldier.

This has been a long journey, and I hope you have found it illuminating. It has taken us through legends of prehistory, the scientific study of fossils in Europe, the explosion of interest in the United States and the unravelling of ideas about the Earth ranging from its age and its origins to the way its continents formed. We have discovered that the present-day view of the Cretaceous period is wrong, and we need to replace its teaching of an arid, largely dry environment with one of extensive shallow water and a warm, balmy climate. Along the way we have discovered some of the rivalries that drove the investigators and have even encountered some that have been born out of this very project. We have discussed the various views about the evolution of life, the classification of dinosaurs, their relationships, and a host of theories about how they died. I have described the species that live with us today (and which were around when the dinosaurs were at their height) and explained the naming of dinosaurs. We have followed the rise and fall of interest in dinosaurs by academia and the public and seen how perfidious science can be. The roaring terrestrial monster created in the 1970s was such a powerful prospect. Palæontologists have assiduously nurtured the notion of the dynamic dinosaur and will do anything to discourage you from thinking of alternatives. Dinosaur specialists adore their pre-eminence and pour scorn on anyone who doubts their comfortable conformity, so one person's resolve to demolish an entire scientific orthodoxy and replace it with a very different view was always destined for a hostile reception.

The first student of the new theory has meanwhile emerged. Emmanuel Calderon is now studying at the University of Reno and was captivated by the concept. For him it was a life-changing experience. He recently wrote to day: 'The aquatic dinosaur theory is why I originally began my education. The more I learn about dinosaurs and life during the Mesozoic, the more I'm convinced about their semi aquatic nature of most of these creatures during this era.' The new theory triggered his resolve to pursue an academic career. 'I didn't really have a plan until now,' he tells me. 'I just never had the confidence to do it.'

There is a twitter account @aquaticdinosaur, you will find more news on facebook.com/aquaticdinosaur, and there are videos on

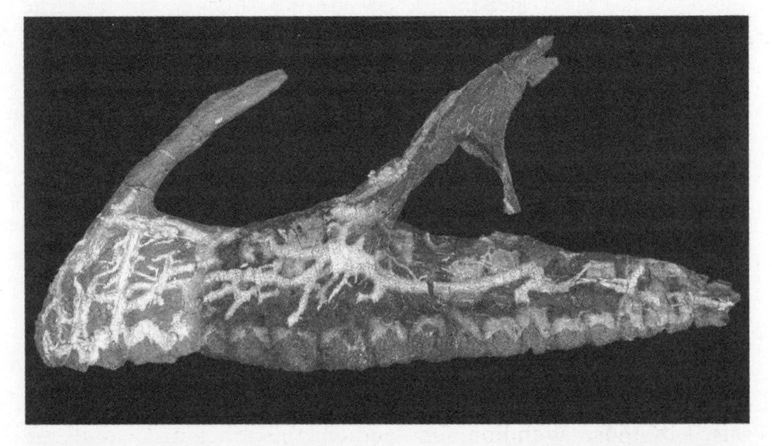

The snout of *Neovenator* from the Isle of Wight was studied by Chris Tijani Barker at Southampton University and published in *Nature Scientific Reports* in 2017. Such complex nerve pathways are typical of aquatic creatures.

YouTube (search for brian ford dinosaurs to start). As soon as this book was announced, the hostility returned. Darren Naish, who has been a devoted critic of the theory, tweeted 'I've recently learnt of Brian J Ford's book "Too Big to Walk: The New Science of Dinosaurs". Ford claims that #dinosaurs were aquatic, that there's a conspiracy to suppress the "truth", and that only he is clever enough to work things out. I call bullshit.' Skorpiodino16 replied at once: 'Wait WTF this sounds hilarious. Do you have a link to an article about it or a link to it?' Chris Dawson responded by tweeting: 'Perhaps it's all a big joke …' and so the tirade starts off again.

In February 2018, in a blog entitled 'Behold the Bronto Baths', Marc Vincent informed his readers that, after my 'peddling utter nonsense about dinosaurs' I had now 'gone and had a book published'. On Facebook he described it as 'very silly' and he ridiculed my proposal of a 'big flood plain [where] the dinosaurs were kept nice and warm all the time. Luxuriating in a swampy bath.' There was a time when I would also have regarded the idea as implausible, but my new theory is predicated upon the existence of vast flooded plains. Science now proves my new theory was right all along. America's great palæoclimatologist, Bill Hay, confirms that I am correct, and the conventional models are wrong.[31] Hay agrees with me that the Cretaceous was indeed wet and warm and he has revealed that 'elevations were less than today and that there were much more extensive wetlands (lakes, meandering rivers, swamps, bogs) on the continents than previously assumed.' His team calculates that up to three-quarters of the land was covered with shallow lakes, precisely what we need to substantiate my theory for the evolution of the giant dinosaurs. When Marc Vincent sarcastically observes: 'Ford's claims about a uniformly warm, aquatic environment made the Mesozoic sound like a spa for dinosaurs' he thought he was being funny, but in fact this is a perfect description of reality.

The theory was debated at Conway Hall in London on May 15, 2018, when Naish reiterated the standard view. Those attending remarked on the lack of objections to the aquatic dinosaur theory raised by those attending. One audience member had wondered whether footprints would persist under water, so I pointed out that in shallow lagoons there are algæ present in silt that confer a glutinous

To illustrate the critique of my Cretaceous climate proposals, the Bronto Baths blog published this beautiful depiction by German illustrator Sara Otterstätter of a group of sauropods luxuriating in their moist, warm environment, with sweat towels round their heads and cushions on which to rest.

consistency which ensures that trackways endure for prolonged periods of time. Human footprints in soft silt dating back hundreds of thousands of years are known.[33] Another mentioned the massive skulls of dinosaurs like *Triceratops*, though these can only be accounted for by an aquatic habitat. On land, supporting such a heavy head would consume inordinate amounts of metabolic energy, whereas – if buoyant in shallow water – a dinosaur would experience a greatly reduced load. The event generated hundreds of hostile remarks, however; Matt Dempsey tweeted about 'the harmful public ravings of a tinfoil hat truther lunatic,' while the BBC's Adam Rutherford proclaimed 'You can't fault the brass balls of this clown [for] his moronic scientifically illiterate ideas.' Truly, the refusal to consider a revolutionary new theory that replaces a widely-held view remains as strong now as it was in Alfred Wegener's time. And the technique remains the same: *ad hominem* – ridicule the person rather than address the issue.

It is the blindness to reasonable alternatives that dogs the heels of those who might wish to innovate in science. The dinosaur palæontologists endlessly repeat the same dogma, even when the evidence clearly shows it is nothing but fake news. One recent example with which we can close is that partial skull of *Neovenator* from the Isle of Wight, which shows all the signs of sense organs on its face of the kind that water-living reptiles are known to need. The scientific evidence, in the words of a newspaper report, shows that: 'it possessed an extremely sensitive snout of a kind previously only associated with aquatic feeders.' This sensible summary fits my theory exactly. Yet the palæontologist who gave the interview, Chris Barker, retorts: 'There is no evidence to suggest the 125-million-year old dinosaur was an aquatic feeder.'[34]

No evidence? He needs to read this book. All the other dinosaur palæontologists should do so too. Barker's response is typical of present-day palæontology. When overwhelming evidence is provided, the dinosaur devotees continue to insist that it does not exist. Yet the scientific evidence is overwhelmingly in support of my theory – the gigantic dinosaurs were obviously aquatic. They can only have evolved in shallow water and there is now no denying this compelling conclusion.

The picture I have presented is of huge herbivorous sauropod dinosaurs buoyant in water, grazing ceaselessly on the luxuriant, verdant vegetation that grew in and around large, limpid lakes and themselves being consumed by smaller populations of grazing therapods that cruised through the thickets like alligators in swamps in a humid, balmy environment. Wherever we look, this theory offers new insights that solve persistent problems. For example, tooth wear on the teeth of herbivorous dinosaurs (like the Hadrosaurs) shows that they masticated abrasive materials. The standard explanation is that they ate plants (like the Equisitales) containing siliceous materials[35] and it has been proposed that they might accidentally have consumed earth. However, the aquatic evolution theory provides the perfect answer: they were consuming alluvial sand as they grazed on marsh plants. This is the only sound and solid answer, and we can now dispense with the previous tentative explanations. Gone is the popular palæontologist's vision of 100-ton monsters pounding across the prairie and elephant-sized carnivores balancing like ballerinas. Now, as each new revelation emerges, the fresh scientific evidence reaffirms my alternative vision. Among the latest was the announcement in April 2018 of dinosaur footprints discovered on the Isle of Skye in Scotland.[36] Most were made by sauropods, with a smaller number left by the carnivorous therapods. What was the habitat like at the time, 170 million years ago? One of the palæontologists, Steve Brusatte, sums up their conclusions. 'This was a subtropical kind of paradise world,' he said.[37] The footprints were 'made in a shallow lagoon.' The era of the terrestrial tyrant is over; the aquatic dinosaur can at last be given its proper place in science. It is the scientific evidence that proves the point.

My findings are even consonant with people who firmly believe in terrestrial dinosaurs. At Manchester University, Bill Sellers has created digital simulations of how a titanosaur like *Argentinosaurus* moved. The results are unconvincing; they show a huge skeleton like a girder moving unsteadily along. To a sceptic, this is obvious evidence of the fact that huge dinosaurs could not walk on dry land. The simulation can be viewed online[38] along with an interview with Sellers.[39] This contains the most convincing substantiation that my view must be correct. After devoting many months of diligent research to recreating the way *Argentinosaurus* might have moved, Sellers concludes: 'This animal is so big … getting up off the ground would have been extraordinarily difficult, if not impossible.' For an ardent advocate of the terrestrial dinosaur to make this admission proves the point. It is true: giant dinosaurs were, simply, too big to walk.

Compiling this research has been a long and exacting task. It has meant diligently demolishing an entire branch of science and establishing an edifice that can stand in its stead. In the end, virtually everything that I have found out about the dinosaurs during 50 years of investigation is summarized in the pages of this book. I hope the various facets of this complex conundrum will have reflected some rays of interest to you, the reader, no matter where your loyalties may lie. You can see now why I hesitated to produce the first edition of this book – but, now that the second edition is in your hands, I am so very glad I did.

NOTES

PREFACE

1. Larry A. Witham (2005) *Where Darwin Meets the Bible, Creationists and Evolutionists in America*, London & New York: Oxford University Press

CHAPTER 1: DINOSAURS AND THE ANCIENTS

1. Kevin Holden Platt (2007) Dinosaur Fossils Part of Longtime Chinese Tonic, *National Geographic News*: July 13
2. http://www.chinahighlights.com/travelguide/article-chinese-dragons.htm
3. Gregory S. Paul (2000) *The Scientific American book of dinosaurs*, New York: St. Martin's Press
4. Magnus Albertus (<1280) *De Causis Proprietatum Elementorum* Book II, tract 3, chapter 5, cited in A.C. Crombie (1959) *The History of Science from Augustine to Galileo*, London: Heinemann
5. Stegosaurus, rhinoceros, or hoax? (2009) http://www.smithsonianmag.com/science-nature/stegosaurus-rhinoceros-or-hoax-40387948/
6. Dennis Swift (2016) https://www.discoveryworld.us/dinosaur-world/acambaro-figurines-from-second-expedition/
7. Charles Di Peso (1953) The Clay Figurines of Acámbaro, Guanajuato, Mexico, *American Antiquity*, 18–4: 388–389
8. Alex Pezzati (2005) Mystery at Acámbaro, Mexico: Did Dinosaurs Coexist with Humans? *Expedition Magazine*, 47 (3): 6–7
9. Javier Cabrera Darquea (1994) *The Message of the Engraved Stones of Ica*, Bolivar: Plaza De Armas

10. Ken Feder (2010) *Encyclopedia of Dubious Archæology*, Westport CT: Greenwood
11. Commonwealth of Australia (2011) Environment Protection and Biodiversity Conservation Act 1999 – inclusion of a place in the National Heritage List – the West Kimberley. *Commonwealth of Australia Gazette* S132: 1–19
12. R.A. Thulborn & M.J. Wade (1984) Dinosaur trackways in the Winton Formation (mid-Cretaceous) of Queensland. *Memoirs of the Queensland Museum*, 21: 413–517
13. B. Chatwin (1987) *The Songlines*, London: Jonathan Cape
14. Commonwealth of Australia (1991) The Injured Coastline: Protection of the Coastal Environment, *Report of the House of Representatives Standing Committee on Environment, Recreation and the Arts*, Canberra: Australian Government Publishing Service
15. J.E. Norman & G.V. Norman (2007) *A Pearling Master's Journey; in the Wake of the Schooner Mist*, Burwood, Victoria: BPA Print Company
16. Steven W. Salisbury, Anthony Romilio, Matthew C. Herne, Ryan T. Tucker & Jay P. Nair (2017) The Dinosaurian Ichnofauna of the Lower Cretaceous (Valanginian–Barremian) Broome Sandstone of the Walmadany Area (James Price Point), Dampier Peninsula, Western Australia, *Journal of Vertebrate Palæontology*, 36 (sup 1): 1–152
17. Paul Arblaster (2004) *Antwerp & the World, Richard Verstegan and the International Culture of Catholic Reformation*, Leuven: University Press
18. Richard Verstegan (1605) *A Restitution of Decayed Intelligence: In Antiquities. Concerning the Most Noble and Renowned English Nation*, Antwerp: John Norton & Iohn Bill
19. Brian J. Ford (2015) The Incredible, Invisible World of Robert Hooke, *The Microscope*, 63 (1): 23–34
20. Robert Hooke (1665) *Micrographia, or, Some physiological descriptions of minute bodies made by magnifying glasses: with observations and inquiries thereupon*, London: James Allestry for the Royal Society
21. Nehemiah Grew (1681) *Musæum Regalis Societatis, or, A catalogue and description of the Natural and Artificial Rarities belonging to the Royal Society and preserved at Gresham Colledge made by Nehemiah Grew; whereunto is subjoyned the Comparative Anatomy of Stomachs and Guts*, London, Thomas Malthus
22. Edward Lhuyd (1699) *Lithophylacii Brittannici Ichnographia*, London: printed for subscribers
23. Robert Plot (1677) *The Natural History of Oxfordshire*, Oxford: Printed at the Theater

24. William Stukeley (1719) An account of the impression of the almost entire sceleton of a large animal in a very hard stone, lately presented to the Royal Society, from Nottinghamshire, *Philosophical Transactions*, 30: 963–968
25. Johann Jacob Scheuchzer (1726) *Homo diluvii testis et theoskopos*, Tiguri: Henrici Byrgklini
26. Bertrand Schultz, Lloyd G. Tanner, Frank Whitmore, Jr., Louis L. Ray & Ellis Crawford (1963) *Paleontologic Investigations at Big Bone Lick State Park, Kentucky: A Preliminary Report*, U.S. Geological Survey Research Paper No. 234
27. Jean-Étienne Guettard (1755) Mémoire sur les encrinites et les pierres étoilées dans lequel on traitera aussi des Entroques, *Mémoire de l'Académie Royale des Sciences*, Paris 1755: 224–263 & 318–354
28. Jean-Étienne Guettard (1780) *Atlas et description minéralogique de la France, entreprize par Ordre du Roi*, Paris: M. Monnet
29. Joshua Platt (1764) An Attempt to Account for the Origin and the Formation of the Extraneous Fossil Commonly Called the Belemnite, *Proceedings of the Royal Society of London*, 54: 38–52
30. John Walcott (1779) *Descriptions and Figures of Petrifications*, Bath: S. Hazard
31. Martinus van Marum (1790) Beschrijving der beenderen van den kop van eenen visch, gevonden in den St Pietersberg bij Maastricht, en geplaatst in Teylers Museum, *Verhandelingen Teylers Tweede Genootschap*, 9: 383–389
32. J. Neumann (1985) Climatic Change as a Topic in the Classical Greek and Roman Literature, *Climatic Change*, 7: 441–454
33. Muninp Intra (2016) *Technological Advancement to contain Climate Change and accompanying natural disasters*, New York: United Nations Commission on Science and Technology for Development
34. Pierre Martel (1744) An account of the glacieres or ice alps in Savoy, in two letters, one from an English gentleman to his friend at Geneva; the other from Pierre Martel, engineer, to the said English gentleman, in C.E. Mathews (1898) *The Annals of Mont Blanc*, London: Unwin
35. Tobias Krüger (2013) *Discovering the Ice Ages, International Reception and Consequences for a Historical Understanding of Climate*, Leiden: Brill Publications
36. James Hutton (1795) *Theory of the Earth, with Proofs and Illustrations*, Edinburgh: Creech
37. Karl Alfred von Zittel, translated by Maria M Ogilvie Gordon (1901) *History of geology and palæontology to the end of the nineteenth century*, London, W. Scott; New York, C. Scribner's Sons, 1901

38. Ignaz Venetz (1833) Mémoire sur les variations de la température dans les Alpes de la Suisse, *Denkschriften der Allgemeinen Schweizerischen Gesellschaft für die gesammten Naturwissenschaften*, 1 (2): 1–3
39. Louis Agassiz (1833–1843) *Recherches sur les poissons fossils*, Neuchâtel: Petitpierre
40. Louis Agassiz (1840) *Études sur les Glaciers*, Neuchâtel: Jent et Gassmann
41. E.P. Evans (1887) The Authorship of the Glacial Theory, *North American Review*, 145 (368): 94–97
42. John Gribbin (1976) *Forecasts, Famines and Freezing*, London: Wildwood House; [and] Nigel Calder (1976) *The Weather Machine and the Threat of Ice*, London: BBC Publications
43. Brian J. Ford (2016) Cloudy with a Chance of Microbes, *The Microscope*, 64 (1): 29–41
44. Walter Broecker (1975) Climatic change; are we on the brink of a pronounced global warming? *Science*, 189 (4201): 460–463
45. Svante Arrhenius (1896) On the Influence of Carbonic Acid in the Air upon the Temperature of the Ground, *The London, Edinburgh, and Dublin Philosophical Magazine and Journal of Science* [fifth series], 41: 237–275
46. Svante Arrhenius (1908) *Worlds in the Making; the Evolution of the Universe*, New York and London: Harper & Brothers
47. Gilbert Norman Plass (1953) The Carbon Dioxide Theory of Climatic Change, *Bulletin of the American Meteorological Society*, 34: 80
48. William W. Hay (2013) *Experimenting on a Small Planet*, Berlin & New York: Springer

CHAPTER 2: EMERGING FROM THE SHADOWS

1. Mrs R. Lee [formerly Mrs T. Edward Bowdich] (1833) *Memoirs of Baron Cuvier*, London: Longman, Rees, Orme, Brown, Green & Longman
2. J. L.N.F. (Georges) Cuvier (1801). Reptile volant [in] Extrait d'un ouvrage sur les espèces de quadrupèdes dont on a trouvé les ossemens dans l'intérieur de la terre, *Journal de Physique, de Chimie et d'Histoire Naturelle*, 52: 253–267
3. James Parkinson (1804–11) *Organic Remains of a Former World*, London: J. Robson, J. White, J. Murray and others
4. J.L.N.F. (Georges) Cuvier (1808) Sur le grand animal fossile des carrières de Maëstricht, *Annales du muséum national d'histoire naturelle (Paris),* 12: 145–176

5. Robert Plot (1677) *The Natural History of Oxfordshire*, Oxford: Printed at the Theater
6. R.T. Gunther (1939) *Dr Plot and the Correspondence of the Philosophical Society of Oxford*, Oxford: privately printed
7. *Genesis* 6:4 – There were giants in the earth in those days; and also after that, when the sons of God came in unto the daughters of men, and they bare [children] to them; *Numbers* 13:33 – And there we saw the giants, the sons of Anak of the giants: and we were in our own sight as grasshoppers, and so we were in their sight
8. C. Plinii (1906) *Naturalis Historiae*, edited by Karl Friedrich Theodor Mayhoff, Leipzig: B.G. Teubneri
9. Richard Brookes (1772) *A New and Accurate System of Natural History*, London: T. Carnan and F. Newbery, Jr.
10. John Aiken & Anna Barbauld (1780) *Eyes or no Eyes; or, the Art of Seeing*, London: Longman
11. Dennis R. Dean (1999) *Gideon Mantell and the discovery of dinosaurs*, Cambridge: University Press
12. George Qvist (1979) Some controversial aspects of John Hunter's life and work, Hunterian ovation, *Annals of the Royal College of Surgeons*, 61: 381–384
13. Johann Gottlob Lehmann (1756) *Versuch einer Geschichte von Flötz-Gebürgen*, Berlin: Klüterschen Buchhandlung
14. Abraham Gottlob Werner (1787) *Kurze Klassifikation und Beschreibung der verschiedenen Gebirgsarten*, Dresden: Waltherischen Buchhandlung
15. Fabien Knoll (2009) Alexander von Humboldt et la bête-main: une contribution du dernier savant universel à la palaéontologie, *Comptes Rendus Palevol*, 8 (4): 427–436
16. Jacques-François Dicquemare (1776) Ostéolithes, *Journal de Physique, de Chimie, d'Histoire Naturelle et des Arts*, 7: 406–441
17. Charles Bacheley (1778) *Notice des pétrifications & autres faits d'histoire naturelle qui se trouvent le long [des] côtes du Pays d'Auge, Collection d'observations sur les maladies et constitutions épidémiques*, Rouen: Imprimerie privilégiée
18. J.L.N.F. (Georges) Cuvier (1800) A quantity of buries found in the rocks in the environs of Honfleur by the late Abbé Bachelet. *Philosophical Magazine* VIII: 290; Ronan Allain (2001) Redescription of *Streptospondylus altdorfensis*, Cuvier's theropod dinosaur, from the Jurassic of Normandy, *Geodiversitas*, 23 (3): 349–367
19. J.L.N.F. (Georges) Cuvier (1808) Sur les ossements fossiles de crocodiles et particulièrement sur ceux des environs du Havre et d'Honfleur, avec des remarques sur les squelettes de sauriens de la

Thuringe, *Annales du Muséum d'Histoire Naturelle de Paris* XII: 73–110
20. Thomas Jefferson (1799) A Memoir on the Discovery of Certain Bones of a Quadruped of the Clawed Kind in the Western Parts of Virginia, *Transactions of the American Philosophical Society*, 4: 246–260
21. Elizabeth Gordon (1894) *The Life and Correspondence of William Buckland*, London: John Murray, p. 91
22. Henry de la Beche & William Conybeare (1821) *Notice of the discovery of a new Fossil Animal, forming a link between the Ichthyosaurus and Crocodile, together with general remarks on the Osteology of the Ichthyosaurus*, London: W. Phillips
23. James Parkinson (1822) *Outlines of Oryctology: an Introduction to the Study of Fossil Remains, especially those found in the English Strata*, London: Sherwood, Neely and Jones
24. William Buckland (1824) Notice on the Megalosaurus or great Fossil Lizard of Stonesfield, *Transactions of the Geological Society of London*, 21 (2): 390–396
25. J.L.N.F. (Georges) Cuvier (1824) *Recherches sur les ossements fossiles, où l'on rétablit les caractères de plusieurs animaux dont les révolutions du globe ont détruit les espèces* (2nd edition), Paris: Déterville
26. Report (1798) *Bath Chronicle*: December 27
27. Hugh Torrens (1995) Mary Anning (1799–1847) of Lyme, *British Journal of the History of Science*, 28: 257–284
28. Charles Lyell (1830) *Principles of Geology*, London: John Murray
29. Ferdinand von Ritgen (1826) Versuchte Herstellung einiger Becken urweltlichter Thiere, *Nova Acta Academiae Caesareae Leopoldino-Carolinae Germanicae Naturae Curiosorum*, 13: 331–358
30. Gideon Mantell (1827) *Illustrations of the Geology of Sussex*, London: Lupton Relfe
31. M. Adams (1808) Some Account of a Journey to the Frozen-Sea, and of the Discovery of the Remains of a Mammoth, *The Philadelphia Medical and Physical Journal*, 3: 120–137
32. J.L.N.F. (Georges) Cuvier, Alexandre Brongniart & Étienne Geoffroy Saint-Hilaire (1812) *Recherches sur les ossemens fossiles de quadrupèdes, où l'on rétablit les caractères de plusieurs espèces d'animaux que les révolutions du globe paroissent avoir détruites*, Paris: Chez Deterville; R.T.J. Moody, E. Buffetaut, D. Naish & D.M. Martill (2010) *Dinosaurs and other extinct Saurians, a historical perspective*, London: Geological Society Special Publication No. 343
33. Charles Kingsley (1863) *The Water-Babies; a Fairy Tale for a Land Baby*, Oxford & New York: Oxford University Press, 1995

34. Friedrich August von Alberti (1834) *Monographie des Bunten Sandsteins, Muschelkalks und Keupers, und die Verbindung dieser Gebilde zu einer Formation*, Stuttgart-Tübingen: Cotta
35. Jean Baptiste Julien d'Omalius d'Halloy (1828) *Description géologique des Pays-Bas* (1831), *Eléments de Géologie* (1833), *Introduction à la Géologie* (1842), *Coup d'oeil sur la géologie de la Belgique* (1843), *Précis élémentaire de Géologie* (1843), *Abrégé de Géologie* (1853) and (1845) *Des Races humaines ou Éléments d'Ethnographie*
36. Rev. William D. Conybeare & William Phillips (1822) *Outlines of the Geology of England and Wales*, London: William Phillips
37. Christopher McGowan (2001) *The Dragon Seekers*, London: Little, Brown
38. Shelley Emmling (2009) *The Fossil Hunter*, London: Palgrave Macmillan
39. Brian J. Ford (2009) On Intelligence in Cells: The Case for Whole Cell Biology, *Interdisciplinary Science Reviews*, 34 (4): 350–365
40. Gideon Mantell (1817) *Outlines of the Mineral Geography of the Environs of Lewes*, Lewes: J Baxter; Gideon Mantell (1818) A Sketch of the Geological Structure of the South-eastern part of Sussex, *The Gleaner's Portfolio*, Lewes: The Sussex Press
41. Gideon Mantell (1822) *The Fossils of the South Downs, or, Illustrations of the Geology of Sussex; the engravings executed by Mrs. Mantell, from drawings by the author*, London: Lupton Relfe
42. Gideon Mantell (1827) *Illustrations of the geology of Sussex: containing a general view of the geological relations of the South-Eastern part of England: with figures and description of the fossils of Tilgate Forest*, London: Lupton Relfe
43. Gideon Mantell (1838) *The Wonders of Geology, or, a Familiar Exposition of Geological Phenomena: being the substance of a Course of Lectures delivered at Brighton*, London: Relfe & Fletcher
44. Jeremy C.T. Fairbank (2004) William Adams and the spine of Gideon Algernon Mantell, *Annals of the Royal College of Surgeons of England*, 86: 349–352
45. Anonymous (1835) Discovery of Saurian Bones in the Magnesian Conglomerate near Bristol, *American Journal of Science and Arts*, 28: 389
46. Henry Riley & Samuel Stutchbury (1836) A description of various fossil remains of three distinct saurian animals discovered in the autumn of 1834, in the Magnesian Conglomerate on Durdham Down, near Bristol, *Proceedings of the Geological Society of London*, 2: 397–399
47. Thomas Hawkins (1834) *Memoirs of Ichthyosauri and Plesiosauri, Extinct Monsters of the Ancient Earth*, London: Relfe & Fletcher

48. Thomas Hawkins (1840) *The Book of the great Sea-dragons, Ichthyosauri and Plesiosauri, Gedolim Taninim of Moses, Extinct Monsters of the Ancient Earth*, London: W. Pickering
49. Richard Owen (1842) Report on British Fossil Reptiles. Part II. *Report of the Eleventh Meeting of the British Association for the Advancement of Science; Held at Plymouth in July 1841*, London: John Murray

CHAPTER 3: THE PUBLIC ERUPTION

1. John Morris (1845), *A Catalogue of British Fossils*, London: John van Voorst
2. Andrew Piddington, director (2002) *The Dinosaur Hunters*, National Geographic Films
3. Johann Andreas Wagner (1861) Neue Beiträge zur Kenntnis der urweltlichen Fauna des lithographischen Schiefers; *Compsognathus longipes* Wagner, *Abhandlungen der Bayerischen Akademie der Wissenschaften*, 9: 30–38
4. Pierre Louis Moreau de Maupertuis (1745) *Vénus Physique*, Paris: La Haye
5. Georges Louis Leclerc, Comte de Buffon (1749) *Histoire des Animaux* and *Théorie de la Terre*, Paris: A. Leleux
6. James Hutton (1794) *An investigation of the principles of knowledge and of the progress of reason, from sense to science and philosophy*, Edinburgh: Strahan & Cadell
7. Richard Joseph Sullivan (1794) *A View of Nature, Letters to a Traveller among the Alps*, London: T. Becket
8. Patrick Matthew (1831) *On Naval Timber and Arboriculture; with critical notes on authors who have recently treated the subject of planting*, Edinburgh: Adam Black; London: Longman, Rees, Orme, Brown & Green
9. Joseph Carroll (2003) *Charles Darwin's On the Origin of Species by means of Natural Selection*, Ontario: Broadview Press, p. 81
10. Anon [Robert Chambers] (1844) *Vestiges of the Natural History of Creation*, London: John Spriggs Morss Churchill
11. Herbert Spencer (1864) *Principles of Biology*, London: Williams & Norgate
12. Letter from Huxley to his sister Elizabeth (Lizzie) dated May 3, 1852, [quoted in] Leonard Huxley (1900) *Life and Letters of Thomas Henry Huxley*, London: Macmillan
13. Brian J. Ford (2011) Darwin, the Microscopist who Didn't Discover Evolution, *The Microscope*, 59 (3): 129–137

14. Charles Darwin (1845) *Journal of Researches into the Natural History of the Countries Visited during the Voyage of the HMS Beagle around the world from 1832–6*, London: John Murray
15. Charles Darwin (1859) *On the Origin of Species by Means of Natural Selection, or the Preservation of Favoured Races in the Struggle for Life*, London: John Murray
16. Brian J. Ford (2009) Charles Darwin and Robert Brown – their microscopes and the microscopic image, *In Focus, Proceedings of the Royal Microscopical Society*, 15: 18–28
17. Robert Hooke (1665) *Micrographia, or, Some physiological descriptions of minute bodies made by magnifying glasses: with observations and inquiries thereupon*, London: James Allestry for the Royal Society
18. Gerald Mayr, Burkhard Pohl & D. Stefan Peters (2005) A Well-Preserved Archæopteryx Specimen with Theropod Features, *Science*, 310 (5753): 1483–1486. doi: 10.1126/science.1120331

CHAPTER 4: GREAT AMERICAN DISCOVERIES

1. Thomas Say (1820) Observations on some Species of Zoophytes, Shells &c, principally Fossil, *American Journal of Science and Arts*, 2 (2): 34–45
2. Joseph Leidy (1856) Notice of remains of extinct reptiles and fishes, discovered by Dr. F.V. Hayden in the Bad Lands of the Judith River, Nebraska Territory, *Proceedings of the Academy of Natural Sciences of Philadelphia*, 8: 72–73
3. Fielding Bradford Meek & Ferdinand V. Hayden (1857) Descriptions of New Species of Acephala and Gasteropoda, from the Tertiary formations of Nebraska Territory, with some general remarks on the Geology of the country about the sources of the Missouri River, *Proceedings of the Academy of Natural Sciences of Philadelphia*, viii: 111–126
4. Cast of *Hadrosaurus*, reference Z.1879.38, Natural Sciences, National Museum of Scotland
5. Edward Drinker Cope (1865) On *Amphibamus grandiceps*, a new batrachian from the Coal Measures, *Proceedings of the Academy of Natural Sciences of Philadelphia* 1865: 134–137 [Note: most sources, such as Wikipedia, wrongly cite this paper as being published by the *Proceedings of the National Academy of Sciences*]
6. Alfred Romer (1964) Cope versus Marsh, *Systematic Zoology*, 13 (4): 201–207
7. Charles Schuchert (1938) Othniel Charles Marsh, *Biographical Memoirs of the National Academy of Sciences*, Washington, DC

8. Christine E. Turner & N.S. Fishman (1991) Jurassic Lake T'oo'dichi': a large alkaline, saline lake, Morrison Formation, eastern Colorado Plateau, *Geological Society of America Bulletin*, 103 (4): 528–558
9. Edward Drinker Cope (1877) On Amphicoe Lias, a genus of Saurians from the Dakota epoch of Colorado, *Proceedings of the American Philosophical Society*, 17: 242–246
10. Othniel C. Marsh (1883) Principal characters of American Jurassic Dinosaurs, restoration of *Brontosaurus*, *American Journal of Science and Art*, 26 (3): 81–85
11. John S. McIntosh & David S. Berman (1975) Description of the Palate and Lower Jaw of the Sauropod Dinosaur Diplodocus (Reptilia: Saurischia) with Remarks on the Nature of the Skull of Apatosaurus, *Journal of Paleontology*, 49 (1): 187–199
12. Report, 1981, Yale brontosaurus gets head on right at last, *New York Times*: October 26
13. Caroline Forster (1996) Species resolution in *Triceratops*: cladistic and morphometric approaches, *Journal of Vertebrate Paleontology*, 16 (2): 259–270. doi:10.1080/02724634.1996.10011313
14. Edward Drinker Cope (1875) *The Vertebrata of the Cretaceous Formations of the West*, Washington D.C.: Government Printing Office
15. William Davis (1876) On the exhumation and development of a large reptile (*Omosaurus armatus*, Owen), from the Kimmeridge Clay, Swindon, Wiltshire, *Geological Magazine*, 3: 193–197
16. Othniel Charles Marsh (1891) Restoration of Stegosaurus, *American Journal of Science*. 3 (42): 179–181
17. Oliver P. Hay (1908) On the Habits and Pose of the Sauropod Dinosaurs, especially of Diplodocus, *The American Naturalist*, 42 (502): 672–681
18. William J. Holland (1910) A Review of Some Recent Criticisms of the Restorations of Sauropod Dinosaurs Existing in the Museums of the United States, with Special Reference to that of *Diplodocus carnegii* in the Carnegie Museum, *The American Naturalist*. 44 (521): 259–283
19. Henry Fairfield Osborn & Charles Craig Mook (1902) Characters and Restoration of the Sauropod genus Camarasaurus, *Proceedings of the American Philosophical Society*, 58: 386–396
20. Kenneth A. Kermack (1951) LXXX, a note on the habits of the Sauropods, *Journal of Natural History* Series 12, 4 (44): 830–883
21. Othniel Charles Marsh (1892) Restorations of Claosaurus and Ceratosaurus, *American Journal of Science and Art*, 44 (262): 343–349
22. Edwin Drinker Cope (1887) *Theology of Evolution*, Philadelphia: Arnold & Company

23. Othniel Charles Marsh (1896) *The Dinosaurs of North America* [from the sixteenth annual report of the U.S. Geological Survey] Washington D.C.: Government Printing Office
24. Camille Flammarion (1864) *Les mondes imaginaires et les mondes réels*, Paris: Flammarion
25. Michael Crichton (2017) *Dragon Teeth*, New York: HarperCollins
26. William R. Wahl, Mike Ross & Judy A. Massare (2007) Rediscovery of Wilbur Knight's Megalneusaurus rex site, new material from an old pit, *Paludicola*, 6 (2): 94–104
27. Brent H. Breithaupt, E.H. Southwell & N.A. Matthews (2008) Brontosaurus giganteus: The 'most colossal animal ever on Earth just found out West' and the discovery of Diplodocus carnegii [In] R. Moody, E. Buffetaut & D.M. Martill (eds.): *Dinosaurs and other extinct saurian, a historical perspective*. Abstracts of the Meeting held on the 6–7 May 2008. The Geological Society of London: History of Geology Group
28. John Bell Hatcher (1901) Diplodocus (Marsh): Its Osteology, Taxonomy, and Probable Habits, with a Restoration of the Skeleton, *Memoirs of the Carnegie Museum*, 1 (1): 1
29. John Bell Hatcher (1900) The Carnegie Museum Paleontological Expeditions of 1900, *Science*, 12 (306): 718

CHAPTER 5: DRIFTING CONTINENTS

1. Abraham Ortelius (1596) *Thesaurus Geographicus* (3rd edition), Antwerp: Plantin
2. Francis Bacon (1620) *Novum Organum Scientiarum*, Leiden: Lugdinum Batavorum
3. Antonio Snider-Pellegrini (1858) *La Création et ses mystères dévoilés*, Paris: Librairies A. Frank et E. Dentu
4. Charles Lyell (1872), *Principles of Geology* (11th edition), London: John Murray, p. 258
5. Robert Mantovani (1889) Les fractures de l'écorce terrestre et la théorie de Laplace, *Bulletin de la Societé des Sciences et Arts de Réunion*: 41–53
6. Alfred Russel Wallace (1891) *Darwinism: An Exposition of the Theory of Natural Selection, with Some of Its Applications*, London: Macmillan
7. William Henry Pickering (1907) The Place of Origin of the Moon – The Volcani Problems, *Popular Astronomy*, 15: 274–287
8. Eduard Suess (1909) *Das Antlitz der Erde*, Oxford: Clarendon Press
9. Frank Bursley Taylor (1910) Bearing of the Tertiary Mountain Belt on the Origin of the Earth's Plan, *Bulletin of the Geological Society of America*, 21: 179–226

10. Alfred Wegener (1912) Die Herausbildung der Großformen der Erdrinde (Kontinente und Ozeane), auf geophysikalischer Grundlage, *Petermanns Geographische Mitteilungen*, 63: 185–195, 253–256 & 305–309 [presented at the annual meeting of the German Geological Society, Frankfurt am Main, January 6, 1912, published as] Alfred Wegener (1912) Die Entstehung der Kontinente, *Geologische Rundschau*, 3 (4): 279–292
11. James D. Dana (1863) *Manual of Geology*, Marlborough, Massachusetts: Bliss Publishing Company
12. Willem van Waterschoot van der Gracht (1928) *Theory of Continental Drift: a symposium on the origin and movement of land masses both intercontinental and intracontinental as proposed by Alfred Wegener*, Tulsa: Symposium of the American Association of Petroleum Geologists
13. Arthur Holmes (1944) *Principles of Physical Geology*, Edinburgh: Thomas Nelson & Sons
14. John Scannella & Jack Horner (2010) Torosaurus Marsh, 1891, is Triceratops Marsh, 1889 (Ceratopsidae: Chasmosaurinae): synonymy through ontogeny, *Journal of Vertebrate Paleontology*, 30 (4): 1157–1168
15. Harry G. Seeley (1887) On the Classification of the Fossil Animals Commonly Named Dinosauria, *Proceedings of the Royal Society of London*, 43: 165–171. doi:10.1098/rspl.1887.0117
16. Elmer S. Riggs (1901) The largest known dinosaur, *Science*, 13 (327): 549–550
17. http://www.angelfire.com/mi/dinosaurs/dinosaurs_trex.html
18. Henry Fairfield Osborn (1912) Integument of the Iguanodont Dinosaur Trachodon, *Memoirs of the American Museum of Natural History*, 1: 33–54
19. Barnum Brown (1916) *Corythosaurus casuarius*, Skeleton, Musculature and Epidermis, *American Museum of Natural History Bulletin*, 38: 709–715
20. Gregory S. Paul (1988) The brachiosaur giants of the Morrison and Tendaguru with a description of a new subgenus, *Giraffatitan*, and a comparison of the world's largest dinosaurs, *Hunteria*, 2 (3): 1–14
21. Arthur Conan Doyle (1879) Gelsemium as a Poison, *The British Medical Journal*, 483, September 20
22. Arthur Conan Doyle (1912) The Lost World, being an account of the recent amazing adventures of Professor George E. Challenger, Lord John Roxton, Professor Summerlee, and Mr. E.D. Malone of the 'Daily Gazette', London: *The Strand Magazine*, and Philadelphia: *Sunday Magazine*

23. Report (1922) Dinosaurs Cavort in Film for Doyle, *New York Times*: June 22
24. Edgar Rice Burroughs (1963) *The Land that Time Forgot*, New York: Ace Books
25. John R. Kielbasa (1998) *Rancho La Brea, Historic Adobes of Los Angeles County*, Pittsburg: Dorrance Publishing Company
26. J.S. Smith & G.R. Brooks (1977) *The Southwest expedition of Jedediah S. Smith: His personal account of the journey to California, 1826–1827*, Glendale, California: A.H. Clark Company
27. http://www.smithsonianmag.com/smart-news/meet-augustynolophus-morrisi-californias-new-state-dinosaur-180965038
28. Roy Chapman Andrews (1926) *On the Trail of Ancient Man*, New York: Garden City Publishing
29. http://www.telegraph.co.uk/news/worldnews/northamerica/10876661/Dinosaur-thief-jailed.html
30. Statement (2014) *Manhattan U.S. Attorney Announces Return To Mongolia Of Fossils Of Over 18 Dinosaur Skeletons*, Department of Justice U.S. Attorney's Office, Southern District of New York, Thursday, July 10 https://www.justice.gov/usao-sdny/pr/manhattan-us-attorney-announces-return-mongolia-fossils-over-18-dinosaur-skeletons
31. Report (2016) *Canadian dealer sentenced for trafficking Chinese fossils at local gem show*, U.S. Immigration and Customs Enforcement, ICE Newsroom, Monday, February 22, https://www.ice.gov/news/releases/canadian-dealer-sentenced-trafficking-chinese-fossils-local-gem-show#wcm-survey-target-id
32. William Lee Stokes (1985) *The Cleveland-Lloyd Dinosaur Quarry: Window to the Past*, Washington D.C.: U. S. Government Printing Office

CHAPTER 6: REPTILE DYSFUNCTION

1. Hitchcock, Edward (1836) Ornithichnology, Description of the foot marks of birds (Ornithichnites) on new Red Sandstone in Massachusetts, *American Journal of Science and Arts*, 29: 307–340
2. http://news.xinhuanet.com/english/2017-06/28/c_136401896.htm
3. David West (2017) *Armoured Dinosaurs*, London: Franklin Watts, Don Lessem (2014) *Armoured Dinosaurs*, Minneapolis: Lerner Publishing; Kenneth Carpenter (2001) *The Armoured Dinosaurs*, Indiana: University Press
4. Celebrating Dinosaur Island (2014) *Biological Journal of the Linnean Society*, 113 (3): 659–896

5. Brian J. Ford (1971) *Nonscience and the Pseudotransmogrificationalific Egocentrified Proclivities Inherently Intracorporated In Expertistical Cerebrointellectualized Redeploymentation with Special Reference to Quasi-Notional Fashionistic Normativity, The Indoctrinationalistic Methodological Modalities and Scalar Socio-Economic Promulgationary Improvementalisationalism Predelineated Positotaxically Toward Individualistified Mass-Acceptance Gratificationalistic Securipermanentalisationary Professionism, or How To Rule The World*, London: Wolfe Publishing
6. Brian J. Ford (1973) *Como se Falsifica la Ciencia; la Nonciencia y las Proclividades Reorientacionales Egocentrificadas Pseudotramogrificacionalificas inheremente a la Redesplegmentacion Exepertistica Cerebrointelecualizada, con Especial Referencia a la Normatividad Modaistica Cuasi-nocional, las Modilidades Metodologicas Adonctrinamientisticas y el Perfeccionamientalismo Escelar, Socioeconomica Promulgacionario Predelineado Positotaxativamenta hacia el Professionalismo, Seguripermanentinicario Gratificstionalistico Individualisticado et la Acepation de las Masas, o Como Regir el Mundio*, Buenos Aires: Granica Editor
7. Brian J. Ford (1982) *The Cult of the Expert*, London: Hamish Hamilton
8. Brian J. Ford (1985) *Der Experten-Kult, vom Maximalen Minimum*, Hamburg and Vienna: Paul Zsolnay Verlag
9. Bede the Venerable (1475) *Historia ecclesiastica gentis Anglorum*, Strasbourg: Heinrich Eggestein
10. Bede the Venerable (c.700) *Jerome, Commentary on the Old Testament book of Isaiah*, Parchment: Codex Sangallensis 254
11. George Bright & John Strype (1684) *Works of Lightfoot*, London: William Rawlins for Richard Chiswell
12. James Ussher (1658) *The Annals of the World*, London: E. Tyler for F. Creek & G. Bedell
13. Edmond Halley (1694) *Some Considerations about the Cause of the Universal Deluge, laid before the Royal Society, on the 12th of December 1694*, London: Royal Society Manuscript RSS 383
14. Georges Louis Leclerc, Comte de Buffon (1774) *Histoire Naturelle: Histoire de la Terre et de l'Homme, Quadrupèdes, Oiseaux, Minéraux, Suppléments*, Paris: De L'Imprimerie Royale
15. Charles Lyell (1830) *Principles of Geology*, London: John Murray
16. Lord Kelvin [William Thomson] (1864) On the Secular Cooling of the Earth, *Transactions of the Royal Society of Edinburgh*, XXIII: 167–169, read April 28, 1862

17. John Joly (1899) An Estimate of the Geological Age of the Earth, *Scientific Transactions of the Royal Dublin Society*, Series ii, vol. vii: 44
18. Bertram B. Boltwood (1907) On the ultimate disintegration products of the radio-active elements. Part II. The disintegration products of uranium, *American Journal of Science*, 23 (134): 77–88
19. Arthur Holmes (1913) *The Age of the Earth*, New York: Harper & Brothers
20. Brian J. Ford (2011) *Secret Weapons, Technology, Science and the Race to Win World War II*, Oxford: Osprey Publishing
21. Harold G. Seeley (1887) On the Classification of the Fossil Animals Commonly Named Dinosauria, *Proceedings of the Royal Society of London*, 43: 165–171
22. Alfred S. Romer (1933) *Vertebrate Paleontology*, Chicago: University Press
23. A.S. Romer (1956) *Osteology of the Reptiles*, Chicago: University Press
24. Gerhard Heilmann (1926) *The Origin of Birds*, London: Witherby Books
25. Gregory S. Paul (2002) *Dinosaurs of the Air: The Evolution and Loss of Flight in Dinosaurs and Birds*, Baltimore: Johns Hopkins University Press
26. Brian J. Ford (1970) *Microbiology and Food*, London: Northwood Publications
27. John Harold Ostrom (1969) *Osteology of Deinonychus antirrhopus, an unusual theropod from the Lower Cretaceous of Montana*, New Haven: Peabody Museum of Natural History
28. Charles R. Knight (1942) Parade of Life through the Ages, *National Geographic Magazine*, 81 (2): 141–184
29. John Harold Ostrom (1970) Stratigraphy and paleontology of the Cloverly Formation (Lower Cretaceous) of the Bighorn Basin area, Wyoming and Montana, *Bulletin of the Peabody Museum of Natural History*, 35: 1–234
30. Digby Johns McLaren (1970) Time, Life and Boundaries, *Journal of Palæontology*, 44 (5): 801–815
31. R.A.F. Grieve (1990) Impact Cratering on the Earth, *Scientific American*, 262: 66–73
32. Walter Alvarez, Luis W. Alvarez, F. Asaro & H.V. Michel (1979) Anomalous iridium levels at the Cretaceous/Tertiary boundary at Gubbio, Italy: Negative results of tests for a supernova origin, in W.K. Christensen & T. Birkelund, *Cretaceous/Tertiary Boundary Events Symposium*, Copenhagen: University Press

33. Walter Alvarez, Luis W. Alvarez, F. Asaro & H.V. Michel (1980) Extraterrestrial cause for the Cretaceous-Tertiary extinction, *Science*, 208: 1095–1108
34. Gerrit L. Verschuur (1996) *Impact! The Threat of Comets and Asteroids*, New York: Oxford University Press
35. Robert T. Bakker (1986) *The Dinosaur Heresies*, New York: William Morrow, pp. 14, 229, 242 & 243
36. Ibid., pp. 233 & 339
37. Robert T. Bakker (1987) The return of the dancing dinosaurs in S.J. Czerkas and C. Olson (eds.), *Dinosaurs Past and Present*, Vol. 1, Seattle: University of Washington Press, pp. 38–69
38. Roger Wilmut (ed.), Graham Chapman, John Cleese, Terry Gilliam, Eric Idle, Terry Jones & Michael Palin (1989) *The Complete Monty Python's Flying Circus: All the Words*, 2: New York: Pantheon Books, pp. 118–120
39. John R. Horner & Robert Makela (1979) Nest of Juveniles Provides Evidence of Family-Structure Among Dinosaurs, *Nature*, 282 (5736): 296–298
40. John Horner & James Gorman (1988) *Digging Dinosaurs: The Search that Unraveled the Mystery of Baby Dinosaurs*, New York: Workman Publishing Co
41. Report (1988) Science Watch: Gregarious Dinosaurs? *The New York Times*, August 16
42. Masaki Matsukawa, Toshikazu Hamuro, Teruo Mizukami & Shoji Fujii (1997) First trackway evidence of gregarious dinosaurs from the Lower Cretaceous Tetori Group of eastern Toyama prefecture, central Japan, *Cretaceous Research*, 18 (4): 603–619
43. Timothy Scott Myers & A. Fiorillo (2009) Evidence for gregarious behavior and age segregation in sauropod dinosaurs, *Palæogeography, Palæoclimatology, Palæoecology*, 274 (1–2): 96–104
44. Leonard Salgado, J. Canudo, A. Garrido & J. Carballido (2012) Evidence of gregariousness in rebbachisaurids (Dinosauria, Sauropoda, Diplodocoidea) from the Early Cretaceous of Neuquén (Rayoso Formation), Patagonia, *Argentine Journal of Vertebrate Paleontology*, 32 (3): 603–613
45. Paul Brinkman (2014) cited in: http://phenomena.nationalgeographic.com/2014/03/24/dinosaur-culture/
46. José F. Bonaparte & Rodolfo A. Coria (1993) Un nuevo y gigantesco sauropodo titanosaurio de la Formacion Rio Limay (Albiano-Cenomaniano) de la Provincia del Neuquen, Argentina, *Ameghiniana*, 30 (3): 271–282

47. Large *Argentinosaurus* walking: https://www.youtube.com/watch?v=TjjZ2Czigmg
48. Jeorge Calvo, J.D. Porfiri, B.J. González-Riga & A.W. Kellner (2007) A new Cretaceous terrestrial ecosystem from Gondwana with the description of a new sauropod dinosaur, *Anais Academia Brasileira Ciencia*, 79 (3): 529–541
49. Richard Monastersky (1997) T. rex Bested by Argentinean Beast, *Science News*, 151 (21): 317–317. doi:10.2307/4018414.
50. David Attenborough (2016) *Attenborough and the giant dinosaur*, Bristol: BBC, January 18

CHAPTER 7: HOW MICROBES MADE THE WORLD

1. Brian J. Ford (1976) *Microbe Power, Tomorrow's Revolution*, London: Macdonald and Jane's
2. Cary Woodruff & John R. Foster, Jr. (2015) The fragile legacy of *Amphicœlias fragillimus* (Dinosauria: Sauropoda; Morrison Formation, Latest Jurassic), *PeerJ PrePrints*, 3: e838v1
3. M.P. Taylor (2009) A re-evaluation of *Brachiosaurus altithorax*, Riggs 1903 (Dinosauria, Sauropoda) and its generic separation from *Giraffatitan brancai* (Janensh 1914), *Journal of Vertebrate Paleontology*, 29 (3): 787–806
4. J.R. Foster (2003) *Paleoecological Analysis of the Vertebrate Fauna of the Morrison Formation (Upper Jurassic), Rocky Mountain Region, U.S.A.*, Albuquerque: New Mexico Museum of Natural History and Science, Bulletin 23
5. Jan Peczkis (1993) Implications of body-mass estimates for dinosaurs, *Journal of Vertebrate Paleontology*, 14 (4): 520–533
6. José L. Carballido, Diego Pol, Alejandro Otero, Ignacio A. Cerda, Leonardo Salgado, Alberto C. Garrido, Jahandar Ramezani, Néstor R. Cúneo & Javier M. Krause (2017) A new giant titanosaur sheds light on body mass evolution among sauropod dinosaurs, *Proceedings of the Royal Society B*, 284: 2017–1219. doi:10.1098/rspb.2017.1219. Published August 9, 2017
7. http://www.newser.com/story/246956/biggest-dino-ever-may-have-been-as-heavy-as-space-shuttle.html
8. http://news.nationalgeographic.com/2017/08/largest-dinosaur-ever-titanosaur-fossil-patagotitan-science/
9. https://phys.org/news/2017-08-patagotitan-mayorum-biggest-dinosaur.html
10. https://usaonlinejournal.com/2017/08/08/meet-patagotitan-the-biggest-dinosaur-ever-found/

11. http://www.itv.com/news/2017-08-09/fossilized-bones-belonged-to-biggest-ever-dinosaur/
12. http://www.bbc.co.uk/mediacentre/latestnews/2016/attenborough-and-the-giant-dinosaur
13. Roger S. Seymour (2009) Raising the sauropod neck: it costs more to get less, *Biology Letters (Mechanics)*. doi:10.1098/rsbl.2009.0096
14. José F. Bonaparte & R.A. Coria (1993) Un neuvo y gigantesco saurópodo titanosaurio de la Formación Rio Limay (Albanio-Cenomaniano) de la Provincia del Neuquén, Argentina, *Ameghiniana*, 30 (3): 271–282
15. Gerardo V. Mazzetta, Per Christiansen & Richard A. Fariña (2004) Giants and Bizarres (*sic*): Body Size of Some Southern South American Cretaceous Dinosaurs, *Historical Biology*, 16 (2–4): 71–78
16. Roger B.J. Benson, Nicolás E. Campione, Matthew T. Carrano, Philip D. Mannion, Corwin Sullivan, Paul Upchurch & David C. Evans (2014) Rates of Dinosaur Body Mass Evolution Indicate 170 Million Years of Sustained Ecological Innovation on the Avian Stem Lineage, *Plos Biology* https://doi.org/10.1371/journal.pbio.1001853
17. P. Dejours, L. Bolis, C.R. Taylor & E.R. Weibel (1987) *Comparative Physiology: Life in Water and on Land*, Padova: IX-Liviana Press, Fidia Research Series
18. David W.E. Hone & Michael J. Benton (2005) The evolution of large size: how does Cope's Rule work? *Trends in Ecology and Evolution*, 20 (1): 4–6
19. Martin Sander, Andreas Christian, Marcus Clauss, Regina Fechner, Carole T. Gee, Eva-Maria Griebeler, Hanns-Christian Gunga, Jürgen Hummel, Heinrich Mallison, Steven F. Perry, Holger Preuschoft, Oliver W.M. Rauhut, Kristian Remes, Thomas Tütken, Oliver Wings & Ulrich Witzel (2011) Biology of the sauropod dinosaurs: the evolution of gigantism, *Biological Reviews Cambridge Philosophical Society*, 86 (1): 117–155
20. Kenneth Kermack (1951) A note on the habits of the Sauropods, *Journal of Natural History* [Series 12], 4 (44): 830–832
21. John Humphrys (2017) Interview with Nick Longrich on the *Today* programme, London: BBC Radio Four, 06:54–06:57 am, May 4
22. Andrew Clarke (2013) Dinosaur Energetics, Setting the Bounds on Feasible Physiologies and Ecologies, *The American Naturalist*, 182 (3): 283–297
23. John M. Grady, B.J. Enquist, E. Dettweiler-Robinson, N.A. Wright & F.A. Smith (2014) Evidence for mesothermy in dinosaurs, *Science*, 344: 1268–1272

24. Kate Littler, Stuart A. Robinson, Paul R. Bown, Alexandra J. Nederbragt & Richard D. Pancost (2011) High sea-surface temperatures during the Early Cretaceous Epoch, *Nature Geoscience*, 4: 169–172
25. Hugh C. Jenkyns, Astrid Forster, Stefan Schouten & Jaap Sinninghe Damsté (2004) High temperatures in the Late Cretaceous Arctic Ocean, *Nature*, 432: 888–892
26. Alan J. Charig & Angela C. Milner (1986) Baryonyx, a remarkable new theropod dinosaur, *Nature*, 324 (6095): 359–361
27. Alan J. Charig & Angela C. Milner (1997) *Baryonyx walkeri*, a fish-eating dinosaur from the Wealden of Surrey, *Bulletin of the Natural History Museum of London*, 53: 11–70
28. Martin G. Lockley & Adrian P. Hunt (1999) *Dinosaur Tracks and Other Fossil Footprints of the Western United States*, Columbia: University Press
29. Myriam R. Hirt, Walter Jetz, Björn C. Rall & Ulrich Brose (2017) The largest animals are not the fastest, *Nature Ecology & Evolution*, 1: 1116–1122. doi:10.1038/s41559-017-0241-4
30. John R. Hutchinson, & M. Garcia (2002) Tyrannosaurus was not a fast runner, *Nature*, 415: 1018–1021
31. Discussion at Morrison Museum (2017): https://www.youtube.com/watch?v=nBHC1-uAePk
32. Sorcha McDonagh (2003) Did dinosaurs do handstands – enormous dinosaurs might have balanced on their front feet when wading in shallow water, *Science News for Students* on www.sciencenewsforstudents.org, October 27
33. Yuong-Nam Lee & Min Huh (2002) Manus-only sauropod tracks in the Uhangri Formation (Upper Cretaceous), Korea and their paleobiological implications, *Journal of Paleontology*, 76 (3): 558–564
34. Michael J. Hawthorne, Rena M. Bonem, James O. Farlow & James O. Jones (2002) Ichnology, stratigraphy and paleoenvironment of the Boerne Lake spillway dinosaur tracksite, south-central Texas, *Texas Journal of Science*, 54 (4): 310
35. Anon (2009) Largest Dinosaur Footprints Ever Found Discovered Near Lyon, France, Science Daily on www.sciencedaily.com *Centre National de la Recherche Scientifique*, Paris
36. J. A. Wilson (2005) Integrating ichnofossil and body fossil records to estimate locomotor posture and spatiotemporal distribution of early sauropod dinosaurs, a stratocladistic approach, *Paleobiology*, 31 (3): 400–423
37. David Fastovsky & David Weishampel (2005) *The Evolution and Extinction of the Dinosaurs* [second edition], Cambridge: University Press

38. Donald Henderson (2004) Tipsy punters: sauropod dinosaur pneumaticity, buoyancy and aquatic habits, *Proceedings of the Royal Society of London* B, 271: S180–S183
39. John R. Cameron, James G. Skofronick & Roderick M. Grant (1999) *Physics of the Body*, Second Edition: 96. Madison, WI: Medical Physics Publishing
40. Noble S. Proctor & Patrick J. Lynch (1993) *Manual of Ornithology*, New Haven, CT: Yale University Press
41. Michael P. Taylor, M.J. Wedel & D. Naish (2009) Head and neck posture in sauropod dinosaurs inferred from extant animals, *Acta Palæontologica Polonica*, 54 (2): 213–220. doi: 10.4202/app.2009.0007
42. Michael J. Ryan, Brenda J. Chinnery-Allgeier, David A. Eberth, Patricia E. Ralrick & Philip J. Currie (2010) A Semi-Aquatic Life Habit for Psittacosaurus [in] Michael J. Ryan, Brenda J. Chinnery-Allgeier & David A. Eberth, *New Perspectives on Horned Dinosaurs*, The Royal Tyrrell Museum Ceratopsian Symposium, Bloomington: Indiana University Press
43. Roger S. Seymour (2009) Raising the sauropod neck: it costs more to get less, *Biology Letters (Mechanics)*, doi: 10.1098/rsbl.2009.0096
44. Tim Radford (2003) Monsters that turned turtle, *The Guardian*, December 10
45. Barney Newman (1970) Stance and gait in the flesh-eating dinosaur *Tyrannosaurus*, *Biological Journal of the Linnean Society*, 2: 119–123. doi: 10.1111/j.1095-8312.1970.tb01707.x
46. Roger S. Seymour (2009) Raising the sauropod neck: it costs more to get less, *Biology Letters (Mechanics)*, doi: 10.1098/rsbl.2009.0096
47. Peter Norton (1995) *Encyclopædia Britannica*, Chicago: Encyclopædia Britannica Inc., 17: 315–350
48. Ernst Freiherr Stromer (1915) Ergebnisse der Forschungsreisen Prof. E. Stromers in den Wüsten Ägyptens. II. Wirbeltierreste der Baharîje-Stufe (Cenoman Unterstes). III. Das Original of Theropoden *Spinosaurus aegyptiacus* ngn sp., *Abhandlungen der Bayerischen Akademie der Wissenschaften*, 18 (3): 1–32
49. William Nothdurft & Josh Smith (2002) *The Lost Dinosaurs of Egypt*, New York: Cosmos Studios
50. Cristiano Dal Sasso, S. Maganuco, E. Buffetaut & M.A. Mendez (2005) New information on the skull of the enigmatic theropod *Spinosaurus*, with remarks on its sizes and affinities, *Journal of Vertebrate Paleontology*, 25 (4): 888–896
51. http://www.cornettedesaintcyr.fr/html/index.jsp?id=5907&lng=&npp=150

52. Cristiano dal Sasso, S. Maganuco & A. Cioffi (2009) A neurovascular cavity within the snout of the predatory dinosaur Spinosaurus, *1st International Congress on North African Vertebrate Palæontology*, Paris: Muséum National d'Histoire Naturelle
53. Romain Amiot, Eric Buffetaut, Christophe Lécuyer, Xu Wang, Larbi Boudad, Zhongli Ding, François Fourel, Steven Hutt, François Martineau, Manuel Alfredo Medeiros, Jinyou Mo, Laurent Simon, Varavudh Suteethorn, Steven Sweetman, Haiyan Tong, Fusong Zhang & Zhonghe Zhou (2010) Oxygen isotope evidence for semi-aquatic habits among spinosaurid theropods, *Geology*, 38 (2): 139
54. Matt Kaplan (2010) Water-dwelling dinosaur breaks the mould, *Nature*, February 19

CHAPTER 8: WADING WITH DINOSAURS

1. Spinosaurus (2011) *Planet Dinosaur, Lost World*, London: BBC productions, September 14
2. Philip Prime (2011) email to the author, September 19, 08:30
3. Brian J. Ford (2012) online video, https://www.youtube.com/embed/kjmNnRFHCzk, *Laboratory News*, March 30
4. Brian J. Ford (2012) A Prehistoric Revolution, *Laboratory News*: 24–26, April 3
5. Original dinosaur image at http://www.madriz.com/sabado-12/
6. Brian J. Ford (2012) A Prehistoric Revolution, *Laboratory News*: 1, 24–26, April 3
7. http://www.telegraph.co.uk/news/science/dinosaurs/9182551/Dinosaurs-must-have-lived-in-water-scientist-claims.html, http://metro.co.uk/2012/04/03/dinosaurs-could-have-ruled-the-water-not-the-earth-claims-scientist-375447/, http://www.dailymail.co.uk/sciencetech/article-2124420/Dinosaurs-DIDNT-rule-earth-The-huge-creatures-actually-lived-water--tails-swimming-aids.html?ito=feeds-newsxml, http://www.ibtimes.co.uk/dinosaurs-lived-underwater-researchers-323260, http://www.foxnews.com/tech/2012/04/03/dinosaurs-lived-underwater-british-scientist-claims.html
8. https://www.autostraddle.com/dinosaur-news-were-your-favorite-prehistoric-beasts-actually-aquatic-13597/
9. Wolfgang Goede (2012) Welt des Wissens: Saurier waren Amphibien, *PM magazine*, April 13
10. http://news.bbc.co.uk/today/hi/today/newsid_9710000/9710630.stm
11. https://svpow.com/2012/04/12/bbc-radio-4-fail-in-their-duty-of-care-to-their-listeners/

12. http://doubtfulnews.com/2012/04/sinking-the-silly-idea-of-aquatic-dinosaurs/
13. https://www.smithsonianmag.com/science-nature/aquatic-dinosaurs-not-so-fast-170096608/
14. Paul Upchurch (2012) email to the author, April 18, 16:02
15. Anshuman J. Das, Denize C. Murmann, Kenneth Cohrn & Ramesh Raskar (2017) A method for rapid 3D scanning and replication of large paleontological specimens, *Plos One*, July 5. doi.org/10.1371/journal.pone.0179264
16. Using Models to teach students how to estimate the Mass of Dinosaurs (2012) https://education.llnl.gov/bep/science/12/tMass.html
17. Charlotte Brassey (2016) Body mass estimation in paleontology: a review of volumetric techniques, *Journal of Paleontology*, Vol. 22: 133–156
18. https://www.labnews.co.uk/features/palaeontology-bites-back-11-05-2012/

CHAPTER 9: COPULATING COLOSSUS

1. Barry Yeoman (2006) Schweitzer's Dangerous Discovery, *Discover* magazine, April 27, http://discovermagazine.com/2006/apr/dinosaur-dna
2. Mary H. Schweitzer, Jennifer L. Wittmeyer & John R. Horner (2005) Gender-specific reproductive tissue in ratites and Tyrannosaurus rex, *Science*, 308 (5727): 1456–1460
3. Henry Fairfield Osborn & Barnum Brown (1906) Tyrannosaurus, Upper Cretaceous carnivorous dinosaur, *Bulletin of the American Museum of Natural History*, 22 (16): 281–296
4. Sandy Fritz (1988) Tyrannosaurus sex – a love tail, *Omni* magazine, February, William A.S. Sarjeant (1993) Lambert Beverly Halstead, his life, his discoveries and his controversies, *Modern Geology*, 18 (Halstead Memorial Volume, 1): 5–59. Vertebrate fossils and the evolution of scientific concepts. *Writings in tribute to Beverly Halstead*. Reading: Gordon & Breach Publishers
5. Edwin H. Colbert, George Geygan & Paul Geygan (1977) *The Dinosaur World*, New York: Stravon Educational Press
6. Robert T. Bakker (1986) *The Dinosaur Heresies*, New York: William Morrow, p. 337
7. Phil Senter (2007) Necks for sex: sexual selection as an explanation for sauropod dinosaur neck elongation, *Journal of Zoology*, 271: 45
8. Nicole Klein, Kristian Remes, Carole T. Gee & P. Martin Sander (2011) *Biology of the Sauropod Dinosaurs, Understanding the Life of Giants*, Indiana: University Press

9. Roger S. Seymour (2009) Raising the sauropod neck: it costs more to get less. *Biology Letters of the Royal Society*, 5: 317–319. doi:10.1098/rsbl.2009.009
10. Nathan P. Myhrvold & Philip J. Currie (1997) Supersonic sauropods? Tail dynamics in the diplodocids, *Paleobiology*, 23 (4): 393–409
11. Shu Li, C. Shih, C. Wang, H. Pang & D. Ren (2013) Forever love: The hitherto earliest record of copulating insects from the Middle Jurassic of China. *PLoS ONE* 8: 11, e78188. doi:10.1371/journal.pone.0078188
12. Martin G. Lockley, Richard T. McCrea, Lisa G. Buckley, Jong Deock Lim, Neffra A. Matthews, Brent H. Breithaupt, Karen J. Houck, Gerard D. Gierliński, Dawid Surmik, Kyung Soo Kim, Lida Xing, Dal Yong Kong, Ken Cart, Jason Martin & Glade Hadden (2016) Theropod courtship: large-scale physical evidence of display arenas and avian-like scrape ceremony behaviour by Cretaceous dinosaurs, *Nature Scientific Reports*, 6, article number: 18952. doi:10.1038/srep18952
13. John A. Long (2012) *The Dawn of the Deed, the Prehistoric Origins of Sex*, Chicago: University of Chicago Press
14. https://www.youtube.com/watch?v=-mv_v4ltSrY
15. Othniel Charles Marsh (1877) A new order of extinct Reptilia (Stegosauria) from the Jurassic of the Rocky Mountains, *American Journal of Science*, 3 (14): 513–514
16. Jon Anderson (1994) The Perplexing Puzzle of Maladroit Mating, *Chicago Tribune*, August 30, http://articles.chicagotribune.com/1994-08-30/features/9408300026_1_large-dinosaurs-landry-two-dinosaurs
17. http://www.dailymail.co.uk/sciencetech/article-2105439/Revealed-How-horny-30-ton-dinosaurs-sex.html#ixzz4n0KiGeTz (February 24, 2012)
18. William W. Hay (2016) *Experimenting on a Small Planet: A History of Scientific Discoveries, a Future of Climate Change and Global Warming*, second edition, Berlin & New York: Springer
19. William W. Hay (2017) Toward understanding Cretaceous climate – An updated review, *Science China Earth Sciences*, 60 (1): pp. 5–19
20. Brian J. Ford (2012) Aquatic Dinosaurs under the Lens, *The Microscope* 60 (3) 123–131. Read online at http: //www.mcri.org/v/830/The-Microscope-Volume-60-Third-Quarter-2012
21. http://timesofindia.indiatimes.com/home/science/95m-year-old-stampeding-tracks-of-dinosaurs-are-swimming-marks/articleshow/18110168.cms?referral=PM
22. http://www.dailymail.co.uk/sciencetech/article-2306230/Researchers-reveal-dino-paddle-T-Rex-used-wade-rivers.html

23. Scott D. Sampson, Lawrence M. Witmer, Catherine A. Forster, David W. Krause, Patrick M. O'Connor, Peter Dodson & Florent Ravoavy (1998) Predatory dinosaur remains from Madagascar: implications for the Cretaceous biogeography of Gondwana, *Science*, 280 (5366): 1048–1051. PMID 9582112. doi:10.1126/science.280.5366.1048
24. William W. Hay (2012) Aquatic Dinosaurs, *The Microscope*, 60 (4) 179
25. Donald Henderson (2012) Aquatic Dinosaurs, *The Microscope* 60, (4) 179–180
26. Donald Henderson (2012) email message to the author, 11:24 pm, December 4
27. Martin G. Lockley & Alan Rice (2008) Did 'Brontosaurus'; [*sic*] ever swim out to sea? Evidence from brontosaur and other dinosaur footprints, *Ichnos*, 1 (2): 81–90. doi.org/10.1080/10420949009386337
28. Lida Xing, Daqing Li, Peter L. Falkingham, Martin G. Lockley, Michael J. Benton, Hendrik Klein, Jianping Zhang, Hao Ran W. Scott Persons & Hui Dai (2016) Digit-only sauropod pes trackways from China – evidence of swimming or a preservational phenomenon? *Science Reports*, 6: 21138. doi: 10.1038/srep21138 PMCID: PMC4758031
29. Brian J. Ford (2014) Die-Hard Dinosaurs, *Mensa Magazine*: 10–14, March
30. Michael S. Potter (2014) Overcoming the scientific dinosaurs, *Mensa Magazine*: letters, April
31. http://scienceviews.com/dinosaurs/dinosaurstatepark.html
32. Walter P. Coombs, Jr. (1980) Swimming Ability of Carnivorous Dinosaurs, *Science*, 207: 1198–1200
33. Walter P. Coombs, Jr. (1975) Sauropod habits and habitats, *Palæogeography, Palæoclimatology, Palæoecology*, 17 (1975): 1–33
34. Terapod Zoology (2009) *Science Blogs*, March 20, http://scienceblogs.com/tetrapodzoology/2009/03/20/junk-in-the-trunk/
35. http://www.smithsonianmag.com/science-nature/did-dinosaurs-swim-47506260/
36. https://www.theguardian.com/science/lost-worlds/2014/jun/16/were-dinosaurs-all-at-sea

CHAPTER 10: TRUTH WILL OUT

1. Nizar Ibrahim N., Paul Sereno, C. Dal Sasso, S. Maganuco, M. Fabbri, D.M. Martill, S. Zouhri, N. Myhrvold & D.A. Iurino (2014) Semiaquatic adaptations in a giant predatory dinosaur, *Science*, 345: 1613–1616

2. https://www.uchicago.edu/features/massive_hunter_prowled_waters_edge/
3. http://press.nationalgeographic.com/2014/09/11/scientists-report-first-semiaquatic-dinosaur-spinosaurus/
4. https://www.youtube.com/watch?v=BhJ2HbnvmXI&t=1s
5. https://www.timeinc.com/experiences/peoples-50-most-beautiful/
6. http://phenomena.nationalgeographic.com/2014/09/11/the-new-spinosaurus/
7. Nizar Ibrahim interview: https://youtu.be/F0op8hbhBwA
8. Spinosaurus interview released by www.telegraph.co.uk on September 11, 2014
9. Donald Henderson (2017) email messages dated November 28, 18:30, and November 29, 20:49
10. Donald Henderson (2018) paper on buoyancy and balance of an immersed *Spinosaurus*, [in] *PeerJ* (online open access journal), in press
11. Video from Singapore Science Centre: https://youtu.be/0KHVAsPa3i0
12. https://blogs.scientificamerican.com/tetrapod-zoology/brian-j-ford-s-aquatic-dinosaurs-2014-edition/ [note: click on 'load comments' for 128 responses]
13. http://www.tested.com/science/43863-why-the-aquatic-dinosaur-theory-doesnt-hold-water/
14. Robert M. DeConto, William W. Hay, Stanley L. Thompson & Jon C. Bergengren (1999) Late Cretaceous Climate and Vegetation Interactions: the Cold Continental Interior Paradox, *Geological Society of America Special Paper*. doi: 10.1130/0-8137-2332-9.391
15. R.A. Spicer & A.B. Herman (2010) The Late Cretaceous environment of the Arctic: A quantitative reassessment based on plant fossils, *Palaeogeography, Palaeoclimatology, Palaeoecology*, 295: 423–442
16. K.L. Bice & R.D. Norris (2002) Possible atmospheric CO_2 extremes of the Middle Cretaceous (late Albian–Turonian), *Paleoceanography*, 17 (1070) 22: 1–16. doi:10.1029/2002PA000778
17. Giuseppe Etiope, T. Fridriksson, F. Italiano, W. Winiwarterd & J. Thelokee (2007) Natural emissions of methane from geothermal and volcanic sources in Europe, *Journal of Volcanology and Geothermal Research*, 165 (1–2): 76–86
18. David M. Wilkinson, E.G. Nisbet & G.D. Ruxton (2012) Could methane produced by sauropod dinosaurs have helped drive Mesozoic climate warmth? *Current Biology*, 22: R292–R293
19. R.A. Rasmussen & M.A.K. Khalil (1983) Global production of methane by termites, *Nature*, 301: 700–702. doi:10.1038/301700a0

20. William W. Hay (2017) Toward understanding Cretaceous climate, an updated review, *Science China – Earth Sciences*, 60 (1): 5–19
21. Elsa Panciroli (2017) A glimpse of when Canada's badlands were a lush dinosaur forest by the sea, *The Guardian*, September 27
22. Brooks Hayes (2017) New Dinosaur discovered in Utah, UTI Science News, April 13
23. Anthony R. Fiorillo (2009) Dental micro wear patterns of the sauropod dinosaurs camarasaurus and diplodocus, *Historical Biology* 1–16. Published online: January 10, 2009. doi.org/10.1080/08912969809386568
24. Frank J. Varriale (2016) Dental microwear reveals mammal-like chewing in the neoceratopsian dinosaur *Leptoceratops gracilis*, *Paleontology and Evolutionary Science* July 6, 2016. PubMed 27441111
25. Marisa DeCandido (2017) UW-Oshkosh professor has new hypothesis about dinosaur fossils in Utah, *NBC26 news*, July 18
26. Lara Sciscio, E.M. Bordy, M. Abrahams, F. Knoll & B.W. McPhee (2017) The first megatheropod tracks from the Lower Jurassic, upper Elliot Formation, Karoo Basin, Lesotho, *Plos One*: October 25, 2017. doi.org/10.1371/journal.pone.0185941
27. Fiann M. Smithwick, Robert Nicholls, Innes C. Cuthill & Jakob Vinther (2017) Countershading and Stripes in the theropod dinosaur *Sinosauropteryx* reveal Heterogeneous Habitats in the Early Cretaceous Jehol Biota, *Current Biology*, 27 (21): pp. 3337–3343. doi: http://dx.doi.org/10.1016/j.cub.2017.09.032
28. Sebastián Apesteguía, Nathan D. Smith, Rubén Juárez Valieri & Peter J. Makovicky (2016) An Unusual New Theropod with a Didactyl Manus from the Upper Cretaceous of Patagonia, Argentina. *PLOS ONE*, 11 (7): e0157793. doi: 10.1371/journal.pone.0157793
29. Henry Osborn (1905) Tyrannosaurus and other Cretaceous carnivorous dinosaurs, *Bulletin of the American Museum of Natural History*, 21 (14): 259–265. hdl:2246/1464
30. Christopher Brochu (2003) Osteology of Tyrannosaurus rex: insights from a nearly complete skeleton and high-resolution computed tomographic analysis of the skull, *Society of Vertebrate Paleontology Memoirs*, 7: 1–138. doi:10.2307/3889334
31. Scientific American, 2007, https://www.scientificamerican.com/article/if-t-rex-fell-how-did-it/
32. John R. Hutchinson & Mariano Garcia (2002) Tyrannosaurus was not a fast runner, *Nature*, 415: 1018–1021 (February 28) doi:10.1038/4151018a

33. Kate Wong (2002) https://www.scientificamerican.com/article/t-rex-not-fleet-of-foot-s/
34. Hannah Osborne (2017) T-Rex Couldn't Run – Doing So Would've Broken Its Legs, *Newsweek*, July 18: https://www.yahoo.com/news/t-rex-couldn-t-run-111009711.html
35. William I. Sellers, Stuart B. Pond, Charlotte A. Brassey, Philip L. Manning & Karl T. Bates (2017) Investigating the running abilities of Tyrannosaurus rex using stress-constrained multibody dynamic analysis, PubMed 28740745, https://peerj.com/articles/3420/
36. Sterling J. Nesbitt, Richard J. Butler, Martín D. Ezcurra, Paul M. Barrett, Michelle R. Stocker, Kenneth D. Angielczyk, Roger M.H. Smith, Christian A. Sidor, Grzegorz Niedźwiedzki, Andrey G. Sennikov & Alan J. Charig (2017) The earliest bird-line archosaurs and the assembly of the dinosaur body plan, *Nature*, 544: 484–487
37. Matthew G. Baron & Paul M. Barrett (2017) A dinosaur missing-link? *Chilesaurus* and the early evolution of ornithischian dinosaurs, *Biology Letters*, 13: 20170220. doi.org/10.1098/rsbl.2017.0220
38. http://www.bbc.co.uk/news/science-environment-40890714
39. Harry G. Seeley (1887) On the Classification of the Fossil Animals Commonly Named Dinosauria, *Proceedings of the Royal Society of London*, 43: 165–171. doi:10.1098/rspl.1887.0117
40. Matthew G. Baron, David B. Norman & Paul M. Barrett (2017) A new hypothesis of dinosaur relationships and early dinosaur evolution, *Nature*, 543: 501–506
41. http://www.bbc.co.uk/news/science-environment-39305750
42. Andrea Cau, Vincent Beyrand, Dennis F.A.E. Voeten, Vincent Fernandez, Paul Tafforeau, Koen Stein, Rinchen Barsbold, Khishigjav Tsogtbaatar, Philip J. Currie & Pascal Godefroit (2017) Synchrotron scanning reveals amphibious ecomorphology in a new clade of bird-like dinosaurs, *Nature*, December 6. doi:10.1038/nature24679
43. M.P. Rowe (2000) Inferring the Retinal Anatomy and Visual Capacities of Extinct Vertebrates, *Palæontologia Electronica*, 3 (1) article 3: 43pp
44. L.M. Witmer (1995) The extant phylogenetic bracket and the importance of reconstructing soft tissues in fossils, in J.J. Thomason (ed.) *Functional Morphology in Vertebrate Paleontology*, Cambridge: University Press, pp. 19–33
45. Gordon L. Walls (1942) *The Vertebrate Eye and Its Adaptive Radiation*, Bloomfield Hills, Michigan: Cranbrook Press
46. Kent A. Stevens (2006) *Binocular Vision in Theropod Dinosaurs*, University of Oregon Press

47. Darla K. Zelenitsky, François Therrien, Ryan C. Ridgely, Amanda R. McGee & Lawrence M. Witmer (2011) Evolution of olfaction in non-avian theropod dinosaurs and birds, *Proceedings of the Royal Society, Biology*, 278: 3625–3634. doi: 10.1098/rspb.2011.0238
48. Press Association (2017) Dinosaurs' sensitive snouts enabled courtship 'face stroking', study suggests, https://www.theguardian.com/science/2017/jun/27/dinosaurs-sensitive-snouts-courtship-face-stroking-study-suggests
49. Chris Barker, Darren Naish, Elis Newham, Orestis L. Katsamenis & Gareth Dyke (2017) Complex neuroanatomy in the rostrum of the Isle of Wight theropod *Neovenator salerii, Scientific Reports* 7, Article number: 3749, doi:10.1038/s41598-017-03671-3
50. http://metro.co.uk/2013/09/22/crocodile-eating-shark-captured-on-camera-by-holidaymaker-in-australia-4075436/

CHAPTER 11: THE LIFE AND DEATH OF DINOSAURS

1. Harry Govier Seeley (1870) On *Ornithopsis*, a gigantic animal of the pterodactyle kind from the Wealden, *Annals of the Magazine of Natural History*, Series 4, 5: 279–283
2. Leon P.A.M. Claessens (2004) Dinosaur gastralia, origin, morphology, and function, *Journal of Vertebrate Paleontology* 24, (1): 89–106
3. http://www.bbc.co.uk/programmes/b08ns0kq
4. Richard McCourt (2017) Dinosaur-era plant found alive in North America for first time [report], *Science Daily*, July 31, https://www.sciencedaily.com/releases/2017/07/170731164122.htm
5. Cesare Emiliani (1992) *Planet Earth: Cosmology, Geology, and the Evolution of Life and Environment*, Cambridge: University Press
6. Darren Naish (2012) There are giant feathered tyrannosaurs now … right? Tetrapod Zoology blog on *Scientific American* website, April 4.
7. Phil R. Bell, Nicolás E. Campione, W. Scott Persons IV, Philip J. Currie, Peter L. Larson, Darren H. Tanke & Robert T. Bakker (2017) Tyrannosauroid integument reveals conflicting patterns of gigantism and feather evolution, *Biology Letters of the Royal Society*, doi: 10.1098/rsbl.2017.0092, June 7.
8. Enrique Peñalver, Antonio Arillo, Xavier Delclòs, David Peris, David A. Grimaldi, Scott R. Anderson, Paul C. Nascimbene & Ricardo Pérez-de la Fuente (2017) Ticks parasitised feathered dinosaurs as revealed by Cretaceous amber assemblages, *Nature online*. doi:10.1038/s41467-017-01550-z
9. Andy Coghlan (2015) Dinosaur blood cells extracted from 75-million-year-old fossil, *New Scientist* website, June 9

10. Rupert T. Gould (1934) *The Loch Ness Monster and Others*, London: Geoffrey Bles
11. Henry H. Bauer (2002) The case for the Loch Ness 'monster', the scientific evidence, *Journal of Scientific Exploration*, 16: 225–246 (a); Ronald Binns (1984) *The Loch Ness Mystery Solved*, London: W.H. Allen & Co (b); Tim Dinsdale (1973) *The Story of the Loch Ness Monster*, London: Allan Wingate (c); Roy P. Mackal (1976) *The Monsters of Loch Ness*, Chicago: Swallow Press (d); Robert H. Rines (1982) Summarizing a decade of underwater studies at Loch Ness, *Cryptozoology*, 1: 24–32 (e); Robert H. Rines, C.W. Wyckoff, H.E. Edgerton & M. Klein (1976) The search for the Loch Ness Monster, *Technology Review*: 78 (5): 25–40 (f); Adrian Shine (2006) *Loch Ness*, Drumnadrochit: Loch Ness Centre (g); Nicholas Witchell (1974) *The Loch Ness Story*, Lavenham: Terence Dalton (h)
12. Claudio Ciofi (2004) *Varanus komodoensis. Varanoid Lizards of the World*, Bloomington & Indianapolis: Indiana University Press
13. Bryan Fry, S. Wroe & W. Teeuwisse (2009) A central role for venom in predation by *Varanus komodoensis* (Komodo Dragon) and the extinct giant *Varanus (Megalania) priscus*, Proceeding of the National Academy of Sciences, 106 (22): 8969–74. doi:10.1073/pnas.0810883106. PMC 2690028 PMID 19451641
14. https://www.youtube.com/watch?v=StcKuJ3q7Xs
15. http://palaeo.gly.bris.ac.uk/Communication/Couch/possible.html
16. R.A.F. Grieve (1990) Impact Cratering on the Earth, *Scientific American*, 262: 66–73
17. Walter Alvarez, Luis W. Alvarez, F. Asaro & H.V. Michel (1979) Anomalous iridium levels at the Cretaceous/Tertiary boundary at Gubbio, Italy: Negative results of tests for a supernova origin [in] W.K. Christensen & T. Birkelund, *Cretaceous/Tertiary Boundary Events Symposium*, Copenhagen: University Press
18. Walter Alvarez (1997) *T. rex and the Crater of Doom*, Princeton: University Press
19. Sergio de Régules (2015) Revisiting the crater of doom, *Physics World*, 28 (9): 33
20. Sean Gulick, Joanna Morgan, Claire L. Mellett, and the expedition 364 scientists (2017) *Expedition 364 Preliminary Report: Chicxulub: Drilling the K-Pg Impact Crater*, International Ocean Discovery Program. http://dx.doi.org/10.14379/iodp.pr.364.2017
21. Gerta Keller, Thierry Adatte, Wolfgang Stinnesbeck, Mario Rebolledo-Vieyra, Jaime Urrutia Fucugauchi, Utz Kramar & Doris Stüben (2004) Chicxulub impact predates the K-T boundary mass extinction, *Proceedings of the National Academy of Sciences of the*

United States of America, 101 (11): 3753–3758. doi: 10.1073/pnas.0400396101

22. http://massextinction.princeton.edu/chicxulub/
23. Rex Dalton (2003) Mass-extinction controversy flares again, core from asteroid crater fuels debate on what wiped out the dinosaurs, *Nature*, doi:10.1038/news030407-7
24. BBC (2002) *The Day the Earth nearly Died*, December 5
25. Gerta Keller (2005) Impacts, volcanism and mass extinction: random coincidence or cause and effect? *Australian Journal of Earth Sciences*, 52: 725–757
26. Paul Barrett (2016) email to the author, May 8, 17:27
27. William Hay (2017) email to the author, November 18, 19:23
28. – (2017) email to the author, November 21, 01:53
29. – (2017) email to the author, November 23, 17:34
30. – (2017) email to the author, December 6, 17:21
31. William Hay, Robert DeConto, Poppe de Boer, Sacha Flögel, Ying Song & Andrei Stepashko (2018) Possible solutions to several enigmas of Cretaceous climate, *International Journal of Earth Sciences* doi:10.1007/s00531-018-1670-2
32. Richard Henderson (2018) The Life Scientific, BBC Radio Four, February 13: 0900–0930h. Online extract: https://youtu.be/AH-y18wMQWo
33. Matthew R. Bennett, Peter Falkingham, Sarita A. Morse, Karl Bates, Robin H. Crompton (2013), Preserving the Impossible: Conservation of Soft-Sediment Hominin Footprint Sites and Strategies for Three-Dimensional Digital Data Capture, *PlosOne*, doi.org/10.1371/journal.pone.0060755
34. https://www.theguardian.com/science/2017/jun/27/dinosaurs-sensitive-snouts-courtship-face-stroking-study-suggests
35. Sara Coelho (2009) Teeth scratches reveal dinosaur menu, Natural Environment Research Council: *PlanetEarth Online*, June 30.
36. Paige Depolo, Stephen Brusatte, Thomas Challands, Davide Foffa, Dugald A. Ross, Mark Wilkinson & Hongyu Yi (2018) A sauropod-dominated tracksite from Rubha nam Brathairean (Brothers' Point), Isle of Skye, *Scottish Journal of Geology*, doi: 10.1144/sjg2017-016, April 2.
37. Nicola Davis (2018) Dinosaur footprints found on Skye, *The Guardian*, April 3.
38. William "Bill" Sellers (2013) digital simulation of *Argentinosaurus*: https://youtu.be/oHm8iZ2yDyk [and] http://www.manchester.ac.uk/discover/news/scientists-digitally-reconstruct-giant-steps-taken-by-dinosaurs/
39. Extract of interview with Bill Sellers (2013) at Manchester University: https://www.youtube.com/watch?v=oHm8iZ2yDyk

ILLUSTRATION CREDITS

All illustrations and photographs are courtesy of the author or in the public domain, except for the following:

Page 3: http://wallpoper.com/wallpaper/dinosaurs-desert-400302
Page 50: Reproduced by permission of the Geological Society of London
Page 219: *China Daily*/www.chinadaily.com.cn
Page 231: Royal Tyrrell Museum, Drumheller, Canada
Page 233: Sedgwick Museum, Cambridge
Page 255: Robert Bakker
Page 257: Still from Episode 31 of *Monty Python's Flying Circus*, 'The All-England Summarize Proust Competition'; aired November 16, 1972: BBC
Page 278: *Radio Times*
Page 281: The Estate of Zdeněk Burian (Jiří Hochman)
Page 293: John Sibbick
Page 302: From Royal Society: 'Tipsy punters: sauropod dinosaur pneumaticity, buoyancy and aquatic habits', *Proceedings of the Royal Society of London* B 271: S180–S183, 2004
Page 305: Nima Sassani/http://paleoking.blogspot.co.uk/
Page 309: Still from Episode 1 of 2-part documentary 'Dinosaur Britain'; aired on ITV1 on August 31, 2015: Maverick TV
Page 318: Still from 'Planet Dinosaur: Lost World'; aired on September 14, 2011: BBC
Page 323: Kevin Ebi/Living Wilderness Nature Photography
Page 349: Ron Embleton, as appeared in *Omni* magazine (1988)
Page 356: Jonathan Poulter

Page 357: Zeljko Zsrdic
Page 377: Anthony Romilio, University of Queensland
Page 393: Walter Coombs
Page 402: Public Affairs, University of Chicago
Page 406: *National Geographic*/NOVA television screen grab
Page 433: As appeared in *Nature*, retouched by Dr Ronald Rines
Page 437: Karen Letch, Adelaide River Cruises

Plate 3: top illustration: Original painting re N01407 at the Tate Gallery, London
Plate 4: top illustration: Still from *Fantasia*, produced by Ben Sharpsteen, released by Walt Disney Productions (1940)
Plate 5: top illustration: Courtesy of Eleanor Kish, Canadian Museum of Nature
Plate 6: bottom photograph: Arterra Picture Library/Alamy
Plate 8: bottom photographs: AP photograph/Francois Mori
Plate 9: middle photograph: San Diego Zoo
Plate 10: top illustration: Sergey Krasovskiy/Public Library of Science
Plate 11: top: Photograph released on Saturday, May 17, 2014, by the Museo Paletontológico Egidio Feruglio, Argentina
Plate 11: bottom: Photograph taken on March 26, 2017 and released by Damian Kelly, University of Queensland
Plate 14: top: Photograph from Enrique Peñalver, Antonio Arillo, Xavier Delclòs, David Peris, David A. Grimaldi, Scott R. Anderson, Paul C. Nascimbene & Ricardo Pérez-de la Fuente in *Nature Communications* volume 8, Article number: 1924 doi:10.1038/s41467-017-01550-z
Plate 14: bottom photograph: Xinhua Agency, China
Plate 15: top right photograph: Philip Lanoue Photography
Plate 15: bottom illustration: Hesham M. Sallam, Eric Gorscak, Patrick M. O'Connor, Iman A. El-Dawoudi, Sanaa El-Sayed, Sara Saber, Mahmoud A. Kora, Joseph J. W. Sertich, Erik R. Seiffert & Matthew C. Lamanna (2018) 'New Egyptian sauropod reveals Late Cretaceous dinosaur dispersal between Europe and Africa', *Nature Ecology & Evolution* volume 2, pages 445–451 doi:10.1038/s41559-017-0455-5

INDEX

Page references in **bold** indicate images.
BJF indicates Brian J. Ford.